Sustainable Agriculture Reviews

Volume 57

Sustainable agriculture is a rapidly growing field aiming at producing food and energy in a sustainable way for humans and their children. Sustainable agriculture is a discipline that addresses current issues such as climate change, increasing food and fuel prices, poor-nation starvation, rich-nation obesity, water pollution, soil erosion, fertility loss, pest control, and biodiversity depletion.

Novel, environmentally-friendly solutions are proposed based on integrated knowledge from sciences as diverse as agronomy, soil science, molecular biology, chemistry, toxicology, ecology, economy, and social sciences. Indeed, sustainable agriculture decipher mechanisms of processes that occur from the molecular level to the farming system to the global level at time scales ranging from seconds to centuries. For that, scientists use the system approach that involves studying components and interactions of a whole system to address scientific, economic and social issues. In that respect, sustainable agriculture is not a classical, narrow science. Instead of solving problems using the classical painkiller approach that treats only negative impacts, sustainable agriculture treats problem sources.

Because most actual society issues are now intertwined, global, and fast-developing, sustainable agriculture will bring solutions to build a safer world. This book series gathers review articles that analyze current agricultural issues and knowledge, then propose alternative solutions. It will therefore help all scientists, decision-makers, professors, farmers and politicians who wish to build a safe agriculture, energy and food system for future generations.

Vinod Kumar Yata
Ashok Kumar Mohanty • Eric Lichtfouse
Editors

Sustainable Agriculture Reviews 57

Animal Biotechnology for Livestock Production 2

Springer

Editors
Vinod Kumar Yata
Animal Biotechnology Centre
ICAR-National Dairy Research Institute
Karnal, India

Ashok Kumar Mohanty
ICAR-Indian Veterinary Research Institute
Mukteswar, India

Eric Lichtfouse
Aix-Marseille University, CNRS, IRD,
INRAE, Coll France, CEREGE
Aix-en-Provence, France

ISSN 2210-4410 ISSN 2210-4429 (electronic)
Sustainable Agriculture Reviews
ISBN 978-3-031-07495-0 ISBN 978-3-031-07496-7 (eBook)
https://doi.org/10.1007/978-3-031-07496-7

This Springer imprint is published by the registered company Springer Nature Switzerland AG
The registered company address is: Gewerbestrasse 11, 6330 Cham, Switzerland

Preface

This book is the second volume of *Animal Biotechnology in Livestock Production*, published in the book series entitled Sustainable Agriculture Reviews. Biotechnology has shown its impact in livestock production, and it will continue to be excel in coming years. This second volume of the book presents the essential concepts and in-depth analysis on animal reproduction and breeding methods.

Chapter 1 focusses the discussion on effects of sexual steroids on stress response and welfare of female ruminants. This chapter also discusses how the behavior and welfare of farm animals could be affected with the application of reproductive biotechnologies.

Chapter 2 provides valuable information on kidney diseases. This chapter includes the discussions on topics such as pathophysiology, molecular biomarkers, and proteomics of kidney diseases.

Chapter 3 covers the updated information on production of genetically modified pigs with the use of CRISPR/Cas9. This chapter focusses the discussion on the production of genetically modified pigs along with pros and cons.

Chapter 4 summarizes various types of anti-nutritional factors and their beneficial and deleterious effects on livestock. This chapter provides information on common factors such as enzymes and chemical compounds found in plant materials used for animal feed.

Chapter 5 summarizes the genetic engineering tools in livestock production. This chapter provides the updated information on biotechnological methods such as molecular gene cloning, diagnostics, vaccines, microarray, marker assisted selection (MAS) in animal breeding, genome editors, role of biotechnology in animal nutrition, artificial insemination, cloning and transgenic animals in livestock production, embryo transfer technology (ETT), and embryo sexing and sperm sexing.

Chapter 6 focuses the discussion on role of specific minerals in female animal reproduction. This chapter covers the discussion on biochemical, enzymatic, and endocrine actions of macromineral (calcium, phosphorus, and magnesium) and micromineral (copper, zinc, and manganese) ions along the hypothalamo-pituitary-ovarian axis.

Chapter 7 discusses genetic selection of livestock such as cattle, sheep, goat, buffalo, and poultry. This chapter also provides a brief overview on statistical models for genomic prediction and whole sequence data.

Dairy cattle production at ICAR-National Dairy Research Institute, India

This book serves as an important reference source for professionals and academicians working in the research area of livestock production. We would like to thank all the authors for their contribution and cooperation. We would like to thank the director of the Indian Council of Agricultural Research (ICAR)-National Dairy Research Institute (NDRI), Karnal, India, for providing institutional support. We would like to extend our thanks to the staff of Springer Nature for their support in publication of this book. We would like to acknowledge the Department of Biotechnology, Government of India, for providing financial support from "DBT-RA Program in Biotechnology & Life Sciences."

Karnal, India Vinod Kumar Yata

Mukteswar, India Ashok Kumar Mohanty

Aix-en-Provence, France Eric Lichtfouse

Contents

About the Editors

Vinod Kumar Yata is an interdisciplinary researcher working at the National Dairy Research Institute, Karnal, India. Previously, he worked as an assistant professor in the Department of Biotechnology, Dr. B R Ambedkar National Institute of Technology Jalandhar, Punjab, India. He received his PhD in biotechnology from the Indian Institute of Technology Guwahati. He specializes in interdisciplinary research which includes nanotechnology, microfluidics, animal biotechnology, cancer biology, and bioinformatics. He has developed a microfluidic device for the separation of live and motile spermatozoa form cattle semen samples. His research interests have been focused on the development of nanocarriers, understanding prodrug enzyme therapy, and targeted drug delivery. He elucidated the structural features and binding interactions of several biomolecules by in silico methods. Vinod has published four books as an editor and one book as an author with Springer Nature. He has published several research papers in peer-reviewed international journals and presented papers at several international conferences.

Ashok Kumar Mohanty is an eminent scientist in animal biotechnology and is currently serving as joint director of ICAR-Indian Veterinary Research Institute, Mukteswar, Uttarakhand, India. His group is involved in various basic and applied research related to animal production systems. His research group has made pioneering contributions to the field of animal biotechnology, with emphasis on gene cloning, expression and functional characterization of animal proteins, proteomics in animal production, cell and molecular biology, and structural biology of proteins. His group has developed a buffalo mammary epithelial cell line for the first time which can be used as a model system to understand lactation biology in animals as well as humans. His team has also developed a pregnancy diagnostic kit for the early detection of pregnancy in cattle and buffalo. His group is also extensively involved in developing low-cost technology for semen sexing in cattle. Ashok has organized a number of national and international workshops and international conferences. He is a recipient of several awards, including DBT Overseas Associateship by the Ministry of Science & Technology, Govt. of India., and Jawaharlal Nehru Award (gold medal) by the Indian Council of Agricultural Research (ICAR), New Delhi, for outstanding postgraduate research in the field of animal biotechnology. He is a fellow of the National Academy of Dairy Sciences (NADS), India, executive member of the Proteomics Society of India, and associate fellow of the National Academy of Veterinary Science, India. He has supervised more than 50 graduates, PhD students, and postdocs. Ashok has published more than 200 peer-reviewed research and review papers. He has also authored eight book chapters in the areas of animal and food biotechnology published by national and international publishers.

Eric Lichtfouse is Professor of Environmental Sciences and Scientific Writing at Aix Marseille University, Xi'an Jiaotong University, and the University of Shanghai for Science and Technology. He has invented carbon-13 dating, a molecular-level method allowing to study the dynamics of organic compounds in temporal pools of complex environmental media. He has discovered temporal pools of individual substances in complex media such as soils. He is chief editor and founder of the journal *Environmental Chemistry Letters* and the book series Sustainable Agriculture Reviews and Environmental Chemistry for a Sustainable World. He is the author of the book *Scientific Writing for Impact Factor Journals*. Eric has awards in analytical chemistry and scientific editing.

Contributors

Salma H. Abu Hafsa Livestock Research Department, Arid Lands Cultivation Research Institute, City of Scientific Research and Technological Applications, Alexandria, Egypt

Alberto Barbabosa-Pliego Facultad de Medicina Veterinaria y Zootecnia, Universidad Autónoma del Estado de México, Toluca, Mexico

Suresh H. Basagoudanavar ICAR-Indian Veterinary Research Institute, Bengaluru, Karnataka, India

Subhasmita Behera ICAR-Indian Veterinary Research Institute, Bengaluru, Karnataka, India

Garima Dadhich Pharmacology and Toxicology Laboratory, Dietetics and Nutrition Technology Division, CSIR-Institute of Himalayan Bioresource Technology, Palampur, India

Mona M. M. Y. Elghandour Facultad de Medicina Veterinaria y Zootecnia, Universidad Autónoma del Estado de México, Toluca, Mexico

Aline Freitas-de-Melo Departamento de Biociencias Veterinarias, Facultad de Veterinaria, Universidad de la República, Montevideo, Uruguay

J. Gadea Physiology of Reproduction Group, Department of Physiology, Universidad de Murcia, Murcia, Spain
Instituto Murciano de Investigación Biosanitaria, IMIB-Arrixaca-UMU, Murcia, Spain

P. S. P. Gupta ICAR-National Institute of Animal Nutrition and Physiology, Bangalore, India

Ayman A. Hassan Animal Production Research Institute, Agricultural Research Centre, Dokki, Gizza, Egypt

Mustafa Hassan Jan ICAR-Central Institute for Research on Buffaloes, Sub-Campus, Nabha, Punjab, India

Shikha Kapil University Institute of Biotechnology, Chandigarh University, Mohali, Punjab, India

Rajiv Kumar Academy of Scientific and Innovative Research (AcSIR), Ghaziabad, Uttar Pradesh, India
Biotechnology Division, CSIR-Institute of Himalayan Bioresource Technology, Palampur, India

Miguel Mellado Department of Animal Nutrition, Autonomous Agrarian University Antonio Narro, Saltillo, Mexico

S. Mondal ICAR-National Institute of Animal Nutrition and Physiology, Bangalore, India

S. Nandi ICAR-National Institute of Animal Nutrition and Physiology, Bangalore, India

S. Navarro-Serna Physiology of Reproduction Group, Department of Physiology, Universidad de Murcia, Murcia, Spain
Instituto Murciano de Investigación Biosanitaria, IMIB-Arrixaca-UMU, Murcia, Spain

J. Parrington Department of Pharmacology, University of Oxford, Oxford, UK

Vikram Patial Pharmacology and Toxicology Laboratory, Dietetics and Nutrition Technology Division, CSIR-Institute of Himalayan Bioresource Technology, Palampur, India
Academy of Scientific and Innovative Research (AcSIR), Ghaziabad, Uttar Pradesh, India

C. Piñeiro-Silva Physiology of Reproduction Group, Department of Physiology, Universidad de Murcia, Murcia, Spain

Madhu Ramesh JNCASR, Jakkur, Bengaluru, Karnataka, India

H. B. Ranjitha ICAR-Indian Veterinary Research Institute, Bengaluru, Karnataka, India

R. Romar Physiology of Reproduction Group, Department of Physiology, Universidad de Murcia, Murcia, Spain
Instituto Murciano de Investigación Biosanitaria, IMIB-Arrixaca-UMU, Murcia, Spain

Abdelfattah Z. M. Salem Facultad de Medicina Veterinaria y Zootecnia, Universidad Autónoma del Estado de México, Toluca, Mexico

Anjum Sherasiya Veterinary World, Wankaner, Gujarat, India

Harpreet Singh Lala Lajpat Rai University of Veterinary and Animal Sciences, Hisar, Haryana, India

S. K. Tripathi ICAR-National Institute of Animal Nutrition and Physiology, Bangalore, India

Rodolfo Ungerfeld Departamento de Biociencias Veterinarias, Facultad de Veterinaria, Universidad de la República, Montevideo, Uruguay

Dhanesh ValiyaValappil ICAR-Indian Veterinary Research Institute, Bengaluru, Karnataka, India

Chapter 1
Impact of Sex Steroids on the Stress Response and Welfare in Female Farm Ruminants

Aline Freitas-de-Melo and Rodolfo Ungerfeld

Abstract The concentrations of sex steroids modify the sensitivity to stressors in mammals: progesterone, reduce the stress response, while progesterone withdrawal increases it. Estrogens appear to increase the sensitivity to stressors, influencing the hypothalamic-pituitary-adrenal axis response. As the concentration of progesterone and estrogens differ according to the physiological status, such as pregnancy, anestrous or the phases of the estrous cycle, it may also modify the sensitivity to stressors. Therefore, it would be important to consider this information for various practices in farm animals as they are frequently subjected to stressful situations, including artificial weaning, shearing, isolation from the group, and transport. Moreover, pharmacological treatments with these hormones are commonly applied to cows, ewes, does, or buffaloes for using different biotechnologies, such as estrous synchronization, estrous induction, or synchronization of the ovulations. Accordingly, it is important to consider that those treatments might modify the sensitivity of females' farm ruminants to human handling.

The chapter summarizes how sexual steroid profiles might change female ruminants' stress response and welfare. The major points included are: a brief description of the natural variation in sex steroid profiles in female ruminants; the main hormones and administration pathways used in the application of reproductive biotechnologies; the different physiological indicators of stress and welfare considered in farm animals; and the main knowledge available on the effects of sex steroids on the stress responses of female ruminants to different farm practices. Lastly, we propose directions in which research is needed to understand better if stress, behavior, and welfare of female farm animals could be affected by the application of reproductive biotechnologies.

Keywords Progesterone · Progestagens · Estrogen · Progesterone withdrawal · Oestrous cycle · Hypothalamic-pituitary-adrenal axis · Glucocorticoids · Sheep · Cattle · Goat

A. Freitas-de-Melo (✉) · R. Ungerfeld
Departamento de Biociencias Veterinarias, Facultad de Veterinaria, Universidad de la República, Montevideo, Uruguay

V. K. Yata et al. (eds.), *Sustainable Agriculture Reviews 57*, Sustainable Agriculture Reviews 57, https://doi.org/10.1007/978-3-031-07496-7_1

Abbreviation

$GABA_A$ aminobutyric acid type A

1.1 Introduction

During the last years, hormonal treatments for reproductive biotechnologies in farm animals increased widely throughout the world. The development of fixed-timed artificial insemination treatments in cattle has a significant impact on the economy and employment of farm animal production (Mapletoft et al. 2018). Most of these treatments include administering sexual steroids, mainly progesterone or progestogens, and different estradiol formulas. Sex steroids are primarily related to reproductive functions, but these hormones also affect metabolism (Ashley et al. 2000; Kalkhoff 1982), immunological status (Bouman et al. 2005), functions in the central nervous system (Genazzani et al. 2002), and can also modulate the stress response (Sze and Brunton 2019). The effects of sex steroids on the stress response are of particular interest in farm animals as they might directly influence their welfare. However, these issues have been mainly studied in laboratory animals and humans, with a paucity of information on farm animals. In ewes, it was reported that the administration of progesterone reduces the stress response at weaning (Freitas-de-Melo et al. 2013), while progesterone withdrawal increases the secretion of cortisol after social isolation (Freitas-de-Melo et al. 2016). Ewes treated with estradiol benzoate respond to an ACTH challenge with greater cortisol concentrations than non-treated ewes (Van Lier et al. 2014).

The concentration of progesterone and estrogens differs according to the physiological status (pregnant, anestrous, or cycling females); thus, it might modify the sensitivity to stressors. Pregnant ewes have slighter behavioral responses than non-pregnant ewes to social isolation or a surprise effect (Viérin and Bouissou 2001) and shearing (Ungerfeld and Freitas-de-Melo 2019). Furthermore, the sensitivity to stressors is greater during the follicular than during the luteal phase (Pinto-Santini and Ungerfeld 2019; Freitas-de-Melo et al., 2022). The reactivity of cows also varies according to the reproductive status, with decreased reactions of pregnant cows to human handling (Freitas-de-Melo et al. 2019).

How female farm animals cope with stressful situations directly affects its' productive, reproductive, and welfare outcomes; thus, the reproductive status and the concentration of different sex steroids should be considered when different managements are applied in these animals. For example, it has been widely known that stress affects cow fertility (see review: Dobson and Smith 2000). However, it is not well studied if the application of hormonal treatments for reproductive biotechnologies might modify the easiness of animal handling, productive efficiency, or animals' welfare.

This chapter aimed to present a brief reminder on how profiles of sexual steroids vary according to the physiological status in female ruminants; the main hormones and administration pathways used in the application of reproductive biotechnologies; the main physiological indicators of stress and welfare; and the knowledge available on the effects of sex steroids during stressful situations in female ruminants. Lastly, we propose directions in which research is needed to understand better if the application of reproductive biotechnologies can affect the sensitivity to stressors, the behavioral responses, and the welfare of female farm animals.

1.2 Natural Variation in Sex Steroid Profiles in Female Ruminants

1.2.1 Estrous Cycle

Sexual steroid concentrations vary naturally throughout the reproductive lives of female ruminants. Before puberty, progesterone concentrations remain at basal levels, and estrogen concentrations may have some oscillations due to the physiological reproductive changes that determine puberty. However, estrogen concentration does not reach concentrations similar to those observed during regular estrous cycles. Since puberty, the estrous cycle is a set of reproductive events that repeats successively. While in sheep, it lasts an average of 17 days, in other ruminants, such as cow, goat, or buffalo, its' length is approximately 21 days. The estrous cycle can be divided into a luteal phase, which in ewes extends from day 2–3 (heat = day 0) of the cycle, to approximately day 13–14; and a follicular phase that goes from luteolysis that occurs on day 13–14 until day 2. In species with estrous cycles of 21 days, the general pattern is the same, maintaining the proportion of the duration of these two phases (Fig. 1.1).

The follicular phase extends from the regression of the corpus luteum to ovulation. During the follicular phase, the growth of the ovulatory follicle ends with the luteinizing hormone surge, ovulation, and the beginning of the follicle luteinization. The ovulatory follicle(s) secrete substantial concentrations of estrogens, which are responsible for triggering the increase in luteinizing hormone secretion, reaching the luteinizing hormone peak during the second half of heat. In turn, these estrogens are responsible for determining the receptive behavior in the female. In summary, estrogen concentrations are high during the follicular phase, increasing from the beginning of this phase until ovulation, when a rather abrupt decrease in concentration begins. The preovulatory luteinizing hormone peak triggers the ovulation of the preovulatory follicle(s), and luteinization of the remaining structure(s), with the subsequent formation of the corpus luteum. As the corpus luteum develops, the amounts of progesterone secreted by it increase. After ovulation, the weight of the corpus luteum increases associated with an important angiogenesis process and an increase in the size of the luteal cells (Smith et al. 1993). The newly formed corpus

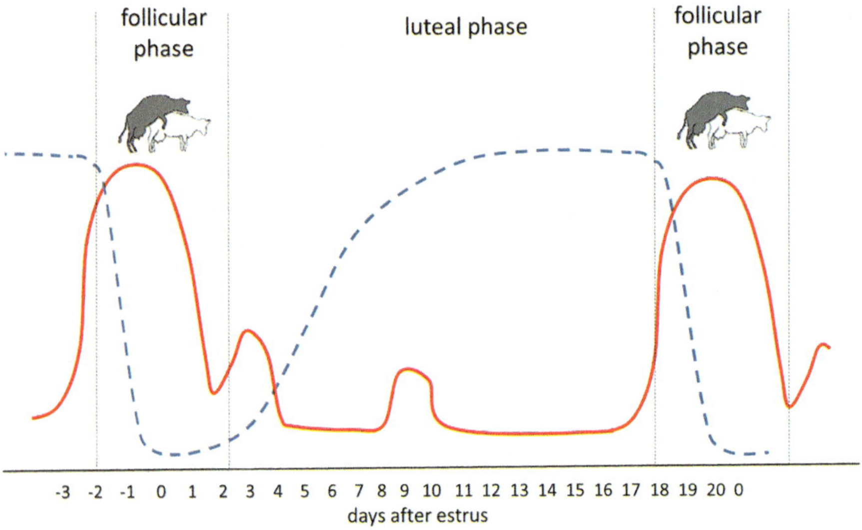

Fig. 1.1 An illustration with the follicular phase on the right and left and the luteal phase in the middle. The follicular phase has an image of cows mating. Progesterone (blue line) and estrogen (red line) profiles during an estrous cycle in cows. The luteal phase is characterized by high concentrations of progesterone and low concentration of estrogens, and during the follicular phase, the opposite hormonal pattern is observed. The line for progesterone is at its peak in the luteal phase and descends in the follicular phase and the one for estrogen is at its peak during the follicular phase and lowest during the luteal phase

luteum receives substantial amounts of blood flow in relation to its size. The production of progesterone increases more than ten times due to the increase of the enzymes that control the precursors of this steroidogenic pathway. In contrast, the production of androgens and estrogens decreases due to the enzymes' loss involved in their synthesis.

1.2.2 Seasonality

According to the species, the estrous cycles are repeated successively in the non-pregnant animal throughout the year in non-seasonal polyestric species, like cows, or only during part of the year, in seasonal polyestric species, like sheep and goat (Fig. 1.2). Although there are variations according to the breed and the latitude where sheep and goats live, the breeding season comprises successive estrous cycles occurring during summer and autumn. The period of the year in which there are no estrous cycles or ovulations is the seasonal anestrus, when progesterone concentration remains at baseline, at subluteal concentrations. During this period, there are small oscillations in the estrogen concentration associated with follicular development, but without reaching the level of a follicular phase.

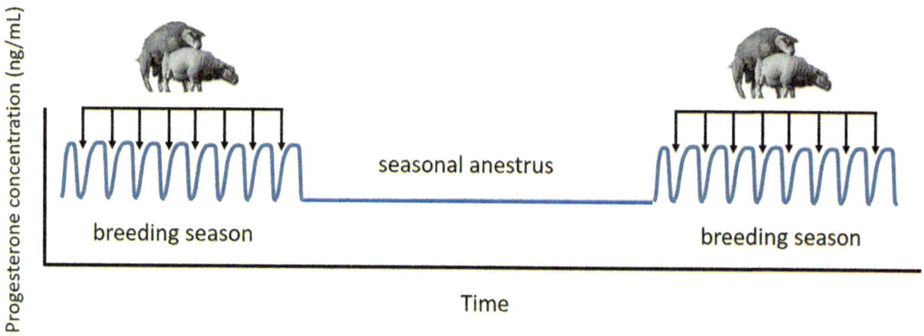

Fig. 1.2 Model of the seasonality of reproduction in sheep. Blue lines show progesterone concentrations, with increases in each estrous cycle. The black arrows indicate when heat occurs, and the estrogen concentration is more remarkable. A graph with time on the x-axis and progesterone concentration on the y-axis. At the beginning and end are a series of continuous peaks for the breeding season with an image of ewes mating, and in between these two series of peaks is a flat line for seasonal anestrus. Arrows are pointing to the dips in the peaks

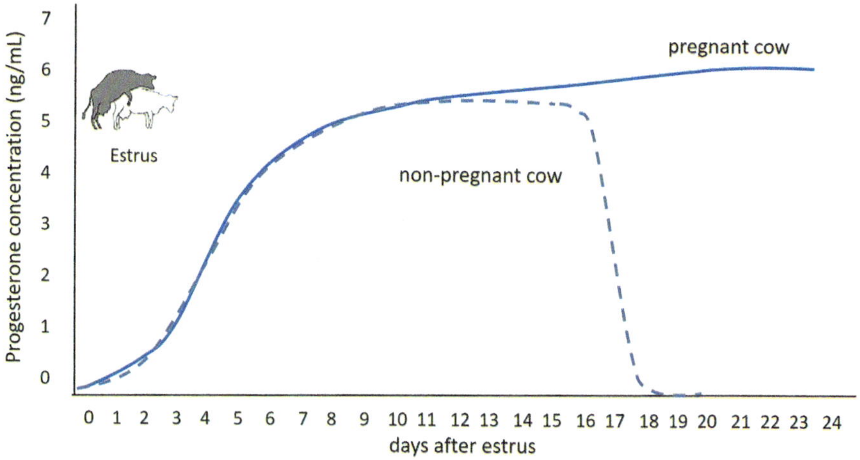

Fig. 1.3 Changes in progesterone concentrations during the estrous cycle (dotted line) or early gestation in cows (continuous line). A graph with days after estrus on the x-axis and progesterone concentration on the y-axis. An image of a cow mating with estrus written below is presented at beginning of the graph. The line for the pregnant cow has an ascending curve that later flatlines and the line for non-pregnant cow has the same ascending curve but dips completely at 16

1.2.3 Gestation and Postpartum Anestrus

Progesterone is the main hormone responsible for the maintenance of pregnancy, so during this period, luteolysis does not occur, maintaining active the corpus luteum (Fig. 1.3). After conception, the embryo produces chemical signals that prevent luteolysis, which in ruminants is mainly interferon tau. The progesterone produced by the corpus luteum is essential to maintain pregnancy during the first half of it in all species. In some species, such as sheep, the placenta also produces progesterone in amounts enough to maintain pregnancy from the mid-gestation until its end. In other species, e.g., cow and goat, the placenta also produces progesterone, but the amount produced is below the threshold necessary to maintain pregnancy *per se*, so progesterone produced by the corpus luteum is essential to maintain the pregnancy. In any case, progesterone concentrations remain elevated throughout pregnancy.

1.3 Exogenous Hormonal Control of Reproduction in Female Ruminants

The control of reproduction for the application of reproductive biotechnologies requires the exogenous administration of sexual steroids. Treatments for applying different biotechnologies, such as estrous synchronization, estrous induction, synchronization of the ovulations, or follicular development for superovulation, require controlling the estrous cycle and/or follicular development. There are several reviews on these topics (Bó et al. 2016; Bó and Baruselli 2014), so this section summarizes the different hormones and routes of administration used in farm ruminants. Table 1.1 summarizes information on the progestogens more frequently used in reproductive management in farm ruminants.

Table 1.1 Progestogens frequently used in reproductive management in farm ruminants

Species	Administration	Hormone	Treatment length
Cattle			
	Intravaginal device	Progesterone	5–7 days
	Solution	Progesterone	Single administration
	Subcutaneous implant	Norgestomet	5–7 days
Small ruminants			
	Intravaginal sponges	Medroxyprogesterone acetate	Short treatments:5–6 days Traditional treatments: 12–14 days in ewes, 15–16 days in does
		Flurogestone acetate	Short treatments:5–6 days Traditional treatments: 12–14 days in ewes, 15–16 days in does
	Intravaginal device	Progesterone	Short treatments:5–6 days Traditional treatments: 12–14 days in ewes, 15–16 days in does
	Solution	Progesterone	Single administration

1.3.1 Progestagens and Progesterone Treatments

The use of treatments based on mimicking a luteal phase requires the administration of sustained amounts of progesterone or progestagens. Progestagens are synthetic molecules with similar reproductive effects to progesterone. In cattle, sheep, and goats progesterone is administered by intravaginal silicone devices impregnated with synthetic progesterone, identical to that natural molecule produced by the corpus luteum. These devices are inserted and remain *in situ* for 5–7 days, depending on the protocol used. After the insertion of these devices, progesterone concentration achieved in the blood increases sharply, remaining at luteal concentrations throughout the treatment (Fig. 1.4). These devices are frequently used more than

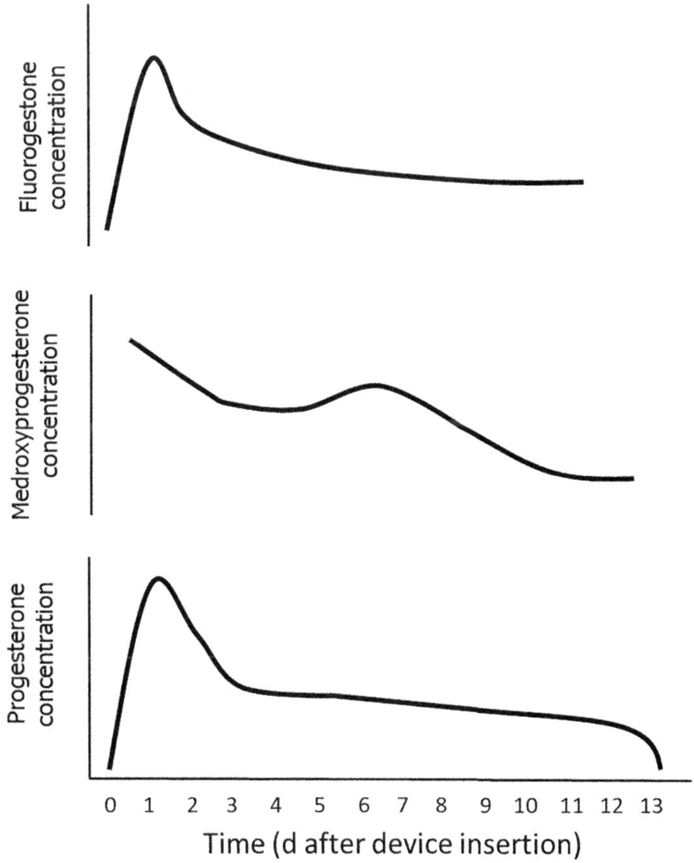

Fig. 1.4 Patterns of fluorogestone, medroxyprogesterone, and progesterone blood concentrations in small ruminants after the insertion of commercial devices. The pattern is based on Gaston-Parry et al. (1988), Greyling et al. (1994), and Rubianes et al. (1998). There are three line graphs with time on the x-axis and progestogen concentration on the y-axis. The graph on top has a steep increase and later dips a little and flatlines. The second graph begins from the top and descends with a small peak in between. The last one has a steep peak which dips a little and flatlines, and finally dips completely

once (Vilariño et al. 2011, 2013; Oliveira et al. 2001). Some authors autoclaved these devices before a second application, increasing progesterone concentration shortly after being inserted, achieving values greater than with used non-autoclaved devices (cattle: Zuluaga and Williams 2008; sheep: Ungerfeld et al. 2013; goats: Alvarez et al. 2013). However, in general, concentrations achieved with used devices are lower than those achieved with new devices (Fig. 1.5).

Similar devices are also used in sheep and goats, but traditional long treatments are frequently used, lasting 12–14 days in sheep and 14–16 days in goats. In this case, progesterone concentrations decrease to subluteal concentrations after 8–10 days of application (Fig. 1.4). During the last period of these treatments, progesterone concentrations are below to those observed during a normal late luteal phase. In these species, intravaginal sponges impregnated with synthetic progestagens, such as medroxyprogesterone and fluorogestone acetate, are more frequently used. There are commercial sponges with different quantities of progestagens, which in general exceed the amount needed for the desired reproductive effects (Ungerfeld et al. 2003). There is scarce information on how the concentrations of progestagens vary in blood after the insertion of the device. In all cases, progestagen concentration decreases throughout the treatment, but it is difficult to know its effects compared to progesterone, as progestogens have a longer half-life and potency. Concerning the main aim of this chapter, there is even less information

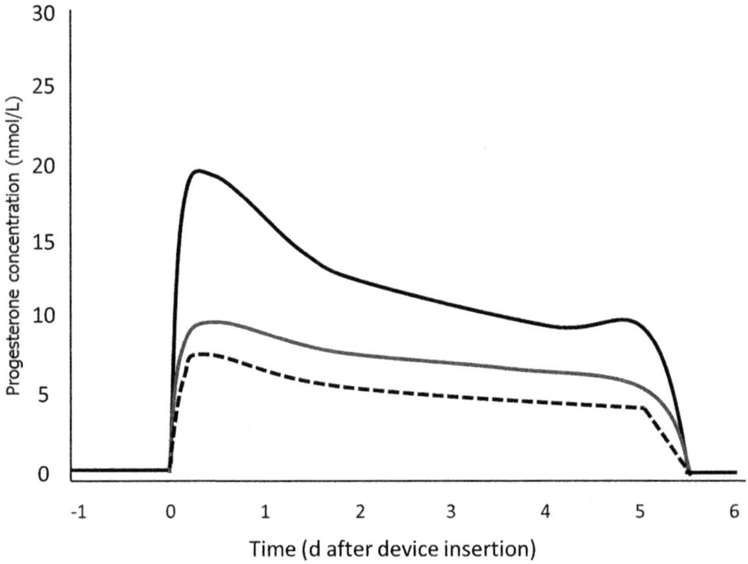

Fig. 1.5 Variations in progesterone concentrations achieved in the blood of goats after the insertion of a new intravaginal silicone device impregnated with progesterone (continuous black line), a device used once before (grey line), or a device used twice before (black-spotted line) (Redrawn from Vilariño et al. 2011). A line graph with the time of the x-axis and progesterone concentration on the y-axis. There are three lines in the graph, and all three of them have a steep peak that descends slowly and finally dips. The peak value of the first line is 20, and the lowest is around 7

about the possible impact of these hormones on the sensitivity to stressors in ewes and does.

Little is known on the effectiveness of injectable progesterone in oil-based formulas (cattle: Andrade et al. 2020; Morotti et al. 2018; Gimenes et al. 2009; goats: Álvarado-Espino et al. 2019a, b), which were developed to substitute the administration through devices. This strategy aims to reduce the hormonal residues in the environment, which can have deleterious effects on native fauna (Sauer et al. 2018; Liu et al. 2015). The effectiveness of injectable progesterone in oil-based is related to the formula used, determining when progesterone concentrations remain over luteal concentrations. The injectable progesterone might be impractical, as in most commercial products, the hormone should be administered several times. The scarce information available suggests that after a single administration of long-acting progesterone prepared in an oil solution decreases after 52 (Cavestany et al. 2008) to 96 h (Corréa Rocha et al. 2011) in cows, or 12–16 h in ewes (Ungerfeld and Freitas-de-Melo, unpublished data).

1.3.2 Estrogens

Although in some countries, the use of estrogens is not allowed, these hormones are widely used, mainly in treatments to control the follicular dynamics in cows. Estrogens are used in treatments to synchronize the ovulation associated with fixed-timed artificial insemination in cows. Although originally follicular wave was synchronized with estradiol-17β (Bó et al. 1994), currently the hormones more commonly used are estradiol benzoate and estradiol cypionate (Monteiro et al. 2015; Sales et al. 2012; Martínez et al. 2000). As it has low solubility in water, estradiol cypionate has a longer half-life than estradiol benzoate or estradiol-17β (Vynchier et al. 1990). In sheep and goats, there are scarce studies using estradiol esters. In some studies, estradiol-17β was initially used to determine if it might synchronize the follicular wave (Ungerfeld et al. 2004), as occurs in cattle. Later, it was used associated with progesterone treatments to synchronize the follicular dynamics for estrous induction (Ungerfeld 2008, 2009) or associated with superovulatory treatments (Souza-Fabjan et al. 2017). More recently, it has been used to synchronize the ovulation for timed artificial insemination in ewes and does (Cosentino et al. 2019). Estradiol benzoate has also been used to promote cervical relaxation in ewes before insemination or embryo collection (Hauschildt Dias et al. 2020).

1.4 Stress Response and Welfare

According to Broom (1986), the welfare of an animal is its state regarding its attempts to cope with its environment. Welfare includes the coordination of different responses, including physiological, behavioral, and immunological responses, and therefore, cannot be evaluated without including several aspects. In livestock cattle, sheep, and goats, females are repeatedly subjected to different challenges, which include human handlings, feeding management, social influences, and the occurrence of pathogens (Broom and Fraser 2015). All these challenges generate physiological, behavioral and immunological responses (Dobson and Smith 2000). As stress implies the modification of the homeostasis of the animal, the coordination of different systems is necessary to return the animal physiology to equilibrium. The responses to the same stressors – considered as any stimulus that triggers a stress response- might differ among different individuals, both in magnitude and duration, depending on their previous experience with that stressor and their temperament (Earley et al. 2010; McEwen and Wingfield 2003). However, other factors such as gender, age, or physiological reproductive status modulate the response to stressors (Freitas-de-Melo and Ungerfeld 2016a; Mormedè et al. 2007; Dallman et al. 2002).

Depending on the intensity of the stressor and the lapse during which it acts, the response can be classified as acute or chronic. When the stressor induces a short response enough to get back to the homeostatic status, it is classified as acute. However, if the stressor continues acting for longer or triggers an intense response with which the animal cannot cope, this is considered a chronic stress response.

1.4.1 Acute Stress Response

There are many different types of acute stressors, but in general, it is assumed that an acute stressor acts from minutes to hours, and the animal should cope with it, returning quickly to the status that it had before the action of the stressor. Acute stressors simultaneously trigger both responding pathways, including the sympathetic autonomic nervous system and the neuroendocrine system (Carrasco and Van de Kar 2003; O'Connor et al. 2021). The sympathetic system is activated immediately after a stressful situation, displaying the main responses a few seconds or even minutes after the perception of the stressor (Charmandari et al. 2005). However, its' action is of short duration, ending quickly. On the other hand, although the neuroendocrine response takes longer to display the main effects, it is maintained during extended periods, from some minutes to hours. Both systems act simultaneously and synergistically, as glucocorticoids are synthesized and released into the blood; this pathway requires more time to be evidenced (Carrasco and Van de Kar 2003; Matteri et al. 2000). The activation of both responses leads to physiological and behavioral changes necessary to cope with the stressor (Mormedè et al. 2007).

The activation of the sympathetic autonomic nervous system provokes the release of catecholamines into the sympathetic synapses, including epinephrine and norepinephrine. The chromaffin cells located in the medulla of the adrenal glands produce both catecholamines and release them into the bloodstream. In general, catecholamines increase the heart and respiratory rates, as well as the body temperature. Catecholamines also stimulate glycogenolysis and lipolysis, increasing glycemia and energy availability. The activation of the sympathetic autonomic nervous system also causes mydriasis, vasodilation in skeletal muscle and peripheral vasoconstriction (Matteri et al. 2000). Therefore, the activation of the sympathetic autonomic nervous system produces a blood redistribution to prioritize the perfusion of the brain, heart, and skeletal muscles. Catecholamines also modify the activity of the animal, increasing its' alertness, vigilance, and animal excitement (Sabban 2010).

When the animal perceives the stressor, the hypothalamus releases corticotrophin-releasing hormone and vasopressin into the portal system. Corticotrophin-releasing hormone acts at the adenohypophysis, stimulating the release of the adrenocorticotrophic hormone into the bloodstream (Matteri et al. 2000). The adrenocorticotrophic hormone stimulates the secretion of glucocorticoids, mainly cortisol in domestic ruminants and corticosterone in rodents, in the cortex of the adrenal gland (Matteri et al. 2000). Glucocorticoids stimulate gluconeogenesis, lipolysis, and the catabolism of proteins, increasing glycemia and energy availability (Kudielka and Kirschbaum 2007). Glucocorticoids also promote an increase in cerebral perfusion, the use of glucose, an increase of blood pressure in most blood vessels, and stimulate heart frequency and cardiac output (Sapolsky et al. 2000). Glucocorticoids also act on the immune system, causing lymphopenia, eosinopenia, and neutrophilia (Griffin 1989). The glucocorticoids exert a negative feedback at the hypothalamus and the pituitary gland, inhibiting the secretion of corticotrophin-releasing hormone, vasopressin, and adrenocorticotrophic hormone (Sabban 2010; McEwen 2007). Therefore, even if the stressor continues acting longer, the high concentrations of glucocorticoids are not maintained, so the decrease in its concentration does not necessarily indicate that the stressful situation ended.

1.4.2 Chronic Stress Response

A chronic stress response is a consequence of a continuous or intermittent but repeated exposure of an animal to stressors during several days (Pacák and Palkovits 2001). Chronic stress response frequently occurs in farm animals, as the general allocating conditions, pathologies, or other types of stressors usually remain active for days or weeks. As it might be expected, a sustained action of glucocorticoids over several days has harmful consequences on the productive and reproductive results, affecting animal health and welfare. If the animal cannot cope with the chronic stressor, it might have several negative consequences, including the animal's death. One consequence of repeated exposure to stressors is the recurrent increase in glucocorticoid concentrations, which negatively affect the normal function of the immune system. Therefore, a chronic stress response might cause immunosuppression, making the animal more susceptible to diseases (Griffin 1989). In this situation, animals lose the appetite, decrease the food conversion, and at the same time, prioritize the catabolic pathways, negatively affecting its' general body condition. Frequently, chronically stressed animals are also more anxious, nervous, and restless, display stereotypic movements, and reduce their sexual behavior (Maniam and Morris 2012; Sapolsky et al. 2000).

Although it is assumed that the stress response is necessary for the animal to cope with stressors, facilitating its adaptation, if the stressor is maintained over time, or the intensity of its effect generates an intense or sustained response, the final result can be harmful to the animal. Therefore, if the reactions displayed by the animal are not enough to recover the homeostasis or the action of the stressor does not stop, the general status of the animal may continuously deteriorate until death. In many husbandry conditions, the stress response cannot modify the stressful situation (Moberg 2001), Therefore, the animal can reestablish its homeostasis only when the stressor disappears.

Some common chronic stressful situations in farm animals are early artificial weaning, transport, and shearing or mixing animals from different groups. Artificial weaning involves several stressors, as the mother-offspring bond is suddenly broken, the physical and social environment is new for the animal, and the animal has to adapt its feeding behavior and digestive processes to the replacement of milk by solid food (Freitas-de-Melo et al. 2022; Freitas-de-Melo and Ungerfeld 2016b). During transportation, the animals are allocated in trucks with high density, provoking injuries and subjecting them to prolonged fasting. Shearing is another stressful situation, as animals are moved from their paddocks, usually by dogs or unknown humans, and located in smaller pens close to the shearing shed. In this condition, they are exposed to high noise levels produced by the shearing machine. During the process, those animals entering to shearing are withdrawn from the group by humans, which implies taking out other group members, which also is a stressful situation for those animals. After shearing, mixing recently shorn animals with those still waiting to be shorn implies the introduction of stressed animals, who are also not easily recognized by non-shorn animals (Ungerfeld et al. 2018). After winter shearing, the stressful situation remains longer with the increase in thermoregulatory demands (Carcangiu et al. 2008; Hargreaves and Huston 1990a, b). Other examples are changes in the social environment, as happens after social grouping when the animals are forced to cohabit with previously unknown individuals (Giriboni et al. 2015), or after social isolation in gregarious species as farm ruminants (Freitas-de-Melo et al. 2016).

In summary, during a chronic stress response, the physiological and behavioral responses are costly with harmful consequences for the animals as the responses cannot modify the cause of the stress. Therefore, it is essential to reduce how the animal perceives the stressor and its response to improve animal welfare.

1.5 Influence Sex Steroids on Stress Response and Animal Welfare

Several studies demonstrate that the concentration of sexual steroids modifies how an animal perceives a stressor and the endocrine stress response itself. In general, it is assumed that progestogens reduce the stress response, inducing calm and relaxed states, while acute progesterone withdrawal and estrogens increase the responses to stressors (Freitas-de-Melo and Ungerfeld 2016a). There are two main metabolites derived from progesterone, allopregnanolone and pregnanolone, which are called neuroactive metabolites, and are responsible for reducing the stress response (Liang and Rasmusson 2018; Barbaccia et al. 2001). These neuroactive metabolites lessen the animal's perception of the stressor, as they present an anxiolytic effect (Wang 2011), and they also act decreasing the response of the hypothalamic-pituitary-adrenal axis (Brunton et al. 2009; Patchev et al. 1996). All these mechanisms reduce the animal's behavioral and physiological stress response (Bitran et al. 1993, 1995). The acute decrease of progesterone concentrations, as occurs immediately after luteolysis or after the withdrawal of progesterone intravaginal devices, increases anxiety and the sensitivity to stressors (Hantsoo and Epperson 2020; Smith et al. 2007). Furthermore, estrogens also increase anxiety and endocrine stress response in rodents (Figueiredo et al. 2007; Jasnow et al. 2006; Morgan and Pfaff 2002). Surprisingly, the effects of the physiological reproductive status and hormonal treatments on stress response are commonly not considered when female farm animals are subjected to stressful situations.

1.5.1 Mechanisms of Action of Progesterone, Progestagens, and Estrogens

Progestagens and estrogens act through genomic mechanisms, modifying the expression of specific genes, stimulating or inhibiting gene transcription, and synthesizing specific proteins (Schumacher et al. 1999). The effect of this pathway requires from minutes to days to be noted (McEwen 1991). Progesterone can also provoke short-term effects, acting through its' neuroactive metabolites. Briefly, progesterone is metabolized mainly in the liver (Pluchino et al. 2009) to various compounds, including dihydro-progesterone, allopregnanolone, and pregnanolone (Corpechot et al. 1993; Seamark et al. 1969). These compounds can cross the blood-brain barrier (Pluchino et al. 2009). Furthermore, both progesterone and dihydro-progesterone can also be metabolized in the brain in allopregnanolone and pregnanolone (Compagnone and Mellon 2000). Therefore, these neuroactive metabolites can arrive from the blood to the central nervous system, or be produced *in situ* (Sze and Brunton 2019), achieving high concentrations in the brain (Paul and Purdy 1992).

Allopregnanolone and pregnanolone can bind to the aminobutyric acid type A ($GABA_A$) and glycine (Gunn et al. 2011; Sarkar et al. 2011; Jiang et al. 2006), which are both brain's inhibitory receptors. Allopregnanolone and pregnanolone can also inhibit excitatory pathways through the nicotine, serotonin, and glutamate receptors (Sedláček et al. 2008; Kaura et al. 2007; Bullock et al. 1997). These neuroactive metabolites are $GABA_A$ positive allosteric modulators, reducing neuronal excitability (Paul et al. 2020; Lambert et al. 2009). The union of these neuroactive metabolites on $GABA_A$ receptor induces quickly anxiolytic, sedative, and analgesic effects (Akk et al. 2007; Wang 2011), exerting a stress-protective influence (Brunton et al. 2009; Ma et al. 2005). Allopregnanolone also decreases the synthesis of mRNA for corticotropin-releasing hormone and the secretion of both adrenocorticotrophic hormone and glucocorticoids in male rats (Brunton et al. 2009). The concentration of allopregnanolone increases along gestation in rats, significantly decreasing just before delivery (Concas et al. 1998). The increase in the concentration of allopregnanolone might be related to the availability of progesterone as a substrate and the increase in the concentration of enzymes that synthesize allopregnanolone (Brunton et al. 2005).

Although it is not wholly well established in the literature, the reduction of stress response could be mediated by the action of progesterone on intracellular receptors. Some studies using mice knock-out for the intracellular receptors of progesterone reported a decline in the stress or anxiety responses, suggesting that progesterone also has a stress-protective effect throughout its receptors (Reddy et al. 2005). However, when rats were treated with progesterone and a progestin receptor antagonist before applying a stressor, the reduction in the stress response was still observed (Bitran et al. 1995). In female ovariectomized rats, the treatment with medroxyprogesterone, a progestin that does not produce the neuroactive metabolites, also reduces the stress response during restrainment (Hassell et al. 2011). It has been suggested that progestagens, progesterone, or dihydro-progesterone could directly join the intracellular receptors located in the amygdala and the bed nucleus of the stria terminalis (Brinton et al. 2008). These brain regions are related to the stress response, fear, or anxiety (Walker et al. 2003), modulating the stress response. Progesterone and medroxyprogesterone differ in their effects on the expression of a subunit of $GABA_A$ receptors in the hippocampus, which have important implications for the modulation of anxiety (Pazol et al. 2009), and in the effects of both hormones in cognition in rats (Frye et al. 2013). Overall, progestagens and the neuroactive metabolites of progesterone may reduce the perception of stressors, and thus, the endocrine stress response.

In contrast, the acute decreases in progesterone concentration and its neuroactive metabolites after chronic exposure can regulate the expression of a specific subunit of $GABA_A$ receptor (Smith et al. 1998, 2007). Progesterone withdrawal leads to an increase in the display of behaviors related to anxiety (Gulinello et al. 2002; Gallo and Smith 1993) and individual risk-taking in rats (Löfgren et al. 2006). In the same direction, the treatment with estradiol benzoate enhances anxiety behavior in three anxiogenic behavior tests in female mice (Morgan and Pfaff 2002). Furthermore, estrogens increase the expression of corticotrophin-releasing hormone mRNA in the central nucleus of the amygdala (Jasnow et al. 2006), and increase the glucocorticoid response to an acute stressor in rodents (Figueiredo et al. 2007; Burgess and Handa 1992).

1.5.2 Action of Sexual Steroids in the Stress Response in Ruminants

There are few studies in ruminants regarding the effects of sex steroids on the stress response, and even a few studies related to the possible mechanisms of action. The concentration of allopregnanolone in the cerebrospinal fluid increases during gestation, reaching a maximum concentration in the last month and decreasing during the first month of lactation (Misztal et al. 2020a). Misztal et al. (2020b) reported that administration of allopregnanolone reduces corticotrophin-releasing hormone and vasopressin mRNA expressions in the paraventricular nucleus of isolated and restrained ewes. Consequently, administration of allopregnanolone to those ewes also reduces the corticotrophin-releasing hormone, adrenocorticotrophic hormone, and cortisol concentrations (Misztal et al. 2020b). Furthermore, there is another possible pathway for the allopregnanolone anxiolytic and sedative effects in this species, as at least in sheep central nervous system-isolated tissue, *in vitro* allopregnanolone binds to the $GABA_A$ receptor (Crossley et al. 2000).

In rodents, the sensitivity to stressors decreases by the end of pregnancy according to the progesterone or allopregnanolone availability (Brunton 2010; Douglas et al. 2005). Although there are scarce studies, pregnant ewes also decrease their sensitivity to stressors. In effect, pregnant ewes have slighter behavioral responses than non-pregnant ewes to social isolation or a surprise test (Viérin and Bouissou 2001) and shearing (Ungerfeld and Freitas-de-Melo 2019). Pregnant ewes also respond with lower cortisol concentrations than non-pregnant ewes to the shearing handling (Ungerfeld and Freitas-de-Melo 2019). Progesterone concentration was negatively related to the intensity of behavioral stress responses (Viérin and Bouissou 2001), suggesting that the inhibitory effect is associated with the progesterone availability. The sensitivity of cows also varies according to the reproductive status, with a lower reaction of pregnant cows to human handling (Freitas-de-Melo et al. 2019).

Recently we observed that estrous ewes stayed more time standing up immobile and alert during social isolation than ewes in their luteal phase (Freitas-de-Melo et al. 2022). Pinto-Santini and Ungerfeld (2019) observed that ewes were probably more sensitive to stressors during the follicular than during the luteal phase, as while during the first there was a clear circadian pattern in cortisol secretion, with a rise during the early morning, but this pattern disappeared during the mid-luteal phase. Furthermore, the surface temperature, total serum protein, globulin, and plasma glucose concentrations were greater in ewes in the follicular phase than in ewes in the luteal phase (Freitas-de-Melo et al. 2022; Pinto-Santini and Ungerfeld 2019). Nevertheless, Kilgour and Szantar-Coddington (1997) did not find any differences between the behavioral response of estrous and non-estrous ewes during social isolation. Furthermore, Orihuela et al. (2002) did not observe differences in the cortisol concentration in diestrous or proestrus ewes after transportation.

The effect of progesterone administration to evaluate ruminants' response to different stressors was studied in some experiments. In particular, long-term treatment with progesterone reduces ewes' responses to weaning (Freitas-de-Melo et al. 2013). Ewes treated with intravaginal devices impregnated with progesterone displayed lesser behavioral changes after weaning. After artificial weaning, treated ewes also present lower globulins concentration than untreated ewes, suggesting a possible protective effect of the treatment. Therefore, to simplify the potential handling of animals, we administered an injection of oil-based progesterone to ewes immediately before weaning, and although it decreased the behavioral response (Freitas-de-Melo and Ungerfeld, unpublished data), the decrease was not as strong as that previously reported with longer treatments (Freitas-de-Melo et al. 2013). On the other hand, the acute withdrawn of progesterone concentrations increase the sensitivity to stressors. In effect, anestrous ewes secrete more cortisol after being isolated 24 h after ending the treatment with progesterone than remaining untreated (Freitas-de-Melo et al. 2016). Similarly, heifer response to human handling is greater after progesterone withdrawal than in untreated animals (Freitas-de-Melo et al. 2019). Although the effects of estrogens on stress response in ruminants were studied to a lesser degree, ewes treated with estradiol benzoate respond to an ACTH challenge with greater cortisol concentrations than untreated ewes (Van Lier et al. 2014).

In this context, although more research is required, there is room to consider treatments with progesterone as practical alternatives to reduce the stress responses to routine handlings in ruminants, thereby improving their well-being. The need for long-term treatments appears as a significant limitation, as it implies the use of intravaginal devices to ensure sustained progesterone concentrations during several days. However, new long-action progesterone formulas might be tested to avoid using devices, which are not allowed in some countries due to the residues in the environment. Table 1.2 summarizes the main effects of progesterone or the reproductive physiological state on the stress response in female ruminants.

1.6 Conclusions

According to our knowledge, there are no studies on how the sensitivity of the animals to stressors is affected by administering sexual steroids during standard reproductive practices. Moreover, although the effectiveness of administering different progestogens (Ungerfeld et al. 1999) or estradiol esters (Melo et al. 2016) on the reproductive responses have been compared, there are no studies relating the possible effects of these treatments with the stress responses. For example, in some studies, it has been reported that the pregnancy rate of ewes is greater when estrus is synchronized with progesterone than with medroxyprogesterone (Santos-Neto et al. 2015). The greater pregnancy rate was related to the characteristics of the intravaginal device or its' effects on follicular dynamics. However, the insertion and withdrawal of the device imply repeatedly moving the animals, taking them to the farm facilities, contacting unknown technicians, and sometimes grouping with other unknown individuals, which are all stressful events. In this sense, as progestagens

Table 1.2 Effects of progesterone or the reproductive physiological state on the stress response in female ruminants

Model	Treatment	Physiological state	Stressor	Physiological response	Behavioral response	References
Ewe	–	Pregnancy	Shearing	↓Cortisol	↓Walking ↑Standing ↑Grazing	Ungerfeld and Freitas-de-Melo (2019)
Ewe	–	Prengancy	Isolation, surprise and presence of a human	–	↓Walking ↓Immobilization ↓Squares crossed ↓Reared against the wall and pawed ↑Time in the central area of the test pen ↓Time near the entrance doors ↑eating	Viérin and Bouissou (2001)
Cow	–	Pregnancy	Human handling and presence	–	↓Flight distance ↓Exit velocity	Freitas-de-Melo et al. (2019)
Ewe	Intravaginal progesterone releasing device for 32 days	Anestrus	Artificial weaning	↑Total protein ↑Globulin	↓Vocalization ↓Pacing	Freitas-de-Melo et al. (2013)
Ewe	Oil-based progesterone immediately before weaning	Anestrus	Artificial weaning	↓ Cortisol	↓Pacing ↓Walking ↑Rumination	Ungerfeld and Freitas-de-Melo (unpublished data)
Heifers	Intravaginal progesterone releasing device for 5 days	Anestrus	Human handling and presence	–	↓Flight distance	Freitas-de-Melo et al. (2019)
Ewes	Intravaginal progesterone releasing device for 14 days + withdrawal	Anestrus	Social isolation	↑Cortisol	No difference was found	Freitas-de-Melo et al. (2016)

(continued)

Table 1.2 (continued)

Model	Treatment	Physiological state	Stressor	Physiological response	Behavioral response	References
Heifers	Intravaginal progesterone releasing device for 5 days + withdrawal	Anestrus	Human handling and presence	–	↑Flight distance	Freitas-de-Melo et al. (2019)
Ewe	–	Follicular vs luteal phase	Social isolation	No difference was found	Tended to ↑vocalization ↑immobilization during the follicular phase	Freitas-de-Melo et al. (2022)
Ewe	–	Follicular vs luteal phase	No stressor	↑cortisol during the follicular phase	–	Pinto-Santini and Ungerfeld (2019)
Ewe		Diestrous vs proestrus	Transportation	No difference was found	–	Orihuela et al. (2002)
Ewe	–	Estrous vs non-estrous	Social isolation	–	No difference was found	Kilgour and Szantar-Coddington (1997)
Ewe	Estradiol benzoate	Anestrus	ACTH challenge	↑cortisol	–	Van Lier et al. (2014)

as medroxyprogesterone are not metabolized to the neuroactive steroids (Pluchino et al. 2009; Bernardi et al. 2006), it is possible that the effects of these hormones on the reduction of stress response also differ, and this might partially explain differences in fertility.

Therefore, considering that the devices are withdrawn close to the moment of artificial insemination, a period in which stress should be avoided to increase the pregnancy rates, the fertility might differ due to differences in the effects of different hormones in the sensitivity to practical handlings. Similarly, different estradiol esters have been used to modify the follicular growth pattern, and thus, the moment in which timed artificial insemination is performed in cattle. Estradiol benzoate is widely used instead of estradiol cypionate, although some studies were done expecting that the last would increase fertility due to its effects at the hypothalamus-pituitary-ovary axis, related to the longer half-life (Melo et al. 2016; Sales et al. 2012). However, the use of estradiol cypionate did not increase the pregnancy rates. Therefore, it should be studied if the sustained concentrations of estradiol during a longer time before the moment in which the cows are grouped, moved to the facilities, handled by the technicians, and inseminated, increase the sensitivity to these stressors, and this might partially explain the lower fertility rate.

It is known that individual temperament affects the results of these treatments (Mello et al. 2020), and it is also known that cows are more responsive to tests used to evaluate temperament after progesterone withdrawal (Freitas-de-Melo et al. 2019), raising the hypothesis that the estradiol ester used might affect the cow perception of stressors, and therefore, their fertility. These examples demonstrate a big room to work in the link between these two types of research areas, which have scarce contribution between them until now. Understanding the animal as an indivisible individual, in which a hormone administrated to produce the desired reproductive effect and also influences other systems, might be the main direction to improve the animals' welfare and results of the reproductive biotechnologies.

References

Akk G, Covey DF, Evers AS, Steinbach JH, Zorumski CF, Mennerick S (2007) Mechanisms of neurosteroid interactions with $GABA_A$ receptors. Pharmacol Ther 116:35–57. https://doi.org/10.1016/j.pharmthera.2007.03.004

Alvarado-Espino AS, Menchaca A, Meza-Herrera CA, Carrillo-Moreno DI, Zúñiga-García S, Arellano-Rodríguez F, Mellado M, Véliz FG (2019a) Ovarian response is not affected by the stage of seasonal anestrus or breed of goats when using a progesterone injection plus human chorionic gonadotropin-based protocol. Anim Reprod Sci 204:60–65. https://doi.org/10.1016/j.anireprosci.2019.03.005

Alvarado-Espino AS, Menchaca A, Meza-Herrera CA, Mellado M, Arellano F, Véliz F (2019b) Use of injectable progesterone and hCG for fixed-time artificial insemination during the non-breeding season in goats. Theriogenology 127:21–25. https://doi.org/10.1016/j.theriogenology.2018.12.035

Alvarez L, Gamboa D, Zarco L, Ungerfeld R (2013) Response to the buck effect in goats primed with CIDRs, previously used CIDRs, or previously used autoclaved CIDRs during the non-breeding season. Livest Sci 155:459–462. https://doi.org/10.1016/j.livsci.2013.05.010

Andrade JPN, Gomez-León V, Andrade FS, Carvalho BP, Lacouth KL, Garcia FZ, Jacob JCF, Sales JNS, Wiltbank MC, Mello MRB (2020) Development of a novel 21-day reinsemination program, ReBreed21, in *Bos indicus* heifers. Theriogenology 155:125–131. https://doi.org/10.1016/j.theriogenology.2020.04.021

Ashley CD, Kramer ML, Bishop P (2000) Estrogen and substrate metabolism. Sports Med 29:221–227. https://doi.org/10.2165/00007256-200029040-00001

Barbaccia ML, Serra M, Purdy RH, Biggio G (2001) Stress and neuroactive steroids. Int Rev Neurobiol 46:243–272. https://doi.org/10.1016/S0074-7742(01)46065-X

Bernardi F, Pluchino N, Pieri M, Begliuomini S, Lenzi E, Puccetti S, Casarosa E, Luisi M, Genazzani AR (2006) Progesterone and medroxyprogesterone acetate effects on central and peripheral allopregnanolone and beta-endorphin levels. Neuroendocrinology 83:348–359. https://doi.org/10.1159/000095400

Bitran D, Purdy RH, Kellogg CK (1993) Anxiolytic effect of progesterone is associated with increases in cortical allopregnanolone and $GABA_A$ receptor function. Pharmacol Biochem Behav 45:423–428. https://doi.org/10.1016/0091-3057(93)90260-z

Bitran D, Shiekh M, McLeod M (1995) Anxiolytic effect of progesterone is mediated by the neurosteroid allopregnanolone at brain $GABA_A$ receptors. J Neuroendocrinol 3:171–177. https://doi.org/10.1111/j.1365-2826.1995.tb00744.x

Bó GA, Baruselli PS (2014) Synchronization of ovulation and fixed-time artificial insemination in beef cattle. Animal 8:144–150. https://doi.org/10.1017/S1751731114000822

Bó GA, Adams GP, Pierson RA, Tribulo HE, Caccia M, Mapletoft RJ (1994) Follicular wave dynamics after estradiol-17β treatment of heifers with or without a progestogen implant. Theriogenology 41:1555–1559. https://doi.org/10.1016/0093-691X(94)90821-Y

Bó GA, de la Mata JJ, Baruselli PS, Menchaca A (2016) Alternative programs for synchronizing and resynchronizing ovulation in beef cattle. Theriogenology 86:388–396. https://doi.org/10.1016/j.theriogenology.2016.04.053

Bouman A, Heineman MJ, Faas MM (2005) Sex hormones and the immune response in humans. Hum Reprod Update 11:411–423. https://doi.org/10.1093/humupd/dmi008

Brinton RD, Thompson RF, Foy MR, Baudry M, Wang J, Finch CE, Morgan TE, Pike CJ, Mack WJ, Stanczyk FZ, Nilsen J (2008) Progesterone receptors: form and function in brain. Front Neuroendocrinol 29:313–339. https://doi.org/10.1016/j.yfrne.2008.02.001

Broom DM (1986) Indicators of poor welfare. Br Vet J 142:524–526. https://doi.org/10.1016/0007-1935(86)90109-0

Broom DM, Fraser AF (2015) Domestic animal behaviour and welfare, 5th edn. CABI Publishing, Wallingford

Brunton PJ (2010) Resetting the dynamic range of hypothalamo–pituitary–adrenal axis stress responses through pregnancy. J Neuroendocrinol 22:1198–1213. https://doi.org/10.1111/j.1365-2826.2010.02067.x

Brunton PJ, Meddle SL, Ma S, Ochedalski T, Douglas AJ, Russell JA (2005) Endogenous opioids and attenuated hypothalamic–pituitary–adrenal axis responses to immune challenge in pregnant rats. J Neurosci 25:5117–5126. https://doi.org/10.1523/JNEUROSCI.0866-05.2005

Brunton PJ, McKay AJ, Ochedalski T, Piastowska A, Rebas E, Lachowicz A, Russell JA (2009) Central opioid inhibition of neuroendocrine stress responses in pregnancy in the rat is induced by the neurosteroid allopregnanolone. J Neurosci 29:6449–6460. https://doi.org/10.1523/JNEUROSCI.0708-09.2009

Bullock AE, Clark AL, Grady SR, Robinson SF, Slobe BS, Marks MJ, Collins AC (1997) Neurosteroids modulate nicotinic receptor function in mouse striatal and thalamic synaptosomes. J Neurochem 68:2412–2423. https://doi.org/10.1046/j.1471-4159.1997.68062412.x

Burgess LH, Handa RJ (1992) Chronic estrogen-induced alterations in adrenocorticotropin and corticosterone secretion, andglucocorticoid receptor-mediated functions in female rats. Endocrinology 131:1261–1269. https://doi.org/10.1210/endo.131.3.1324155

Carcangiu V, Vacca GM, Parmeggiani A, Mura MC, Pazzola M, Dettori ML, Bini PP (2008) The effect of shearing procedures on blood levels of growth hormone, cortisol and other stress haematochemical parameters in Sarda sheep. Animal 2:606–612. https://doi.org/10.1017/S1751731108001651

Carrasco GA, Van de Kar LD (2003) Neuroendocrine pharmacology of stress. Eur J Pharmacol 463:235–272. https://doi.org/10.1016/S0014-2999(03)01285-8

Cavestany D, Fernández D, Salazar E, Sánchez A, Leyton L, Crespi D (2008) Determinación de niveles de progesterona en sangre luego de la administración parenteral de progesterona en vacas Holando ovariectomizadas o ciclando. In: Jornadas Uruguayas de Buiatría, XXXVI, Paysandú, Uruguay, pp 218–219

Charmandari E, Tsigos C, Chrousos G (2005) Endocrinology of the stress response. Annu Rev Physiol 67:259–284. https://doi.org/10.1146/annurev.physiol.67.040403.120816

Compagnone NA, Mellon SH (2000) Neurosteroids: biosynthesis and function of these novel neuromodulators. Front Neuroendocrinol 21:1–56. https://doi.org/10.1006/frne.1999.0188

Concas A, Mostallino MC, Porcu P, Follesa P, Barbaccia ML, Trabucchi M, Purdy RH, Grisenti P, Biggio G (1998) Role of brain allopregnanolone in the plasticity of g–aminobutyric acid type A receptor in rat brain during pregnancy and after delivery. Proc Natl Acad Sci USA 95:13284–13289. https://doi.org/10.1073/pnas.95.22.13284

Corpechot C, Young J, Calvel M, Wehrey C, Veltz JN, Touyer G, Mouren M, Prasad VVK, Banner C, Sjijvall J, Baulieu EE, Robel P (1993) Neurosteroids: 3 alpha-hydroxy-5 alphapregnan-20-one and its precursors in the brain, plasma, and steroidogenic glands of male and female rats. Endocrinology 133:1003–1009. https://doi.org/10.1210/endo.133.3.8365352

Corréa Rocha D, Beskow A, Mc Manus Pimentel CM, Costa Mattos R, Macedo Gregory R (2011) Níveis séricos de progesterona em vacas ovariectomizadas tratadas com MAD4 com diferentes concentracões e vías de administracão. Acta Sci Vet 39:1–6

Cosentino IO, Balaro MFA, Arashiro EKN, Santos JDR, Carvalho ABDS, Pérez-Clariget R, Ungerfeld R, Brandão FZ (2019) Hormonal protocols for early resynchronization of ovulation in ewes: the use of progestagens, eCG, and inclusion of early pregnancy diagnosis with color Doppler ultrasound. Theriogenology 133:113–118. https://doi.org/10.1016/j.theriogenology.2019.04.033

Crossley KJ, Walker DW, Beart PM, Hirst JJ (2000) Characterization of $GABA_A$ receptors in fetal, neonatal and adult ovine brain: region and age related changes and the effects of allopregnanolone. Neuropharmacology 39:1514–1522. https://doi.org/10.1016/s0028-3908(99)00222-1

Dallman ME, Viau VG, Bhatnagar S, Gomez F, Laugero K, Bell ME (2002) Corticosteroids, stress, and sugar: energy balance, the brain, and behavior. In: Pfaff W et al (eds) Hormones, brain, and behaviour. Academic, San Diego, pp 571–632

Dobson H, Smith RF (2000) What is stress, and how does it affect reproduction? Anim Reprod Sci 60:743–752. https://doi.org/10.1016/s0378-4320(00)00080-4

Douglas AJ, Meddle SL, Toschi N, Bosch OJ, Neumann ID (2005) Reduced activity of the noradrenergic system in the paraventricular nucleus at the end of pregnancy: implications for stress hyporesponsiveness. J Neuroendocrinol 17:40–48. https://doi.org/10.1111/j.1365-2826.2005.01272.x

Earley B, Buckham-Sporer K, Gupta S, Pang W, Ting S (2010) Biologic response of animals to husbandry stress with implications for biomedical models. Anim Physiol 2:25–42. https://doi.org/10.2147/OAAP.S9354

Figueiredo HF, Ulrich YM, Choi DC, Herman JP (2007) Estrogen potentiates adrenocortical responses to stress in female rats. Am J Physiol Endocrinol Metab 292:E1173–E1182. https://doi.org/10.1152/ajpendo.00102.2006

Freitas-de-Melo A, Ungerfeld R (2016a) Progesterona y respuesta de estrés: mecanismos de acción y sus repercusiones en rumiantes domésticos. Revisión Rev Mex Cien Pec 7:185–199

Freitas-de-Melo A, Ungerfeld R (2016b) Destete artificial en ovinos: respuesta de estrés y bienestar animal. Revisión Rev Mex Cien Pec 7:361–375

Freitas-de-Melo A, Banchero G, Hötzel MJ, Damián JP, Ungerfeld R (2013) Progesterone administration reduces the behavioural and physiological responses of ewes to abrupt weaning of lambs. Animal 7:1367–1373. https://doi.org/10.1017/S1751731113000621

Freitas-de-Melo A, Damián JP, Hötzel MJ, Banchero G, Ungerfeld R (2016) Progesterone pretreatment increases the stress response to social isolation in ewes. Hormones 15:81–87. https://doi.org/10.14310/horm.2002.1625

Freitas-de-Melo A, Orihuela A, Magri G, Cruz BD, Rubio I, Corro M, Alonso MA, Ungerfeld R (2019) Physiological reproductive status and progesterone concentration affect the results of tests to measure temperament traits in female beef cattle. Livest Sci 221:39–43. https://doi.org/10.1016/j.livsci.2019.01.010

Freitas-de-Melo A, Garcia Kako Rodriguez M, Crosa C, Ungerfeld R (2022) Social Stress during the Estrus or Luteal Phase in Sheep. J Appl Anim Welf Sci 1–9. https://doi.org/10.1080/10888705.2021.2021408

Frye CA, Koonce CJ, Walf AA (2013) Progesterone, compared to medroxyprogesterone acetate, to C57BL/6, but not 5α-reductase mutant, mice enhances object recognition and placement memory and is associated with higher BDNF levels in the hippocampus and cortex. Neurosci Lett 551:53–57. https://doi.org/10.1016/j.neulet.2013.07.002

Gallo M, Smith S (1993) Progesterone withdrawal decreases latency to and increases duration of electrified prod burial: a possible rat model of PMS anxiety. Pharmacol Biochem Behav 46:897–904. https://doi.org/10.1016/0091-3057(93)90219-J

Gaston-Parry O, Heasman K, Nemorin JKE, Robinson TJ (1988) A radioimmunoassay for fluorogestone acetate (FGA) and its application to the measurement of plasma FGA and progesterone in ewes treated with FGA-impregnated intravaginal sponges. Austr J Biol Sci 41:57–67. https://doi.org/10.1071/BI9880057

Genazzani AR, Monteleone P, Gambacciani M (2002) Hormonal influence on the central nervous system. Maturitas 43:11–17. https://doi.org/10.1016/s0378-5122(02)00144-5

Gimenes LU, Fantinato Neto P, Arango JSP, Ayres H, Baruselli PS (2009) Follicular dynamics of Bos indicus, Bos taurus and Bubalus bubalis heifers treated with norgestomet ear implant associated or not to injectable progesterone. Anim Reprod 6:256 (abstract)

Giriboni J, Lacuesta L, Damián JP, Ungerfeld R (2015) Grouping previously unknown bucks is a stressor with negative effects on reproduction. Trop Anim Health Prod 47:317–322. https://doi.org/10.1007/s11250-014-0722-2

Greyling JPC, Kotzé WF, Taylor GJ, Hagendijk WJ (1994) Synchronization of oestrus in sheep: use of different doses of progestagen outside the normal breeding season. S Afr J Anim Sci 24:33–36

Griffin JFT (1989) Stress and immunity: a unifying concept. Vet Immunol Immunopathol 20:263–312. https://doi.org/10.1016/0165-2427(89)90005-6

Gulinello M, Gong QH, Smith SS (2002) Progesterone withdrawal increases the alpha4 subunit of the GABA(A) receptor in male rats in association with anxiety and altered pharmacology-a comparison with female rats. Neuropharmacology 43:701–714. https://doi.org/10.1016/s0028-3908(02)00171-5

Gunn BG, Brown AR, Lambert JJ, Belelli D (2011) Neurosteroids and GABA(A) receptor interactions: a focus on stress. Front Neurosci 5:article 131. https://doi.org/10.3389/fnins.2011.00131

Hantsoo L, Epperson CN (2020) Allopregnanolone in premenstrual dysphoric disorder (PMDD): evidence for dysregulated sensitivity to GABA-A receptor modulating neuroactive steroids across the menstrual cycle. Neurobiol Stress 12:100213. https://doi.org/10.1016/j.ynstr.2020.100213

Hargreaves AL, Hutson GD (1990a) The stress response in sheep during routine handling procedures. Appl Anim Behav Sci 26:83–90. https://doi.org/10.1016/0168-1591(90)90089-V

Hargreaves AL, Hutson GD (1990b) An evaluation of the contribution of isolation up-ending and wool removal to the stress response to shearing. Appl Anim Behav Sci 26:103–113. https://doi.org/10.1016/0168-1591(90)90091-Q

Hassell J, Miryala CSJ, Hiegel C, Uphouse L (2011) Mechanisms responsible for progesterone's protection against lordosis-inhibiting effects of restraint: I. Role of progesterone receptors. Horm Behav 60:219–225. https://doi.org/10.1016/j.yhbeh.2011.05.006

Hauschildt Dias J, Pupin MA, Saloni Duarte G, Lopes Brair V, Carvalho de Paula CJ, Paula de Sousa MA, Tavares Pereira Batista RI, Souza-Fabjan JMG, Franco Oliveira ME, Fonseca JF (2020) Successful transcervical uterine flushing can be performed without or reduced dose of oestradiol benzoate in cervical relaxation protocol in Dorper ewes. Reprod Domest Anim 55:844–850. https://doi.org/10.1111/rda.13692

Jasnow AM, Schulkin J, Pfaff DW (2006) Estrogen facilitates fear conditioning and increases corticotropin-releasing hormone mRNA expression in the central amygdala in female mice. Horm Behav 49:197–205. https://doi.org/10.1016/j.yhbeh.2005.06.005
Jiang P, Yang CX, Wang YT, Xu TL (2006) Mechanisms of modulation of pregnanolone on glycinergic response in cultured spinal dorsal horn neurons of rat. Neuroscience 141:2041–2050. https://doi.org/10.1016/j.neuroscience.2006.05.009
Kalkhoff RK (1982) Metabolic effects of progesterone. Am J Obstet Gynecol 142:735–738. https://doi.org/10.1016/s0002-9378(16)32480-2
Kaura V, Ingram CD, Gartside SE, Young AH, Judge SJ (2007) The progesterone metabolite allopregnanolone potentiates $GABA_A$ receptor-mediated inhibition of 5-HT neuronal activity. Eur Neuropsychopharmacol 17:108–115. https://doi.org/10.1016/j.euroneuro.2006.02.006
Kilgour RJ, Szantar-Coddington MR (1997) The arena test and cortisol response of sheep as indirect selection criteria for the improvement of lamb survival. Anim Reprod Sci 46:97–108. https://doi.org/10.1016/s0378-4320(96)01591-6
Kudielka BM, Kirschbaum C (2007) Sex differences in HPA axis responses to stress: a review. Biol Psychol 69:113–132. https://doi.org/10.1016/j.biopsycho.2004.11.009
Lambert JJ, Cooper MA, Simmons RDJ, Weir CJ, Belelli D (2009) Neurosteroids: endogenous allosteric modulatorsof $GABA_A$ receptors. Psychoneuroendocrinology 34:48–58. https://doi.org/10.1016/j.psyneuen.2009.08.009
Liang JJ, Rasmusson AM (2018) Overview of the molecular steps in steroidogenesis of the GABAergic neurosteroids allopregnanolone and pregnanolone. Chronic Stress 2:2470547018818555. https://doi.org/10.1177/2470547018818555
Liu SS, Ying GG, Liu YS, Yang YY, He LY, Chen J, Liu WR, Zhao JL (2015) Occurrence and removal of progestagens in two representative swine farms: effectiveness of lagoon and digester treatment. Water Res 77:146–154. https://doi.org/10.1016/j.watres.2015.03.022
Löfgren M, Johansson IM, Meyerson B, Lundgren P, Bäckström T (2006) Progesterone withdrawal effects in the open field test can be predicted by elevated plus maze performance. Horm Behav 50:208–215. https://doi.org/10.1016/j.yhbeh.2006.03.002
Ma S, Shipston MJ, Morilak D, Russell JA (2005) Reduced hypothalamic vasopressin secretion underlies attenuated adrenocorticotropin stress responses in pregnant rats. Endocrinology 146:1626–1637. https://doi.org/10.1210/en.2004-1368
Maniam J, Morris MJ (2012) The link between stress and feeding behaviour. Neuropharmacology 63:97–110. https://doi.org/10.1016/j.neuropharm.2012.04.017
Mapletoft RJ, Bó GA, Baruselli PS, Menchaca A, Sartori R (2018) Evolution of knowledge on ovarian physiology and its contribution to the widespread application of reproductive biotechnologies in South American cattle. Anim Reprod 15:1003–1014. https://doi.org/10.21451/1984-3143-AR2018-0007
Martínez MF, Kastelic JP, Adams GP, Janzen E, McCartney DH, Mapletoft RJ (2000) Estrus synchronization and pregnancy rates in beef cattle given CIDR-B, prostaglandin and estradiol, or GnRH. Can Vet J 41:786–790
Matteri RL, Carroll JA, Dyer CJ (2000) Neuroendocrine responses to stress. In: Moberg GP, Mench JA (eds) The biology of animal stress: basic principles and implications for animal welfare. CABI Publishing, Cambridge, pp 43–76
McEwen BS (1991) Non-genomic and genomic effects of steroids on neural activity. Trends Pharmacol Sci 12:141–147. https://doi.org/10.1016/0165-6147(91)90531-v
McEwen BS (2007) Physiology and neurobiology of stress and adaptation: central role of the brain. Physiol Rev 87:873e904. https://doi.org/10.1152/physrev.00041.2006
McEwen BS, Wingfield JC (2003) The concept of allostasis in biology and biomedicine. Horm Behav 43:2–15. https://doi.org/10.1016/s0018-506x(02)00024-7
Mello BP, Maturana Filho M, Lemes KM, Gonçalvez RL, Mendes Lollatod JP, Zanella AJ, Vasconcelos Ferreira TF, Pugliesif G, Madureira EH, Gonella-Diaza A, Membrive CMB (2020) Importance of temperament in the pregnancy by timed insemination in bovine females *Bos taurus indicus*. Livest Sci 240:104104. https://doi.org/10.1016/j.livsci.2020.104104

Melo LF, Monteiro PLJ, Surjus RS, Drum JN, Wiltbank MC, Sartori R (2016) Progesterone-based fixed-time artificial insemination protocols for dairy cows: gonadotropin-releasing hormone versus estradiol benzoate at initiation and estradiol cypionate versus estradiol benzoate at the end. J Dairy Sci 99:9227–9237. https://doi.org/10.3168/jds.2016-11220

Misztal T, Czauderna MR, Młotkowska P, Misztal A, Marciniak E (2020a) Temporal changes in the cerebrospinal fluid allopregnanolone concentration and hypothalamic-pituitary-adrenal axis activity in sheep during pregnancy and early lactation. Livest Sci 231:103871. https://doi.org/10.1016/j.livsci.2019.103871

Misztal T, Młotkowska P, Marciniak E, Misztal A (2020b) Allopregnanolone reduces neuroendocrine response to acute stressful stimuli in sheep. J Endocrinol 244:201–211. https://doi.org/10.1530/JOE-19-0376

Moberg GP (2001) Biological response to stress: implications for animal welfare. In: Moberg GP, Mench JA (eds) The biology of animal stress: basic principles and implications for animal welfare. CABI Publishing, Wallingford, pp 1–21

Monteiro PLJ Jr, Borsato M, Silva FLM, Prata AB, Wiltbank MC, Sartori R (2015) Increasing estradiol benzoate, pretreatment with gonadotropin-releasing hormone, and impediments for successful estradiol-based fixed-time artificial insemination protocols in dairy cattle. J Dairy Sci 98:3826–3839. https://doi.org/10.3168/jds.2014-9040

Morgan MA, Pfaff DW (2002) Estrogen's effects on activity, anxiety, and fear in two mouse strains. Behav Brain Res 132:85–93. https://doi.org/10.1016/S0166-4328(01)00398-9

Mormedè P, Andanson S, Aupérin B, Beerda B, Guémené D, Malmkvist J, Manteca X, Manteuffel G, Prunet P, van Reenen CG, Richard S, Veissier I (2007) Exploration of the hypothalamic-pituitary-adrenal function as a tool to evaluate animal welfare. Physiol Behav 92:317–339. https://doi.org/10.1016/j.physbeh.2006.12.003

Morotti F, de Campos JT, Lunardelli PA, Costa CB, Bergamo LZ, Rigo Barreiros TR, dos Santos GMG, Seneda MM (2018) Injectable progesterone in timed artificial insemination programs in beef cows. Anim Reprod 15:17–22. https://doi.org/10.21451/1984-3143-2017-AR928

O'Connor DB, Thayer JF, Vedhara K (2021) Stress and health: a review of psychobiological processes. Annu Rev Psychol 72:663–688. https://doi.org/10.1146/annurev-psych-062520-122331

Oliveira MAL, Guido SI, Lima PF (2001) Comparison of different protocols used to induce and synchronize estrus cycle of Saanen goats. Small Rumin Res 40:149–153. https://doi.org/10.1016/S0921-4488(00)00222-4

Orihuela A, Sanchez-Mejorada H, Toledo M (2002) Effect of short transport during di-oestrus and pro-oestrus on cortisol levels and oestrous behaviour of sheep. J Agric Sci 138:93–96. https://doi.org/10.1017/S0021859601001733

Pacák K, Palkovits M (2001) Stressor specificity of central neuroendocrine responses: implications for stress-related disorders. Endocr Rev 22:502–548. https://doi.org/10.1210/edrv.22.4.0436

Patchev VK, Hassan AH, Holsboer DF, Almeida OF (1996) The neurosteroid tetrahydroprogesterone attenuates the endocrine response to stress and exerts glucocorticoid-like effects on vasopressin gene transcription in the rat hypothalamus. Neuropsychopharmacology 5:533–540. https://doi.org/10.1016/S0893-133X(96)00096-6

Paul SM, Purdy RH (1992) Neuroactive steroids. FASEB J 6:2311–2322. https://doi.org/10.1096/fasebj.6.6.1347506

Paul SM, Pinna G, Guidotti A (2020) Allopregnanolone: from molecular pathophysiology to therapeutics. A historical perspective. Neurobiol Stress 12:100215. https://doi.org/10.1016/j.ynstr.2020.100215

Pazol K, Northcutt KV, Patisaul HB, Wallen K, Wilson ME (2009) Progesterone and medroxyprogesterone acetate differentially regulate α4 subunit expression of GABAA receptors in the CA1 hippocampus of female rats. Physiol Behav 97:58–61. https://doi.org/10.1016/j.physbeh.2009.01.021

Pinto-Santini L, Ungerfeld R (2019) The phase of the estrous cycle modifies the endocrine, metabolic and behavior rhythms in ewes. Physiol Behav 204:324–335. https://doi.org/10.1016/j.physbeh.2019.03.011

Pluchino N, Cubeddu A, Giannini A, Merlini S, Cela V, Angioni S, Genazzani AR (2009) Progestogens and brain: an update. Maturitas 62:349–355. https://doi.org/10.1016/j.maturitas.2008.11.023

Reddy DS, O'Malley BW, Rogawski MA (2005) Anxiolytic activity of progesterone in progesterone receptor knock-out mice. Neuropharmacology 48:14–24. https://doi.org/10.1016/j.neuropharm.2004.09.002

Rubianes E, de Castro T, Kmaid S (1998) Estrous response after a short progesterone priming in seasonally anestrous goats. Theriogenology 49:356 (abstract)

Sabban EL (2010) Catecholamines and stress. In: Soreq H et al (eds) Stress: from molecules to behavior: a comprehensive analysis of the neurobiology of stress responses. Wiley Blackweel, Weinheim, pp 19–36

Sales JNS, Carvalho JBP, Crepaldi GA, Cipriano RS, Jacomini JO, Maio JRG, Souza JC, Nogueira GP, Baruselli PS (2012) Effects of two estradiol esters (benzoate and cypionate) on the induction of synchronized ovulations in *Bos indicus* cows submitted to a timed artificial insemination protocol. Theriogenology 78:510–516. https://doi.org/10.1016/j.theriogenology.2012.02.031

Santos-Neto PCD, García-Pintos C, Pinczak A, Menchaca A (2015) Fertility obtained with different progestogen intravaginal devices using Short-term protocol for fixed-time artificial insemination (FTAI) in sheep. Livest Sci 182:125–128. https://doi.org/10.1016/j.livsci.2015.11.005

Sapolsky RM, Romero LM, Munck AU (2000) How do glucocorticoids influence stress response? Integrationing, permissive, suppressive stimulatory and preparative actions. Endocr Rev 2:55–89. https://doi.org/10.1210/edrv.21.1.0389

Sarkar J, Wakefield S, MacKenzie G, Moss SJ, Maguire J (2011) Neurosteroidogenesis is required for the physiological response to stress: role of neurosteroidsensitive $GABA_A$ receptors. J Neurosci 31:18198e210. https://doi.org/10.1523/JNEUROSCI.2560-11.2011

Sauer P, Stará A, Golovko O, Valentová O, Bořík A, Grabic R, Kroupová HK (2018) Two synthetic progestins and natural progesterone are responsible for most of the progestagenic activities in municipal wastewater treatment plant effluents in the Czech and Slovak republics. Water Res 137:64–71. https://doi.org/10.1016/j.watres.2018.02.065

Schumacher M, Coirini H, Robert F, Guennoun R, El-Etr M (1999) Genomic and membrane actions of progesterone: implications for reproductive physiology and behavior. Behav Brain Res 105:37–52. https://doi.org/10.1016/s0166-4328(99)00081-9

Seamark RF, Nancarrow CD, Gardiner J (1969) Progesterone metabolism in ovine blood: the formation of 3α-hydroxypregn-4-en-20-one and other substances. Steroids 15:589–604. https://doi.org/10.1016/S0039-128X(70)80086-1

Sedlâèek M, Koøínek M, Petroviè M, Cais O, Adamusová E, Chodounská H, Vyklický L Jr (2008) Neurosteroid modulation of ionotropic glutamate receptors and excitatory synaptic transmission. Physiol Res 97:49–57

Smith MF, McIntush A, Smith GW (1993) Mechanisms associated with corpus luteum development. J Anim Sci 72:1857–1872. https://doi.org/10.2527/1994.7271857x

Smith S, Gong Q, Hsu F, Markowitz R, Ffrench-Mullen J, Li X (1998) GABA(A) receptor alpha4 subunit suppression prevents withdrawal properties of an endogenous steroid. Nature 392:926–930. https://doi.org/10.1038/31948

Smith SS, Shen H, Gong QH, Zhou X (2007) Neurosteroid regulation of $GABA_A$ receptors: focus on the 4 and subunits. Pharmacol Therap 116:58–76. https://doi.org/10.1016/j.pharmthera.2007.03.008

Souza-Fabjan JMG, da Rosa RM, Balaro MFA, Pinto PHN, dos Santos GB, Arashiro EKN, da Fonseca JF, Ungerfeld R, Brandão FZ (2017) Effect of different hormonal combinations on follicular wave emergence and superovulatory response in sheep. Theriogenology 103:24–29. https://doi.org/10.1016/j.theriogenology.2017.07.036

Sze Y, Brunton PJ (2019) Sex, stress and steroids. Eur J Neurosci 52:2487–2515. https://doi.org/10.1111/ejn.14615

Ungerfeld R (2008) Response of anestrous ewes pretreated with a single dose of oestradiol-17beta or progesterone and oestradiol-17beta to the introduction of rams and ewes in oestrus. N Z Vet J 56:36–39. https://doi.org/10.1080/00480169.2008.36802

Ungerfeld R (2009) The induction of oestrus in ewes during the non-breeding season using pre-used CIDRs and oestradiol-17beta treatment. Small Rumin Res 84:129–131. https://doi.org/10.1016/j.smallrumres.2009.06.011

Ungerfeld R, Freitas-de-Melo A (2019) Stress and behavioural responses to winter shearing differ between pregnant and non-pregnant ewes. Physiol Behav 210:112653. https://doi.org/10.1016/j.physbeh.2019.112653

Ungerfeld R, Pinczak A, Forsberg M, Rubianes E (1999) Response of Corriedale ewes to the ram effect after primings with medroxyprogesterone, fluorogestone, or progesterone in the non-breeding season. Acta Vet Scand 40:299–305. https://doi.org/10.1186/BF03547009

Ungerfeld R, Suárez G, Carbajal B, Silva L, Laca M, Forsberg M, Rubianes E (2003) Medroxyprogesterone primings and response to the ram effect in Corriedale ewes during the non-breeding season. Theriogenology 60:35–45. https://doi.org/10.1016/s0093-691x(02)01302-x

Ungerfeld R, Dago AL, Rubianes E, Forsberg M (2004) Response of anestrous ewes to the ram effect after follicular wave synchronization with a single dose of estradiol-17beta. Reprod Nutr Dev 44:89–98. https://doi.org/10.1051/rnd:2004010

Ungerfeld R, Gamboa D, Alvarez L (2013) Response of ewes primed with CIDRs, previously used CIDRs, or previously used and autoclaved CIDRs to the ram effect during the non-breeding season. Anim Reprod 10:704–707

Ungerfeld R, Freitas-de-Melo A, Damián JP, Fernández-Werner A, Kremer R (2018) Behavioural and physiological changes in pregnant ewes after winter shearing. Small Rumin Res 169:57–61

Van Lier E, Carriquiry M, Meikle A (2014) Sex steroid modulation of cortisol secretion in sheep. Animal 8:960–967. https://doi.org/10.1017/S1751731114000780

Viérin M, Bouissou MF (2001) Pregnancy is associated with low fear reactions in ewes. Physiol Behav 72:579–587. https://doi.org/10.1016/S0031-9384(01)00416-4

Vilariño M, Rubianes E, Menchaca A (2011) Re-use of intravaginal progesterone devices associated with the short-term protocol for timed artificial insemination in goats. Theriogenology 75:1195–1200. https://doi.org/10.1016/j.theriogenology.2010.11.030

Vilariño M, Rubianes E, Menchaca A (2013) Ovarian responses and pregnancy rate with previously used intravaginal progesterone releasing devices for fixed-time artificial insemination in sheep. Theriogenology 79:206–210. https://doi.org/10.1016/j.theriogenology.2012.10.007

Vynchier L, Debackere M, De Kruif A, Coryn M (1990) Plasma estradiol-17b concentrations in the cow during induced estrus and after injection of estradiol-17b benzoate and estradiol-17b cypionate-a preliminary study. J Vet Pharmacol Therap 13:36–42. https://doi.org/10.1111/j.1365-2885.1990.tb00745.x

Walker DL, Toufexis DJ, Davis M (2003) Role of the bed nucleus of the stria terminalis versus the amygdala in fear, stress, and anxiety. Eur J Pharmacol 463:199–216. https://doi.org/10.1016/s0014-2999(03)01282-2

Wang M (2011) Neurosteroids and $GABA_A$ receptor function. Front Endocrinol 2:1–23. https://doi.org/10.3389/fendo.2011.00044

Zuluaga JF, Williams GL (2008) High-pressure steam sterilization of previously used CIDR inserts enhances the magnitude of the acute increase in circulating progesterone after insertion in cows. Anim Reprod Sci 10:30–35. https://doi.org/10.1016/j.anireprosci.2007.06.006

Chapter 2
Insights from Proteomics in Kidney Disease Diagnosis and Various *In Vitro* and *In Vivo* Experimental Models

Vikram Patial, Garima Dadhich, and Rajiv Kumar

Abstract Kidney disease is emerging as a significant cause of mortality and public health burden worldwide. Diabetes, hypertension, obesity, and drug toxicity are the leading causes of kidney damage. The biomarkers and experimental models of kidney disease play an important role in diagnosis, prognosis, and follow-up treatment. Therefore, detailed knowledge about molecular markers and experimental models could be helpful to understand kidney diseases. Proteomic technology contributes to clinical research, particularly biomarker discovery and prognosis of kidney disease by estimating protein expression. Here, we review various kidney biomarkers which are efficient and associated with early detection of chronic kidney disease, diabetic nephropathy and acute kidney injury compared to surrogate markers. Also, we highlight the recent progress of proteomic quantification techniques like gel-based and gel-free techniques primarily used for biomarker discovery. Various *in vitro* cell line models used for screening and molecular studies were explained. Further, the different rodent and alternative models like zebrafish and drosophila used in kidney research are also discussed in detail. In conclusion, this chapter aims to review molecular biomarkers, the role of proteomics in kidney diseases, and different *in vitro* and *in vivo* models for understanding disease pathogenesis and therapeutics development.

V. Patial (✉)
Pharmacology and Toxicology Laboratory, Dietetics and Nutrition Technology Division, CSIR-Institute of Himalayan Bioresource Technology, Palampur, India

Academy of Scientific and Innovative Research (AcSIR), Ghaziabad, Uttar Pradesh, India
e-mail: vikrampatial@ihbt.res.in

G. Dadhich
Pharmacology and Toxicology Laboratory, Dietetics and Nutrition Technology Division, CSIR-Institute of Himalayan Bioresource Technology, Palampur, India

R. Kumar (✉)
Academy of Scientific and Innovative Research (AcSIR), Ghaziabad, Uttar Pradesh, India

Biotechnology Division, CSIR-Institute of Himalayan Bioresource Technology, Palampur, India
e-mail: rajiv@ihbt.res.in

V. K. Yata et al. (eds.), *Sustainable Agriculture Reviews 57*, Sustainable Agriculture Reviews 57, https://doi.org/10.1007/978-3-031-07496-7_2

Keywords Animal models · Biomarkers · Kidney disease · Proteomics

Abbreviations

ADMA	Asymmetric dimethylarginine
DOCA	Deoxycorticosterone acetate
ELISA	Enzyme-linked immunosorbent assays
eNOS	endothelial nitric oxide synthase
GeLC	Gel electrophoresis Liquid Chromatography-Mass Spectrometry
GFR	Glomerular Filtration Rate
IL-18	Interleukin 18
IL-1β	Interleukin-1β
KIM-1	Kidney Injury Molecule 1
LC-MS	Liquid Chromatography-Mass Spectrometry
LDH	Lactate dehydrogenase
L-FABP	Liver-fatty acid-binding protein
MCP-1	Monocyte chemoattractant protein-1
NAG	N-acetyl-β-D-glucosaminidase
NGAL	Neutrophil gelatinase-associated lipocalin
RBP4	Retinol-binding protein 4
α_1M	α_1-microglobulin
α-GST/π-GST	α/π-Glutathione s-transferase
β_2M	β-microglobulin

2.1 Introduction

Kidney disease is an important contributor to health burden and mortality globally. More than 2.5 million people receive kidney replacement therapy, and the number is projected to double by 2030. The disease screening programme in countries like USA, Australia, and Norway have shown kidney diseasemarkers in more than 10% of the adult population (Bikbov et al. 2020). Three million people in UK alone are diagnosed with kidney ailments; however, more than one million people are undiagnosed (Outtandy et al. 2019). Chronic kidney disease is the 12th most common cause of death worldwide. In India, areas like Goa and Odisha were reported to have high chronic kidney disease incidence of unknown etiology (Jayasekara et al. 2015). According to the global burden of disease, approximately 1.5 million people died from kidney disease in 2015 alone, which was almost 32% higher from 2005 (Wang et al. 2016).

The term "Kidney disease" is used for any abnormality or changes that occur in kidney function. The abrupt change in glomerular filtration rate, increased serum creatinine, and decreased urine output over 6 to 24 h is termed as acute kidney injury (Peerapornratana et al. 2019). It is mainly characterized by azotemia and oliguria. Acute kidney injury was defined by Acute Dialysis Quality Initiative (ADQI) as risk, injury, failure, loss of kidney function, end stage kidney disease (RIFLE), and later adopted by the acute kidney injury network (Lopes and Jorge 2013). Sepsis, ischemia, and drug toxicity are the leading causes of acute kidney injury. Recent studies suggest that acute kidney injury may lead to the development of chronic kidney disease, which requires kidney replacement therapy (Peerapornratana et al. 2019). Chronic kidney disease is characterized by kidney impairment that leads to decreased glomerular filtration rate (less than 60 ml/min/1.73m^2), elevated albuminuria, and serum creatinine. In addition, kidney injury causes deterioration in kidney function, resulting impaired urine output and higher connective tissue deposition in the kidney to cause fibrotic disorders (Levin et al. 2013). End-stage kidney disease is the last stage when the kidney cannot work and identified by a glomerular filtration rate of less than 15 ml/min (Sgambat et al. 2019). In the last few decades, an increasing prevalence of diseases such as obesity, diabetes and hypertension predisposed individuals to chronic kidney disease. Chronic kidney disease has been recognized as a significant risk factor for the higher rate of cardiovascular disease and mortality. Therefore, its early detection and prevention is a health-care priority in both developed and developing countries (Matovinović 2009).

The prognosis of kidney disease is greatly facilitated by the detection of biomarkers. Biomarkers are protein/enzymes released by damaged kidney tissue, and their presence directly reflects pathological condition. Further, the identification of proteins, enzymes and transcription factors in pathological conditions will increase our understanding about the molecular regulation of disease. The use of proteomics for disease diagnosis has recently been gathered pace (Wu and Fenton 2018). Proteomic approaches are applied for qualitative and quantitative estimation of biomarkers in different samples like plasma, urine, or tissue that will be useful for early diagnosis and treatment of kidney disease. Experimental models have a critical role in studying the pathology and pharmacology of kidney diseases. The *in vitro* and *in vivo* models provide the best insights into the mechanism of kidney disease and identify potential therapeutic targets, which further help clinical science for understanding disease and develop medications. Cell lines, zebrafish, and drosophila provides great alternatives to rodent and primates in screening the compound and studying the molecular pathways in renal diseases (Bao et al. 2018). This chapter provides an overview of various biomarkers for early diagnosis of kidney disease, proteomic technologies for their identification and validation, and different non-transgenic experimental models to study kidney diseases based on published research articles.

2.2 Pathophysiology of Kidney Disease

To understand the pathophysiology of kidney disease, the renal structure, physiology, and various mechanism of kidney injury must be taken into consideration. The kidney is a bean-shaped organ located in the upper abdominal area. Broadly kidneys are divided into two regions, *viz.* cortex and medulla. The cortex is mainly composed of the glomeruli, proximal and distal tubules, whereas collecting ducts and loop of Henle from the medullary part (Wang et al. 2017a, b). The nephron is the kidney's functional unit and its development in human occur during the 36th week of gestation period. The number of nephrons varies in an individual with an average of 0.2 to more than 2.5 million. Although when body mass increases the nephron hypertrophy occurs like in the case of obesity, pregnancy, and kidney transplantation (Bertram et al. 2011). The primary functions of nephron include glomerular filtration, tubular reabsorption, and secretion. Glomerulus consists of bundles of capillaries, mesangial cells, and podocytes enclosed by bowman's capsule, forming kidney corpuscles. It filters the blood plasma and allows waste products to pass through tubules (Fig. 2.1) (Wang et al. 2017a, b). The direct damage to glomerular (mesangial, podocytes, and endothelial cells), and tubular cells lead to impaired kidney function. The change in kidney blood flow due to cytokines and chemokines, alteration in glomerular filtration rate (due to disturbance in electrostatic barrier), drugs, and genetic factors are the primary reasons for the kidney injury (Matovinović 2009). The kidney injury can be acute or chronic, depending upon the duration of the disease process. Acute kidney injury causes sudden kidney perfusion changes, resulting in reduced glomerular filtration rate and blood flow (Basile et al. 2011). The significant pathological changes include inflammation, cytokines activation, tubular, epithelial, and endothelial cell injury (Bellomo et al. 2017). Ischemia is the leading causes of acute kidney injury that results in tubular cell damage due to ATP depletion and redistribution of Na^+/K^+ ATPase pump from basolateral to apical membrane of proximal tubular cells. This further results

in cytoskeleton structure disturbance, loss of cell-cell interaction, brush border damage, and cell swelling (Shaw et al. 2018; Chawla and Kimmel 2012). The kidney blood flow reduction due to kidney injury induces conversion of prorenin to renin and angiotensinogen activation. Angiotensinogen converting enzyme convert angiotensinogen into angiotensin I and activates angiotensin II, leading to vasoconstriction and high blood flow. Angiotensin II causes the secretion of aldosterone, leading to impairment of glomerular barrio function by reducing the expression of podocytes protein responsible for maintaining glomerular barrier function (Fig. 2.2) (Benigni et al. 2004; Lim 2014). The altered glomerular filtration rate and pressure lead to glomerular hypertension and nephron hypertrophy. This further increases the shear stress of podocytes resulting in podocyte detachment, glomerulosclerosis, and kidney injury (Romagnani et al. 2017). Ischemic kidney injury also causes some epigenetic changes like DNA methylation, acetylation and mRNA expression contributing to kidney impairment (Rodriguez-Romo et al. 2015).

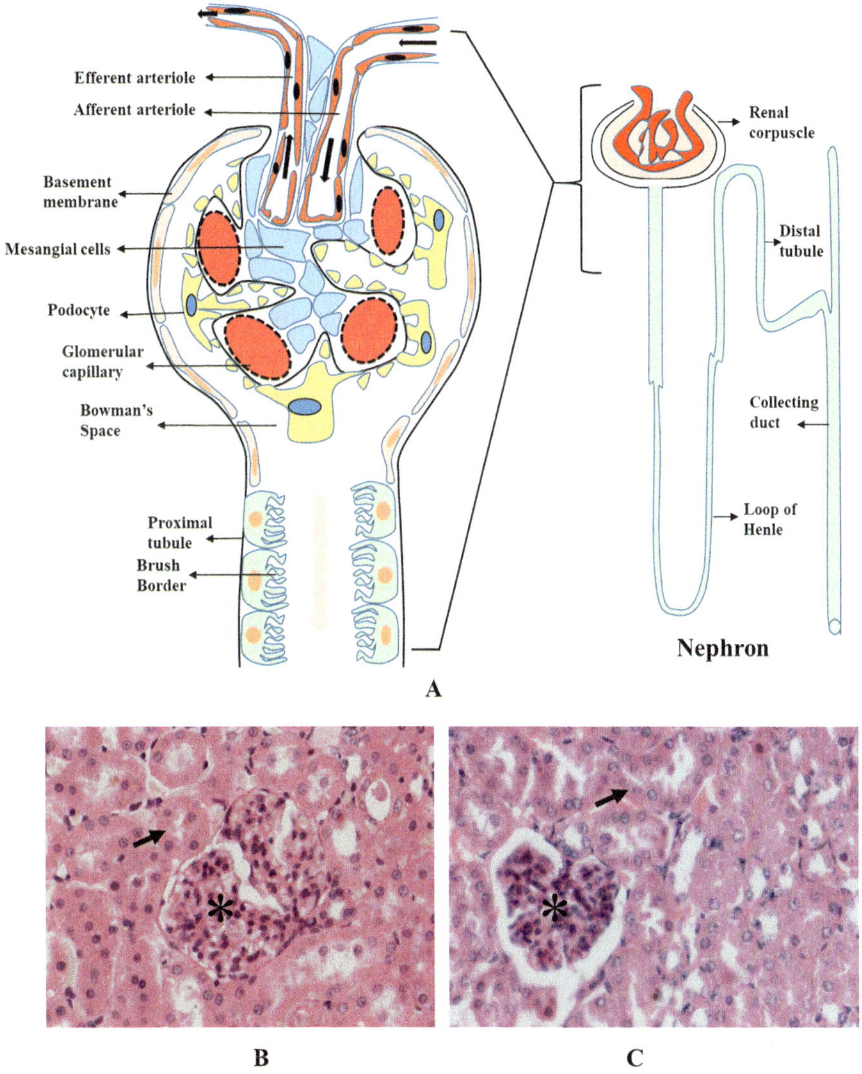

Fig. 2.1 (**a**) Showing different parts of the nephron and glomerulus. The blood enters the glomerulus through the afferent arteriole and leaves through the efferent. The glomerular capillaries, podocytes and mesangial cell together form a filtration bed in the glomerulus. The glomerulus is followed by proximal convoluted tubules with brush border epithelial cells, the loop of Henle, distal convoluted tubules and collecting ducts; (**b**) Histology of rat kidney showing normal glomerulus (asterisk) and proximal convoluted tubule with brush border epithelium (arrow); (**c**) Mouse kidney showing normal glomeruli (asterisks) and proximal convoluted tubules (arrow)

If the kidney function abnormalities remain for more than 3 months, the condition is known as chronic kidney disease. Chronic kidney disease is identified by albuminuria, nephron loss, impaired globular filtration rate, hypertrophy of nephrons, and kidney fibrosis (Romagnani et al. 2017). Its development is associated with three factors: first is cause-specific, second is related to fibrosis and dysfunction, and third is related to progressive damage to nephrons (Zhong et al. 2017). The tubules and interstitium form approximately 90% of the kidney volume. In proteinuria and tubulointerstitial damage, excessive protein reabsorption in proximal tubules is increased via lysosomal processing that causes lysosomal rupture resulting in tubular toxicity. Kidney injury activates the pro-inflammatory, and profibrotic factors contributing to kidney fibrosis development. Chemokine secretion in proteinuria may induce secretion of TGF-β, and intercellular adhesion molecule. The degree of tubulointerstitial fibrosis is directly linked to the level of expression of adhesion molecule (Fig. 2.2) (Nogueira et al. 2017).

Hypertension is one of the primary reasons for kidney damage. Increased blood pressure level for the long term can cause epithelial damage resulting in reduced

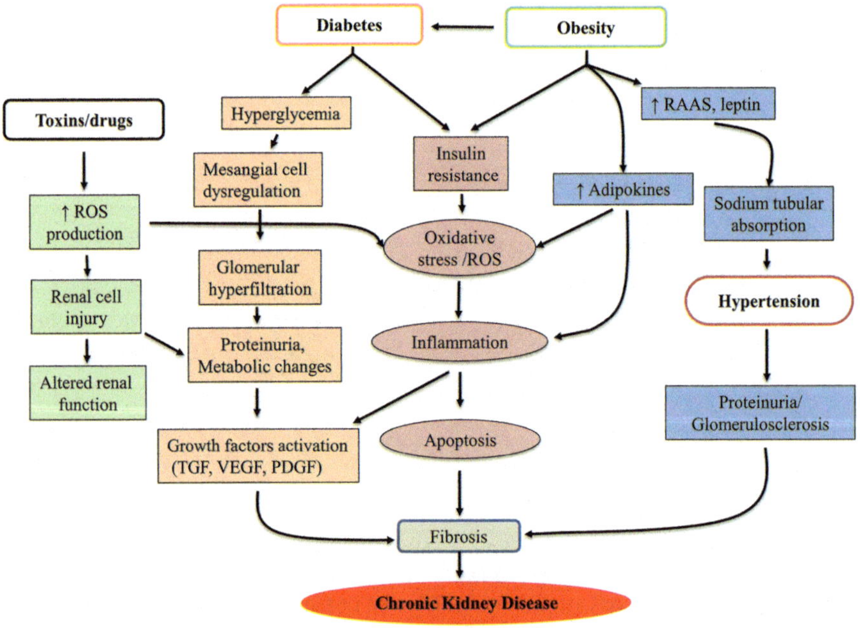

Fig. 2.2 Schematic representation of the pathogenesis of kidney diseases. The hyperglycemia and insulin resistance in diabetes leads to glomerular function changes, oxidative stress, inflammation, and activation of growth factors leading to fibrotic changes. Obesity increases adipokine levels and oxidative stress in the renal tissue. The activation of RAAS increases tubular absorption of sodium, leading to hypertension and pathological changes in the kidneys. Toxin/drugs cause direct injury to the renal cells and cause impairment of renal function. PDGF: Platelet-derived growth factor; RAAS: Renin-angiotensin-aldosterone system; ROS: Reactive oxygen species; TGF: Transforming growth factor; VEGF: Vascular endothelial growth factor

globular filtration rate arterial stenosis, arterial stiffening, and kidney ischemia (Fenoglio et al. 2019). Nephron loss increases the glomerular workload leading to glomerular hypertension, which further activates chronic kidney disease pathways. Similarly, the diabetic condition causes microvascular complications in the kidney leading to albuminuria, proteinuria, and decreased glomerular filtration rate (Orasanu and Plutzky 2009; Ioannou 2017). In diabetic nephropathy (DN), proximal tubules are unable to reduce the glucose transport rate due to hyperglycemia leading to hypertrophy, hyperfiltration, elevated reactive oxygen species generation, and angiotensinogen production (Fig. 2.2) (Vallon 2011). Most glomerular diseases are caused by a defective immune system, metabolism, and stress-related injuries. Proliferative glomerular disease is characterized by immunoglobulins accumulation, leading to glomerular injury and inflammation. The non-proliferative glomerular disease occurs without any glomerular inflammation, and immunoglobulin accumulation as in case diabetic, metabolic, and toxins related damage (Fine and Norman 2008). Some inherited diseases like nail-patella syndrome, thin membrane, and partial lipodystrophy also affect the glomerular function.

2.3 Molecular Biomarkers of Kidney Diseases

Biomarkers are defined as 'characteristic' or 'measurable/quantifiable biological parameters' that differentiate between normal biological processes to pathogenic and other clinical conditions (Provenzano et al. 2020). Molecular biomarkers represent disease-associated proteins for various pathophysiology of kidney, including acute kidney failure, chronic kidney disease, autoimmune kidney disease, and kidney cancer. Biomarker's most important feature is the ease of availability, sensitivity, specificity, reproducibility, and non-invasive and easy-to-perform in the clinical laboratory. The application of molecular biomarkers in clinical practice is vital for the early diagnosis of disease, leading to a better prognosis with improved clinical outcome. Besides, biomarkers were also used to differentiate disease subtype, etiology, predicting the disease severity, and drug development. Although no clear guidelines are available for molecular biomarkers identification; however, the two most common approaches; deductive and inductive were used for kidney biomarkers identification. Deductive approaches include protein study of specific diseases like inflammation, while inductive methods include studying a tissue at a particular condition using different genomic and proteomic techniques followed by bioinformatics analysis. In the recent past, numerous studies have been reported for a better understanding of biomarkers and their role in the early diagnosis of kidney disease (Vaidya et al. 2008; Nguyen and Devarajan 2008; Lopez-Giacoman and Madero 2015; Dobrek and Thor 2016; Alicic et al. 2017; Malhotra and Siew 2017). Here, we discuss prominent biomarkers associated with kidney disease (Table 2.1).

Measurement of glomerular filtration rate is considered to be a direct assessment of kidney functions. In clinical practices, estimation of glomerular filtration rate is usually estimated through measuring alternate markers like serum creatinine (a

Table 2.1 Various molecular markers of kidney disease

Biomarkers	Tissue	Renal disease	Biological function	References
NGAL	Serum, urine	AKI,CKD,DN	Indicate renal injury	Mitsnefes et al. (2007), Nguyen and Devarajan (2008), Bolignano et al. (2009), Kohl et al. (2020)
Cys-C	Serum, urine	CKD, DN, AKI	Glomerular and tubular dysfunction	Vaidya et al. (2008), Kim et al. (2013), Lopez-Giacoman and Madero (2015)
KIM-1	Urine	CKD, DN, AKI	Tubular dysfunction	Vaidya et al. (2005) Uwaezuoke (2017)
NAG	Plasma, urine	CKD, DN, AKI	Tubular injury	Vaidya et al. (2008)
L-FABP	Plasma, urine	CKD, DN, AKI	Renal injury	Vaidya et al. (2008), Bonventre et al. (2010), Lopez-Giacoman and Madero (2015)
β_2M	Plasma and urine	CKD,DN,AKI	Tubular injury	Lopez-Giacoman and Madero (2015), Vaidya et al. (2008)
α_1M	Urine	AKI	Tubular injury	Hart (2005)
RBP4	Urine	CKD, AKI	Tubular dysfunction	Lopez-Giacoman and Madero (2015)
IL-18	Urine	CKD, AKI, DN	Tubular damage	Miyauchi et al. (2009)
Clustrin	Urine	CKD, AKI, DN	Tubular damage	Vaidya et al. (2008)
ADMA	Plasma, urine	CKD,DN	Kidney disease	Raptis et al. (2013)
MCP-1	Urine	CKD, DN	Renal injury	Wada et al. 2000 and Lopez-Giacoman and Madero (2015)
α-GST/π-GST	Urine	AKI	Renal injury	Harrison et al. (1989)
NHE3	Urine	AKI	Tubular injury	du Cheyron et al. (2003)
Fetuin A	Urine	AKI	Tubular damage	Zhou et al. (2006)
TIMP-2 and IGFBP7	Urine	AKI	G_1 cell-cycle arrest	Kashani et al. (2013)
Osteopontin	Urine	AKI	Kidney disease	Vaidya et al. (2008)
Clustrin	Urine	AKI	Tubular injury	Vaidya et al. (2008)
FGF-23	Urine	CKD		Isakova et al. (2011)

Abbreviations: *AKI* Acute Kidney Injury, *CKD* Chronic Kidney Disease, *DN* Diabetic Nephropathy, *NGAL* Neutrophil gelatinase-associated lipocalin, *Cys-C* cysteine protease inhibitor, *KIM-1* Kidney Injury Molecule 1, *NAG* N-acetyl-β-D-glucosaminidase, *L-FABP* Liver-fatty acid-binding protein 1, β_2M β-microglobulin, α_1M α_1-microglobulin, *RBP4* Retinol-binding protein 4, *IL-18* Interleukin 18, *ADMA* Asymmetric dimethylarginine, *MCP-1* Monocyte chemoattractant protein-1, *α-GST/π-GST* α/π-Glutathione s-transferase, *NHE3* Sodium–hydrogen exchanger isoform, *TIMP-2* Tissue Inhibitor of Metalloproteinases-2, *IGFBP7* Insulin-Like Growth Factor-Binding Protein-7, *FGF-23* fibroblast growth factor-23

113 kDa metabolic waste product of muscle-derived from creatine degradation), albuminuria (excess albumin excretion in urine), blood urea nitrogen, and urine output. Prediction of kidney function through increased serum creatinine has some limitations as its concentration may be influenced by patient muscle mass, physical activity, diet, time log between kidney injury that results in low predictive value (Uchino 2010). In kidney disease, albuminuria and glomerular filtration rate are inversely proportional, i.e., higher albuminuria lowers the glomerular filtration rate. Increased albuminuria results due to diabetes, hypertensive nephrosclerosis, and glomerular disease. Excess albuminuria directly affects kidney glomeruli and tubules or induces inflammation that results in kidney damage. However, in contrast to previous report MacIsaac et al. (2014) found that microalbumin concentration is moderately associated with kidney disease progress. Besides, many diabetic kidney disease cases are associated with non-proteinuria (Tsalamandris et al. 1994). Blood urea nitrogen usually increases as glomerular filtration rate decreases; however, urea production may be related to a protein diet and tissue breakdown. Due to insensitivity, non-specificity, and delayed response, these surrogate markers are not sufficient and reliable for early diagnosis of kidney disease. Therefore, identifying novel, sensitive, and specific biomarkers for prior and accurate diagnosis of kidney disease is essential, particularly for reflecting intrinsic organ injury.

Identifying kidney biomarkers in the urine mainly depends on the change in glomerular permeability or tubular resorption, and increased synthesis/release of a biomarker in response to tissue damage. Alteration in glomerular permeability and tubular damage is the leading cause of kidney disease. Tubular protein like neutrophil gelatinase-associated lipocalin (NGAL) is associated with damaged glomerular and tubular epithelial cells of the kidney. An increased concentration of neutrophil gelatinase-associated lipocalin in blood/urine is used to detect acute kidney injury, kidney transplant, and kidney disease associated with diabetes, hypertension, and atherosclerosis (Mitsnefes et al. 2007; Nguyen and Devarajan 2008; Bolignano et al. 2009; Kohl et al. 2020). Cystatin C (13.4-kDa) an indicator of glomerular and tubular damage. An increased concentration of cystatin C in urine has been used to predict chronic kidney disease and diabetic nephropathy (Lopez-Giacoman and Madero 2015; Kim et al. 2013). However, in patients with kidney transplants and cardiac surgery, an increase of 50% urinary cystatin C level was reported that represents a potential biomarker for acute kidney injury (Vaidya et al. 2008).

Kidney injury molecule 1 (KIM-1) is a transmembrane protein usually detected in urine after tubulointerstitial impairment. An increased level of KIM-1 in urine has been reported in chronic kidney disease and diabetic nephropathy (Uwaezuoke 2017). Likewise, its levels significantly increases in acute kidney injury patients with ischemic and cardiac surgery (Vaidya et al. 2005). A lysosomal enzyme in the proximal tubule, N-acetyl-β-D-glucosaminidase (NAG) has been associated with altered urine levels in tubular damage in acute kidney injury, chronic kidney disease and diabetic nephropathy (Vaidya et al. 2008). However, the non-specificity of NAG for acute kidney injury limits its use as biomarkers. Liver-fatty acid-binding protein 1 (L-FABP; intracellular binding protein of fatty acid, highly expressed in tubular damage) has been used as next-generation biomarkers to diagnose kidney disease

(Vaidya et al. 2008; Bonventre et al. 2010; Lopez-Giacoman and Madero 2015). β-microglobulin (β_2M), a 11.8 kDa light chain of major histocompatibility class 1 protein is normally filtered through the glomerulus and reabsorbed by proximal tubular cells. Its increased concentration in urine indicates tubular dysfunctions and act as an early marker for kidney disease (Lopez-Giacoman and Madero 2015). However, rapid degradation in urine with a pH less than 6.0 at room temperature limits its use as biomarkers for acute kidney injury. Similarly, increased urinary α_1-microglobulin (α_1M) level has been used as a sensitive biomarker for proximal tubular damage in acute kidney injury (Hart 2005).

Other protein/enzymes associated with tubular damage like retinol-binding protein 4 (RBP4; a 21kD protein involves in vitamin A transport from liver to other tissue), Interleukin 18 (IL-18; an inflammatory cytokine), and Clusterin; 75 kD-disulfide-linked heterodimeric glycoprotein have been used for detection of chronic kidney disease, acute kidney injury and diabetic nephropathy (Vaidya et al. 2008; Miyauchi et al. 2009; Lopez-Giacoman and Madero 2015). Further, C reactive protein belongs to a protein family known as pentraxin, is associated with endothelial injury and impaired vasodilation in the kidney leading to glomerular damage and progressive kidney function impairment. Another protein monocyte chemoattractant protein-1 (MCP-1; a member of chemotactic cytokines affects kidney structure and function), asymmetric dimethylarginine (ADMA) and fibroblast growth factor-23 also has been used to diagnosse chronic kidney disease and diabetic nephropathy (Wada et al. 2000; Bonventre et al. 2010; Isakova et al. 2011; Raptis et al. 2013; Lopez-Giacoman and Madero 2015). Various other glomerular proteins like transferrin (serum iron-binding protein), immunoglobulin G, laminin (components of glomerular basement membrane), type IV collagen (components of the glomerular basement membrane and mesangial matrix), fibronectin1 (intrinsic component of the glomerular extracellular matrix), resistin (Serum marker for reduced glomerular filtration), prostaglandin D2 synthase (Increased serum level indicates impaired kidney function), ceruloplasmin, glycosaminoglycans, lipocalintype, podocalyxin, vascular endothelial growth factor, human cartilage glycoprotein-40, and nephrin has been used for early diagnosis of chronic kidney disease (Dobrek and Thor 2016). Furthermore, proteins associated with cell cycle regulation (G_1 cell-cycle arrest) including insulin-like growth factor-binding protein-7, tissue inhibitor of metalloproteinases-2, α/π-glutathione s-transferase (α-GST/π-GST), sodium–hydrogen exchanger isoform, fetuin-A, osteopontin, clusterin have been identified as early acute kidney diseasebiomarkers (Harrison et al. 1989; du Cheyron et al. 2003; Zhou et al. 2006; Vaidya et al. 2008; Kashani et al. 2013).

One of the most crucial reasons for the progression of any disease is oxidative stress. Oxidative stress leads to generation of reactive oxygen species, and at higher concentrations, they produce unwanted modification of protein, lipid, and DNA. Of a particular interest, protein oxidative products and protein carbonyls were used as molecular markers as they significantly increase in chronic kidney disease and end stage renal disease patients (Himmelfarb et al. 2000; Mitrogianni et al. 2009; Machowska et al. 2016). Mitrogianni et al. (2009) reported that the carbonylation of albumin gradually increases with chronic kidney disease progression. Further, lipid peroxidation products; malondialdehyde, 4-hydroxynonenal, thiobarbituric acid reactive substances, and isoprostanes such as 8-iso-prostaglandin F2α were used as biomarkers for chronic kidney disease (Sugiura and Wada 2009). Moreover, urinary 8-oxo-7,8-dihydro-2-deoxyguanosine; an oxidative modification of DNA is a typical biomarker used for diabetic nephropathy. Recently, Michele et al. reviewed in detail about molecular biomarkers involved in chronic kidney disease. They discussed the serum level of oxidative stress-induced biomarkers involved in tissue remodelling, and metabolism; myeloperoxidase, matrix metalloproteinase (matrix metalloproteinase-2, −8 and − 9), endopeptidases, tissue inhibitor of metalloproteinases-1, C-reactive protein and fibroblast growth factor-23 were increased significantly. Using these biomarkers in clinical practice, diagnosis and prognosis of kidney diseases at an early stage becomes more accurate than surrogate markers. In addition, these biomarkers will also increase our molecular understanding about pathophysiology of kidney damage that could be utilised for developing novel therapy to treat kidney diseases.

2.4 Proteomics in Kidney Disease

Proteome; analogue of genome, may be defined as "total protein complement of a cell, tissue or of an organism at a particular time, condition or environment" (Pennington et al. 1997). The emergence of mass spectrometry with better specificity and sensitivity, enrichment of less abundant proteins and depletion of abundant proteins with different pre-fractionation steps, computer tools for data processing, peptide sequencing, and protein identification has paved the way for analysis of protein and peptides at large scale in high-throughput mode (Matthiesen et al. 2004). Proteomic techniques have been used for the identification and quantification of protein that will be essential to inference the biological functions. Mostly two complementary approaches are common in proteomics; top-down or bottom-up. In top-down approach, isolated proteins were first fractionated, and intact protein was trypsin digested, followed by data acquisition and analysis by mass spectrometry (Han et al. 2006). This approach allows identification of protein, post-translational modifications, and protein isoforms. In contrast, the bottom-up approach includes characterization of protein through peptides generated from protein digestion (Yates 2004). The bottom-up approach is generally called shotgun proteomics when it is performed on a mixture of proteins (Yates 1998). In general, shotgun proteomic experiment includes trypsin digestion of protein mixture to generate peptides, resulting peptides mixture was fractionated and identified by mass spectrometry.

In the past few decades, significant progress has been made to explore the mechanism of kidney disease progression and biomarker discovery for disease diagnosis, prognosis, and their follow-up treatment through proteomic approaches (Mischak et al. 2007; Chen et al. 2018). In this context, body fluids (blood and urine) are seem to be ideal for the identification of new biomarkers as they contained secreted/leaked proteins associated with a disease or altered state of tissue or organ (Lescuyer et al. 2007). Among different body fluids (blood/plasma/serum), proteome is highly complex in which peptide concentration can vary ten order of magnitude. However, urine proteome is comparatively less complex, and the pathophysiological state of kidney is indicated by change in urinary proteome. Considering urine for proteomic study has several advantages as it collected noninvasively in large volume, stored at 37 °C for several hours without any proteolytic degradation and it devoid highly abundant plasma proteins that create hindrance in the extraction process. In this section, we discussed proteomic approaches involved in biomarkers discovery for kidney disease and also highlighting the advantage and its limitation.

A large number of proteomic techniques like capillary electrophoresis, two-dimensional gel electrophoresis, surface-enhanced laser desorption/ionization, liquid chromatography-mass spectrometry (LC-MS), and among others have been used for the identification of kidney biomarkers (Mischak et al. 2015; L'Imperio et al. 2016; Magalhães et al. 2016; Van et al. 2017). Capillary electrophoresis coupled with mass spectrometry is one of the important proteomic technique for fractionation of protein with good resolution in the presence of electric field using capillary in a single step (Maier and Schmitt-Kopplin 2016). It has been used for the identification of biomarker peptides where sample complexity is the limiting factor. In 2010, a human urinary peptidome panel developed CKD273 classifier; composed of 273 specific peptide markers for diabetic kidney disease (Good et al. 2010). Most of the peptide biomarker of CKD273 classifier is derived from collagens, serum albumin, uromodulin, membrane-associated progesterone receptor component 1, sodium/potassium-transporting ATPase γ chain, haemoglobin α chain, α1-antitrypsin, and fibrinogen α chain proteins. CKD273 classifier has been used to diagnose diabetic nephropathy (Zürbig et al. 2012), hypertensive nephropathy (Ovrehus et al. 2015), and progression to end-stage renal disease (Pontillo et al. 2017). In addition, it also predicts the progress of chronic kidney disease from non-albuminuria to macroalbuminuria (Roscioni et al. 2013).

Two dimensional gel electrophoresis coupled with mass spectrometry is another powerful tool to fractionate and visualize thousands of proteins from a complex biological sample. The protein fractionation takes place using two fundamental properties of protein; isoelectric point and molecular weight. After protein fractionation gels were stained with Coomassie brilliant blue/silver stain), and analysed through different software e.g. PDQuest. Further differential protein spots were excised from the gel, trypsin digested, and identified by mass spectrometry. This technique has several advantages like protein quantification, detection of post-translational modification, and protein isoforms and wide application in clinical pathology for biomarker discovery (Park et al. 2006). Ferlizza et al. (2015) reported a reference map of urinary proteome of 23 healthy cats and 17 chronic kidney disease associated cats and found an increased abundance of retinol-binding protein, cystatin M and apolipoprotein-H while a decreased abundance of uromodulin and cauxin that could be a candidate biomarker for chronic disease disease. Recently performed urinary proteome of 22 healthy dogs and 28 dogs with chronic kidney disease using gel electrophoresis revealed uromodulin could be a putative biomarker for chronic kidney disease (Ferlizza et al. 2020). However, two dimensional gel electrophoresis approach is limited by poor reproducibility, less coverage, and unable to detect low abundant proteins. These limitations were overcome by the use of differential gel electrophoresis in which proteins isolated from different treatments were labelled with a specific fluorophore, pooled and run in a single gel. The main advantage of differential gel electrophoresis is less number of replicate gel that increase the reproducibility of the experiment. Fluorophore has four-fold higher linear response than Coomassie brilliant blue/silver stain that results in detection of low abundant proteins (Tannu and Hemby 2006; Sui et al. 2010). Recently, Romanova et al. (2020) examined serum sample of 26 adult chronic kidney disease patients and 10 healthy adults using differential gel electrophoresis and identify 46 differentially expressed proteins, among which 21 proteins were further quantified by multiple reaction monitoring and enzyme-linked immunosorbent assays (ELISA) suggesting these 21 proteins could be an early biomarker for chronic kidney disease progression (Romanova et al. 2020).

In order to identify low abundant protein, another approach has been developed known as multidimensional protein identification technology that allows identification proteins in complex mixtures without prior separation of protein. In this approach, protein lysate was first digested with trypsin followed by separation of peptides using strong cation exchange chromatography and later on by C18 reversed-phase. The resulting peptides were analyzed by mass spectrometry for protein identification. Another proteomic technique known as GeLC has been developed for complex samples in which proteins were first fractionated by one dimensional gel electrophoresis, fractionated protein bands were chopped into slices, digested, extracted and analyzed by reversed phase nanoLC-MS that gives better protein identification. However, above described shotgun approaches perform quantification at peptide level and lack information on the intact protein. In addition, these techniques have poor reproducibility in liquid chromatography separation, variable ion signal, and strong ion suppression effect that makes it challenging. However, these limitations were overcome by chemical tagging or by metabolic labelling of protein during growth of biological material. Zhang et al. performed proteomic profiling of serum samples with acute kidney allograft rejection using isobaric tag for relative and absolute quantification and identified 109 differential proteins that will be helpful for the identification of allograft rejection (Zhang et al. 2020). Further, a label-free proteomics analysis has been developed, which involves the comparison of normalized relative peptide peak intensities between multiple LC-MS datasets using different software that results identification and quantification of peptides and proteins. This technique reduces time for spectra search and simultaneously increases the throughput of biomarker discovery. Recently studied urinary proteome of acute kidney injury patients using gel-free proteomic approach, identified 1810 proteins in which 168 were differential proteins. Further validation through ELISA showed that annexin A5, neutrophil gelatinase-associated lipocalin, and protein S100-P are promising biomarkers for acute kidney injure (Jung et al. 2020).

Once the biomarkers were identified from proteomic techniques quantification/ validation was performed through affinity-based assay like ELISA and western blotting. ELISA is a highly sensitive and specific method for absolute quantification of proteins; however, prior to ELISA molecular structure of protein must be known for purification of protein and generation of antibody to validate its specificity. In addition, it has low dynamic range that needs optimum dilution of antibody for cross-reactivity of antigen for the standard curve (Vaidya et al. 2008). Further, mass spectrometry based absolute quantification of protein with high sensitivity has been developed for validation of biomarker. The two methods are multiple reaction monitoring mode or selected reaction monitoring, which are not discussed in this chapter (Aebersold et al. 2013; Picotti and Aebersold 2012; Nilsson et al. 2010). Therefore, recent development in proteomic techniques offers great promise towards the identification of kidney biomarkers for early disease diagnosis; however, there is a need to develop proteomic techniques used for validation of biomarkers specificity.

2.5 Experimental Model for Studying Kidney Diseases

2.5.1 In Vitro *Models*

The cell lines are one of the reliable methods to study kidney diseases as they can closely mimic the cells' physiological condition in the body. Cell lines also provide an advantage over animal models for cost and ethical constraints. In the primary cell culture, cells are directly separated and propagated from the tissue. It is still used as a useful tool in drug screening and toxicity studies; however, they have limitation of their growth capacity. The secondary cell line or immortalized cells overcome this gap and divide indefinitely (Faria et al. 2019). Various types of glomerular and tubular cell lines are used to study kidney disease as *in vitro* models.

2.5.1.1 Podocytes and Mesangial Cell Lines

Podocytes are the glomerular cells having a role in the filtration and maintenance of the slit diaphragm. These cells are used as *in vitro* models to study the disease condition and can be isolated from rat, mouse and human (Ni et al. 2012). The altered fatty acid metabolism in diabetes and obesity affects the podocytes. Palmitic acid treatment of the cultured podocytes was reported to alter the gene expression profile of stearoyl-CoA desaturase-1, stearoyl-CoA desaturase-2, and carnitine palmitoyltransferase and induced cell apoptosis, which mimics the altered fatty acid metabolism in the diabetic nephropathy (Sieber et al. 2013). Conditionally immortalized human podocyte's treatment with glucose (5 nM and 25 nM), angiotensin II and methylglyoxal (an advanced glycated end-product precursor) mimic the conditions of diabetic nephropathy. Methylglyoxal (1.5 mM) treatment increased the oxidative stress, decreased the expression of nephrin-like protein 1 and Tjp1/ZO-1, which are tight junction proteins and maintains the podocyte integrity and slit diaphragm (He et al. 2020). Mesangial cells are perivascular pericytes in glomerulus involved in mesangial matrix synthesis, maintenance of glomerular hemodynamics, and filtration process. IP15 is a human stable immortalized mesangial cell line with the ability to contract with angiotensin II (L'Azou et al. 2007). Spontaneously transformed rat mesangial cell lines treatment with 25 mM D-glucose resulted in enhanced caspase-3 activity, decreased Bax: Bcl-2 ratio, cell apoptosis, inhibition of nuclear factor kappa B, and transforming growth factor beta activation. High glucose treatment of mesangial cells induces their proliferation via toll-like receptor 4 dependent inhibition of hydrogen sulfide synthesis (Khera et al. 2006). Both podocytes and mesangial cells are the important cells of glomerulus, and their cell lines are very useful in studying the glomerular function in renal disease.

2.5.1.2 Proximal Tubular Epithelial Cell Lines

Proximal tubular epithelial cells perform the primary function of reabsorption of numerous solute present in the glomerular filtrate. These cells are highly prone to injury in kidney toxicity or disease conditions and thus are used as a potential *in vitro* model system to study kidney diseases. Human kidney cells immortalized by transduction with human papillomavirus 16 E6/E7 genes are the most commonly used tubular cell line. Human kidney 2 cells treated with Tenofovir for 72 h showed cell cytotoxicity, oxidative stress, mitochondrial dysfunction, upregulation of tumor necrosis factor-alpha and caspases (3 and 9). These cells mimic the various conditions related to Tenofovir induced nephrotoxicity (Murphy et al. 2017). The human renal proximal tubule epithelial cells immortalized by the human telomerase reverse transcriptase form the RPTEC/TERT1 cell line. These are extensively used to study human kidney disease conditions, including kidney cancer (Simon-Friedt et al. 2015). Human embryonic kidney 293 is another human embryonic kidney cell derived line used for *in vitro* studies. The cell's treatment with nephrotoxic molecule ochratoxin A (25 μM) showed the impairment of miRNA biogenesis correlated to cancer and signal transduction pathways (Zhao et al. 2016).

Normal rat kidney-52E epithelial cells are used to study kidney disease and for toxicological drug screening. The treatment of epithelial cells with transforming growth factor beta-1 activates the various pathway of kidney fibrosis. It induces cells' transition into spindle-like cells and activates Smad2/3 signalling pathway with up-regulation of alpha smooth muscle actin, fibronectin, and integrin-linked kinase expressions (Wei et al. 2015). Similarly, treatment of these cells with high glucose concentration (30 mM) mimics the physiological conditions of diabetic nephropathy. Glucose treatment induces oxidative stress and activates the pro-inflammatory cytokines (TNF-α and IL-1β) and fibrogenic factors (TGF-β1) in the kidney epithelial cells (Sharma et al. 2019). Caki-1 is a human clear cell renal cell carcinoma (ccRCC) line with the morphology of epithelial cells, and used to study renal cancer. Other cells are porcine kidney cells (LLC-PK1) and Madin-Darby canine kidney cells (MDCK) (Barnett et al. 2018). Together, the previous reports suggest that *in vitro* kidney disease models are vital for renal molecular studies, drug screening, and safety assessment. The range of *in vitro* renal models allows researchers to select the most suitable model to answer their scientific questions.

2.5.2 In Vivo *Rodent Models of Kidney Disease (Non-transgenic)*

2.5.2.1 Acute Kidney Injury Models

Animal models are widely used for understanding the human kidney disease as they mimic the human condition to a more considerable extent. Various type of non-transgenic *in vivo* models are enlisted in Table 2.2.

Ischaemia-Reperfusion (I/R) Model

Ischaemia-reperfusion is one of the primary reasons of acute kidney injury. It triggers robust oxidative damage and inflammatory response in the tubular cells, leading to lipid peroxidation, mitochondrial damage, and cell death. Ischemia-reperfusion induced acute kidney injury can be developed in both rat and mouse. This model involves a surgical procedure in which midline dorsal incision is used to locate the kidneys and bilateral clamping of kidney pedicles for 30 minutes. The kidneys are observed for color change before removing the clamps. The renal ischaemia-reperfusion results in a sharp increase in the serum creatinine and blood urea nitrogen. Kidney histopathology showed cast formation, severe tubular dilation, loss of proximal tubular brush border, cell apoptosis, and interstitial inflammation (LiGong et al. 2020; Zheng et al. 2019). This model is a robust model to study the pathophysiology of I/R induced acute kidney damage.

Acetaminophen

Acetaminophen is a widely used antipyretic and analgesic drug. It is a safe drug at therapeutic doses; however, overdosing of the drug causes hepatotoxicity and nephrotoxicity. At higher doses, harmful metabolites like N-acetyl-p-benzoquinone imine are produced from acetaminophen leading to severe oxidative damage and kidney toxicity. A single oral dose of paracetamol (500 mg/kg BW) administered to rats resulted in elevated kidney injury markers (serum urea and creatinine), lipid peroxidation, glutathione depletion, and reduced activities of antioxidant enzymes (super oxide dismutase, catalase, and glutathione peroxidase) (Kandemir et al. 2017). Mice treated with acetaminophen for 24 h at a dose of 300 mg/kg (intraperitoneal) increased the creatinine, blood urea nitrogen, and cystatin C levels. Acetaminophen treatment also elevated the markers of tubular injury *viz.* kidney injury molecule−1 and neutrophil gelatinase-associated lipocalin by 5 and 347-fold

Table 2.2 Various non-transgenic rodent models of kidney diseases

Agent	Model/strain	Method	Characteristics	References
Acute kidney injury				
Acetaminophen	Sprague-Dawley (SD) rats	500 mg/kg BW and 2 g/kg, single oral dose	↑ Serum urea and creatinine, lipid peroxidation, oxidative stress and tubular damage	Kandemir et al. (2017)
	BALB/c mice	300 mg/kg BW intraperitoneal, single dose	↑ Creatinine, BUN, cystatin C, NGAL and KIM-1	Hua et al. (2018)
	SD rat	2 g/kg BW single dose	↑ Urea, creatinine, MDA and TNF-α	Haidara et al. (2020)
Adriamycin	SD rat	2 mg/kg BW intravenous for 20 days	Lipid peroxidation and oxidative stress	Okuda et al. (1986)
	BALB/c	10 mg/kg BW intravenous, single dose	Proteinuria, glomerulosclerosis, tubular atrophy and fibrosis	Tan et al. 2013
Cisplatin	SD rat	7 mg/kg BW intraperitoneal, and 6 mg/kg BW, single dose	↑ BUN, creatinine and TNF-α Oxidative stress, tubular degeneration and altered urine output	El Amir et al. (2019)
Cyclophosphamide	Albino rats	75 mg/kg BW, intraperitoneal (six doses)	Altered urine output, serum Na+, serum kidney injury markers and tubular damage	El-Shabrawy et al. (2020)
	Swiss albino mice	CP (200 mg/kg BW intraperitoneal) two doses	↑ BUN, creatinine, LDH and Tubular degeneration, inflammation and apoptosis	Sharma et al. (2017)
Cyclosporine A	SD rat	7.5 mg/kg,15 mg/kg and 20 mg/kg BW subcutaneous	↓ GFR rate, ↑ creatinine levels, endothelial dysfunction and tubular atrophy	Bing et al. (2006) Huang et al. 2018

(continued)

Table 2.2 (continued)

Agent	Model/strain	Method	Characteristics	References
Ischaemia-reperfusion model	SD rat	Bilateral clamping of renal pedicles for 30 minutes	↑ BUN and creatinine, tubular injury, cast formation, loss of brush border and apoptosis	Gong et al. (2020)
	C57bl/6 mice		Oxidative stress, tubular degeneration, cell apoptosis, inflammation and fibrosis	Zheng et al. (2019)
Mercuric chloride	SD rat	1 mg/kg and 3.5 mg/kg BW (intraperitoneal)	Proximal tubular damage, mitochondrial swelling, vacuolization and apoptosis	Stacchiotti et al. (2003)
		2 mg/kg BW subcutaneous	↑ BUN, creatinine, ↓ GSH and GPx, glomerular and tubular degeneration, inflammation	Gao et al. (2016)
Renal fibrosis				
Carbon tetrachloride	Wistar rat	1.5 mL/kg BW of 30% CCl_4 in olive oil along with 1 gm of BSA as a 3 mL solution in NS	Connective tissue proliferation, protein cast and lymphocytic infiltrations in peritubular area	Limbu et al. (2017)
	Swiss albino mice	0.5 ml/ kg BW /1 ml olive oil for 4 weeks	↑ Collagen deposition and α-SMA upregulation, oxidative stress, ↑ TNF-α and IL-6	Ma et al. (2020)
		1.5 ml/kg BW (1:1 v/v with olive oil) intraperitoneal twice a week for 15 days	Lipid peroxidation, oxidative stress, ↑cytokines and apoptosis	Safhi (2018)

(continued)

Table 2.2 (continued)

Agent	Model/strain	Method	Characteristics	References
Vanadate	Wistar rat	0.125 mg/mL for 12 weeks	↓fluid intake, urine output, body weight, and urinary creatinine ↑ MDA, cystatin C and KIM-1	Ścibior et al. (2014)
		0.9 mg/kg BW subcutaneous for 16 days	↑ collagen deposition and inflammation	Al-Bayati et al. (1989)
Folic acid	CD1 mice	240 mg/kg BW intraperitoneal	Folic acid crystals deposition, tubular obstruction, necrosis and interstitial fibrosis	Yuan et al. (2003)
	C57BL/6 mice	250 mg/kg BW, intraperitoneal, single dose	Impairment of renal function, lipid peroxidation, oxidative stress, inflammation and fibrosis	Li et al. (2020) and Hsu et al. (2020)
Nephrectomy	129/Sv and C57BL/6JRj	5/6 nephrectomy of left kidney through left laparotomy	↑ Serum creatinine and urinary albumin/creatinine ratio glomerulosclerosis, tubular damage and fibrosis	Hamzaoui et al. (2020)
	C57BL/6 mice	Upper and lower poles of left kidney were ligated	↑ BUN, creatinine and proteinuria ↑ α-SMA and severe renal fibrosis	Tan et al. (2019)
Unilateral ureteral obstruction	C57BL/6 mice	Ligation of left ureter	↑ BUN and creatinine Pyelonephrosis, renal papillary necrosis, tubular degeneration and fibrosis	Gu et al. (2019, 2020)

(continued)

Table 2.2 (continued)

Agent	Model/strain	Method	Characteristics	References
Chronic kidney disease				
Adenine	Wistar rat	0.25% adenine in diet for 16 weeks	↑ Body weight BUN, creatinine, LDH activity, and potassium concentration glomerulosclerosis, and tubular atrophy	Diwan et al. (2013)
	SD rats	Dietary adenine (0.75%) for 3 weeks and unilateral nephrectomy	Hypocalcemia, hyperphosphatemia, anemia and ↓ GFR Tubular degeneration, inflammation and interstitial fibrosis	Abellán et al. (2019)
High calorie diet	Wistar rat	Special diet (carbohydrate 56%, 30% fat, and 14% protein) for 28 weeks	Renal vasorelaxation, ↓ catalase activity and catalase mRNA transcripts IL-1β and endothelial dysfunction	Rangel Silvares et al. (2019)
	Wistar rat	High fructose diet (35%) for 16 weeks	↑ BUN, creatinine, lipid peroxidation, amyloid deposition, atherosclerosis and vacuolar degeneration of tubular epithelial cells	Bratoeva et al. (2017)
	Rcc:Han WIST rats	Cafeteria diet for 14 weeks	↓ Plasma antioxidant capacity, cytoplasmic antioxidant enzymes and induction of apoptosis	La Russa et al. (2019)
	C57BL/6 mice	HFD (60% calorie from fat) for 16 weeks	↑ Weight gain, blood glucose, BUN and creatinine, albuminuria, oxidative stress Glomerular and tubular degeneration and apoptosis	Sun et al. (2020)

(continued)

Table 2.2 (continued)

Agent	Model/strain	Method	Characteristics	References
DOCA	SD rat	Uninephrectomy and DOCA (50 mg pellet) and 0.9% NaCl and 0.2% KCl in drinking water for 4 to 16 weeks	↑ Albumin/creatinine ratio, urinary KIM-1 marker excretion, ↓ GFR, accumulation of glomerular extracellular matrix, collagen and mesangial proliferation	Wang et al. (2017a, b)
	SD rat	DOCA (3.3 mg/day BW subcutaneous) along with 1% NaCl +0.2% KCl in the drinking water	Higher blood pressure with proteinuria and glomerulosclerosis	Polichnowski et al. (2017)
Streptozotocin	SD rats	50 mg/kg BW intraperitoneal, single dose	↑ BUN and creatinine, lipid peroxidation and ↓ antioxidant parameters (SOD, CAT and GPx), mesangial proliferation, thickening of basement membrane, glomerular hypertrophy and degeneration	Qi et al. (2020)
	SD rats	60 mg/kg BW intraperitoneal, single dose	↑ Blood glucose, urinary volume, urinary protein excretion, lipid peroxidation and eNOS immunoreactivity	Nakhoul et al. (2006)

(continued)

Table 2.2 (continued)

Agent	Model/strain	Method	Characteristics	References
Lupus nephritis	(NZB X NZW)F1 (B/W) lupus-prone mice MRL/lpr lupus mice and BXSB mice strain	Spontaneous models	Deposition of immune complexes in subendothelium or mesangium, mesangial proliferation, glomerulonephritis, and tubular degeneration	McGaha and Madaio (2014)
	BALB/c mice	0.5 ml Pristane (2,6,10,14 tetramethylpentadecane) intraperitoneal	Moderate proteinuria and proliferative glomerulonephritis	Satoh et al. (1995)
Thy-1 nephritis model	Brown Norway/ RijHsd rat	Single intravenous injection of mouse or rabbit anti-thymocyte monoclonal antibody (OX-7 1 mg/kg bw)	Proteinuria, hematuria, mesangiolysis, mononuclear infiltration and endothelial damage	Westerweel et al. (2012)

Abbreviations: *α-SMA* Alpha smooth muscle actin, *BW* Body weight, *BUN* Blood Urea nitrogen, *CAT* Catalase, *DOCA* Deoxycorticosterone acetate, *eNOS* endothelial nitric oxide synthase, *GFR* Glomerular Filtration Rate, *GPx* Glutathione Peroxidase, *GSH* Glutathione, *IL-6* Interleukin-6, *IL-1β* Interleukin--1β, *KIM-1* Kidney Injury Molecule 1, *LDH* lactate dehydrogenase, *MDA* Malondialdehyde, *NGAL* Neutrophil gelatinase-associated lipocalin, *SOD* Superoxide dismutase, *TNF- α* Tumor necrosis factor-alpha

at mRNA levels, respectively (Hua et al. 2018). In another study, a single dose of acetaminophen (2 g/kg) to rats resulted in elevated kidney injury markers in serum, alterations in the ultrastructure of the proximal convoluted tubule, lipid peroxidation and high tumor necrosis factor-alpha (Haidara et al. 2020). Acetaminophen induced renal injury model is used for long time and one of the preferred models to study acute kidney injury till date.

Adriamycin

It is an anti-neoplastic and antibiotic drug used to study chronic glomerular disease. Rats treated with adriamycin for short term manifest glomerular lesions, fibrosis, and kidney damage. A dose of 2 mg/kg by intravenous route in Sprague Dawley rats was given for twenty days to induce kidney disease (Okuda et al. 1986). Adriamycin causes oxidative stress and lipid peroxidation in the glomerular epithelial cells. A 10 mg/kg dose of adriamycin to C57/balbc mice *via* tail vein resulted in proteinuria, podocytes injury, glomerular damage, tubular atrophy, and fibrosis. The adriamycin-induced kidney damage model mimics the human disease of focal segmental glomerulosclerosis (Tan et al. 2013). The adriamycin toxicity is mainly attributed to the oxidative damage caused to cellular components including the plasma membrane, lipids, and mitochondria.

Cisplatin

Cisplatin is a common anticancer agent used for the treatment of solid tumors. Kidney damage is one of the side effects of this drug; therefore, this is used to develop kidney injury models in biomedical research. Cisplatin-induced acute kidney injury involves the induction of oxidative stress, kidney inflammation, vascular and tubular injury. Cisplatin was reported to impair kidney functions, induce tubular damage, kidney cast formation, and increased neutrophil gelatinase-associated lipocalin expression in kidneys of mice (Wu et al. 2020). Cisplatin (7 mg/kg BW intraperitoneal, single dose) reported elevating the kidney injury markers (blood urea nitrogen and creatinine) and tumor necrosis factor-alpha in the serum of SD rats. The kidney tubule showed vacuolar degeneration and necrosis on microscopic evaluation (El Amir et al. 2019). In another study, cisplatin treatment (6 mg/kg BW, single dose) increased the serum urea, creatinine, and urine glucose levels in rats. The drug also altered the urine output and osmolality excretion rate (Hosseinian et al. 2016). Overall, cisplatin triggers complex molecular pathways involving tubular cell injury, inflammatory response, and cell death.

Cyclophosphamide

Cyclophosphamide is a widely used agent for a range of neoplastic conditions, autoimmune diseases, and organ transplantation. Nephrotoxicity is one of the primary side effects of this drug. Cyclophosphamide gets metabolized in the liver into highly reactive compound acrolein, which is responsible for oxidative damage and cell toxicity (Mahmoud 2014). Cyclophosphamide (200 mg/kg intraperitoneal) administration for two consecutive days resulted in elevated creatinine and blood urea nitrogen in mice. The drug resulted in proximal tubular swelling, glomerular degeneration, and elevated pro-inflammatory cytokines (Interleukin-1β and tumor necrosis factor-alpha). Cyclophosphamide also induced cell apoptosis by altering caspase 3/9 activity and Bax/Bcl-2 ratio (Sharma et al. 2017). In a recent study, nephrotoxicity was induced by administering cyclophosphamide (75 mg/kg bw, intraperitoneal) on days 3, 4, 5, 19, 20, and 21 of the trial. The agent altered the urine volume, serum Na^+ and serum kidney injury markers. The kidney tissue showed pathological changes with lipid peroxidation, elevated pro-inflammatory cytokines and apoptotic changes (El-Shabrawy et al. 2020). The induction of oxidative stress, inflammation, cell injury, and apoptosis in the kidney tissue by cyclophosphamide makes it a suitable model to study drug-induced renal damage.

Cyclosporine A

Cyclosporine A is an immunosuppressant agent used to increase the efficacy of organ transplantation. It is a calcineurin inhibitor, and its mechanism of action is linked to the side effect of nephrotoxicity. Cyclosporine A interacts with the renal tubular cell membrane, and long-term use leads to tubulointerstitial nephropathy and kidney fibrosis. The drug is used to study interstitial fibrosis in rodent models; however, there are limitations of cost, hepatic dysfunction, and higher doses in animal studies than in clinical trials (Zhao et al. 2015; Nogueira et al. 2017). Cyclosporine A is given by subcutaneous route (7.5 mg/kg and 15 mg/kg) in rats for 28 days resulted in decreased glomerular filtration rate with an elevated creatinine level. Morphological investigation showed tubular atrophy with endothelial dysfunction (Bing et al. 2006). In another study on rats, cyclosporine A (30 mg/kg) treatment elevated serum blood urea nitrogen and creatinine levels and reduced the endogenous creatinine clearance. Cyclosporine A also induced kidney cell injury by causing vacuolar degeneration in glomerular cells (Huang et al. 2018). The short-term administration of cyclosporine A produces acute renal cell damage, however, long term use may lead to the development of chronic tubule-interstitial nephropathy.

Mercuric Chloride ($Hgcl_2$)

Mercuric chloride is used to induce kidney damage by triggering oxidative stress in the organ. The toxicity mechanism involves disturbing membrane potential, precipitation of proteins, alteration in protein synthesis, and intracellular calcium homeostasis (Zalups 2000; Dhanapriya et al. 2016). $HgCl_2$ at 1 mg/kg and 3.5 mg/kg BW (intraperitoneal) doses induced nuclear and cytoplasmic changes in the proximal tubular cells. The ultrastructural pathology showed nucleolar segregation, mitochondrial swelling, vacuolization, loss of brush border, necrosis and apoptosis of tubular epithelial cells (Stacchiotti et al. 2003). Similarly, rats treated with a single dose of HgCl2 (2 mg/kg) subcutaneously showed elevated kidney serum injury markers (Blood urea nitrogen and creatinine), depletion of glutathione level, and decreased glutathione peroxidase activity. The kidneys appeared enlarged and whitish in color. Microscopically, glomerular degeneration, swelling of proximal tubule epithelial cell, granular degeneration, infiltration of inflammatory cells, and extensive necrosis were reported (Gao et al. 2016). The $HgCl_2$ induced kidney damage is a classical model to study acute kidney injury. Overall, various analgesic, anticancer, immunosuppressive drugs, and chemicals induced acute kidney injury model offers a great opportunity to explore the pathophysiology and molecular mechanisms of acute kidney injury.

2.5.2.2 Kidney Fibrosis Models

Carbon Tetrachloride

Carbon tetrachloride is a potent toxin to induce oxidative injury in various physiological conditions. The metabolism of CCl_4 in the liver leads to the generation of trichloromethyl radical (CCl_3•) and trichloromethyl peroxyl radical (CCl_3O_2•) which causes oxidative tissue injury by binding with the macromolecules (lipids, proteins, and DNA) (Chhimwal et al. 2020). The administration of carbon tetrachloride (0.5 ml/kg BW/1 ml olive oil) to mice for 4 weeks induced kidney fibrosis by deposition of collagen and upregulation of α-smooth muscle actin. It also promoted oxidative stress and inflammatory response by activation of proinflammatory cytokines (Interleukin-6 and tumor necrosis factor-alpha) (Ma et al. 2020). A combination of carbon tetrachloride and bovine serum albumin (1.5 mL/kg of 30% carbon tetrachloride in olive oil and 1 g of bovine serum albumin as a 3 mL solution in NS) was reported to induce kidney fibrosis in female rats. The combination was found to develop the early stages of kidney fibrosis (Limbu et al. 2017). In another study on Swiss albino mice, carbon tetrachloride treatment at 1.5 ml/kg (1:1 v/v with olive oil) intraperitoneally twice a week for 15 days resulted in lipid peroxidation and disruption of antioxidant defense in the kidney tissue. It also increased the cytokines' levels with the induction of apoptosis in the kidney cells (Safhi 2018). Free radical generation, lipid peroxidation, inflammation, and increased fibrous tissue deposition are the main steps involved in carbon tetrachloride-induced renal fibrosis.

Vanadate and Folic Acid

Vanadate is pro-oxidant, can alter the oxidative stress levels in the cells, and activates oxidative damage mechanisms. An exposure of sodium metavanadate (0.125 mg V/mL) to rats for 12 weeks decreased the fluid intake, urine output, body weight, and urinary creatinine excretion rate. The relative kidney weight and the malondialdehyde levels were also found elevated. The study suggested that cystatin C and KIM-1 might be the most appropriate makers to evaluate sodium metavanadate-induced alterations in kidney function (Ścibior et al. 2014). Vanadate-induced kidney fibrosis is basically dose-dependent. A dose of 0.9 mg/kg for 16 days *via* the subcutaneous route to rats resulted in kidney fibrosis and inflammation. Pathological and other biochemical parameters confirmed the kidney injury. After 12 days of administration, cellular proliferation was observed in the cortex and medulla, and fibrotic changes appeared in the glomerular tuft and interstitium. The collagen deposition in the kidney was reported highest after 25th day of the trial (Al-Bayati et al. 1989).

Folic acid, also known as vitamin B, is water-soluble and helps the body in the production of new cells. The administration of folic acid to (240 mg/kg by intraperitoneal route) mice reported depositing folic acid crystals and obstruction in the tubules within 2 weeks leading to acute tubular necrosis and nephrotoxicity. The animals were reported to develop interstitial fibrosis by the end of fourth week (Yuan et al. 2003). In another study, the folic acid treatment in mice (250 mg/kg BW, i.p., single dose) resulted in kidney function impairment, lipid peroxidation, and reduction in the kidney antioxidant parameters. Besides, severe glomerular inflammation, tubular swelling, kidney fibrosis, and apoptosis were also noticed. (Li et al. 2020; Hsu et al. 2020). Deposition of folic acid crystal is the main triggering point for folic acid-induced renal fibrosis.

Nephrectomy and Unilateral Ureteral Obstruction Models

These methods are commonly used in the study of kidney disease; however, there is disadvantage of performing the surgery, loss of kidney tissue and mortality. Gender and strain (Sprague Dawley are resistant then Wistar rats) also influence kidney disease progression in the experimental model. Hamzaoui et al. (2020) developed a kidney disease model in 129/Sv and C57BL/6JRj mice by performing 5/6 nephrectomy of the left kidney through left laparotomy. A significant increase was observed in the creatinine and urinary albumin/creatinine ratio after 12 weeks. Histopathological evaluation showed glomerulosclerosis, tubular damage, perivascular, and interstitial fibrosis, vascular thickening and inflammation in the kidney tissue. Recently a highly efficient method was developed in C57BL/6 mice to reduce the mortality and infection in the 5/6 nephrectomy model of kidney fibrosis. In this method, the right kidney was removed surgically, and after 1 week the upper and lower poles of the left kidney were ligated directly to cause necrosis of ligated poles. The method results in high levels of serum blood urea nitrogen, creatinine, and proteinuria after 12 weeks. The tissue also showed increased expression of smooth muscle alpha-actin and severe kidney fibrosis (Tan et al. 2019).

Unilateral ureteral obstruction is a model of urinary path obstruction resulting in damage of kidney structures. It leads to kidney enlargement and hydronephrosis. Unilateral ureteral obstruction model is the most frequently used model to study non-immunological tubulointerstitial fibrosis (Grande et al. 2010). In this model, the left kidney of mouse is exposed surgically, and ligation of left ureter is performed. Unilateral ureteral obstruction results in an elevation in the serum kidney injury markers (blood urea nitrogen and creatinine). The kidney tissue shows typical pathology of obstructive nephropathy with pyelonephrosis, renal papillary necrosis, severe tubular dilatation and atrophy, interstitial inflammation, and fibrosis (Gu et al. 2019). Similarly, Gu et al. (2020) showed that Unilateral ureteral obstruction in mice resulted in proinflammatory activation cytokines and kidney fibrosis along with impairment of kidney function. This is a frequently used model; however, the drawback is that a complete obstruction rarely occurs in humans.

2.5.2.3 Chronic Kidney Disease Models

Adenine

Adenine administration in the diet (0.25% to 0.75%) was used as a potential model to study chronic kidney disease in rodents. Adenine is converted to 2,8-dihydroxyadenine, which deposit as crystals in the proximal tubular cells, leading to tubular degeneration and kidney damage. Administration of 0.25% adenine in the diet for 16 weeks in rats mimics the slow progression of chronic kidney disease in humans. The rats showed weight gain with increased blood urea nitrogen, creatinine, lactate dehydrogenase (LDH) activity, and potassium concentration. Microscopically, the kidney showed glomerulosclerosis, loss of proximal tubular brush border epithelium and tubular atrophy (Diwan et al. 2013). In another method, dietary adenine (0.75%) for 3 weeks combined with unilateral nephrectomy in rats was used to develop chronic kidney disease. The rats develop hypocalcemia, hyperphosphatemia, and anemia along with altered glomerular filtration rate. Histopathology of kidneys showed tubular degeneration and dilatation, adenine crystal deposits, mononuclear cell infiltration and interstitial fibrosis (Abellán et al. 2019). The adenine-induced model is also preferred to study the chronic kidney disease associated secondary hyperparathyroidism and hyperphosphatemia related to mineral bone disease.

Streptozotocin

Streptozotocin is a N-acetyl glucosamine analogue transported to pancreatic beta cells by GLUT-2 receptors, causing toxicity and used to develop diabetic nephropathy in rodents. Diabetic nephropathy is the highest cause of chronic kidney disease and end-stage kidney disease. At high doses, streptozotocin can also lead to nonspecific kidney toxicity; therefore, single or multiple low doses are used to induce diabetic nephropathy in mice. Streptozotocin at a dose of 60 mg/kg BW i.p. was reported to induce diabetic nephropathy in Sprague Dawley rats. The rats showed elevated blood glucose, urinary volume, and urinary protein excretion. The kidney tissue also showed increased lipid peroxidation and endothelial nitric oxide synthase immunoreactivity (Nakhoul et al. 2006). Similarly, streptozotocin (50 mg/kg BW i.p.) administration to male Sprague Dawley rats elevated the kidney injury markers (blood urea nitrogen and creatinine), lipid peroxidation and antioxidant parameters (superoxide dismutase, catalase and glutathione peroxidase). Histopathologically, kidney tissue showed glomerular hypertrophy, mesangial proliferation, thickening of the basement membrane, glomerular epithelial cell swelling, and infiltration of inflammatory cells (Qi et al. 2020). Streptozotocin is very effective in inducing the diabetic condition in rodent; therefore, this model is highly preferred to study diabetic nephropathy in experimental animals.

Deoxycorticosterone Acetate Salt

Deoxycorticosterone acetate (DOCA), a synthetic mineralocorticoid derivative, and high salt intake are used to induce hypertension and kidney damage in rodents. This model mimics the human disease condition to a greater extent. The uninephrectomized rats receiving deoxycorticosterone acetate (DOCA; 50 mg pellet) and 0.9% NaCl plus 0.2% KCl in drinking water develop increased systolic blood pressure and kidney and heart weights after 4 to 16 weeks. The DOCA salt treatment decreased GFR and increased the urinary albumin/creatinine ratio and urinary kidney injury molecule-1 marker excretion. Microscopically, kidney tissue showed accumulation of extracellular matrix in glomeruli, mesangial proliferation, and collagen deposition in the kidney tissue (Wang et al. 2017a, b). Similarly, Sprague Dawley rats administered with DOCA (3.3 mg/day s.c.) and 1% NaCl plus 0.2% KCl in the drinking water developed higher blood pressure with proteinuria after 6 weeks. The kidney pathology showed the development of glomerulosclerosis in rats (Polichnowski et al. 2017). Hypertension is one of the primary reasons for chronic kidney disease, as high blood pressure causes deleterious effects on kidney vasculature. DOCA salt-induced model provides an opportunity to study the pathophysiology of hypertension-induced chronic kidney disease in experimental animals.

Diet-Induced Chronic Kidney Disease Models

A metabolic syndrome is a group of complications that includes obesity, diabetes, hypertension, and dyslipidemia. Patients with metabolic syndrome are having more risk of kidney injury (Nashar and Egan 2014). An obese person is more prone to chronic kidney injury, although the exact mechanism is not clear. A study reported that a few weeks of weight gain leads to glomerular cell proliferation and structural changes in the kidney (Laurentius et al. 2019). Lifestyle and high-calorie diet are significant factors for the development of metabolic syndrome and related complications. There are many diets induced rodent model which are used to study the kidney damage.

A high-fat diet contains a high-fat percentage, increasing the calorie intake and leading to obesity and related complications. Feeding of high-fat diet (60% calorie from fat) to C57BL/6 mice for 16 weeks resulted in excessive weight gain, elevated blood glucose levels, and kidney function impairment. The mice showed increased accumulation of blood urea nitrogen, creatinine and albuminuria. The kidneys showed marked degeneration in the glomeruli and tubules, leading to damaged glomerular filtration barrier and apoptosis of tubular epithelial cells. The intake of high-fat diet triggers lipogenic pathways in the kidney tissue and increases kidney triglyceride and cholesterol contents. It also increases oxidative stress and induction of pro-apoptotic pathways in the kidney tissue (Sun et al. 2020). A daily intake of high-fat diet (carbohydrate 56%, 30% fat, and 14% protein) for 28 weeks in Wistar rats altered kidney vasorelaxation, decreased catalase activity, and catalase mRNA transcripts. The high-fat diet fed rats showed upregulation of interleukin-1β and endothelial dysfunction in the kidneys. The study emphasized that vascular changes as a result of high caloric intake play an important role in kidney dysfunction (Rangel Silvares et al. 2019). Similarly, a high fructose diet is also used to induce kidney damage. High fructose ingestion affects blood pressure, adenosine triphosphate depletion, and uric acid generation, leading to glomerular hypertension, tubulointerstitial damage, and inflammation (Johnson et al. 2010). Male Wistar rats exposed to a high fructose diet (35%) for 16 weeks resulted in obesity, diabetes, and kidney impairments. The fructose-rich diet elevated the serum kidney injury markers (blood urea nitrogen and creatinine) and malondialdehyde levels in the kidney tissue. Microscopically, kidney tissue showed amyloid deposition, atherosclerosis, vacuolar degeneration of tubular epithelial cells, and eosinophilia in the distal tubules (Bratoeva et al. 2017). Recently, La Russa et al. (2019) showed that ad libitum feeding of cafeteria diet (cookies, snakes, chocolates, potato chips) for 14 weeks used to develop obesity and associated kidney disease in rats. This induced severe prooxidant effects by reducing the plasma antioxidant capacity, cytoplasmic antioxidant enzymes, and induction of apoptosis in the kidney cells.

Lupus and Thy-1 Nephritis Models

Kidney damage resulting from the autoimmune reaction is one of the comorbidities related to Systemic Lupus Erythematosus. The disease is characterized by deposition of immune complexes in sub-endothelium or mesangium, mesangial proliferation, glomerulonephritis, deposition of casts in the tubules and tubular degeneration (McGaha and Madaio 2014). There are mice strains that show spontaneous development of the lupus nephritis *viz.* (NZB X NZW) F1 (B/W) lupus-prone mice, deoxyribonuclease 1 (Dnase1)-deficient mice, MRL/lpr lupus mice, and BXSB mice strain. However, intraperitoneal administration of pristane (2,6,10,14 tetramethylpentadecane) to mice is one of the potential inducible lupus nephritis models. BALB/c mice injected with pristine were reported to develop moderate proteinuria and proliferative glomerulonephritis after six months (Satoh et al. 1995). The Thy-1 antigen is mainly present on thymocytes, however, also expressed on the glomerular mesangial cells. Thy-1 nephritis is an experimental mesangio-proliferative glomerulonephritis model, induced by the single intravenous injection of mouse or rabbit anti-thymocyte monoclonal antibody (Yang et al. 2010). The disease condition is characterized by marked proteinuria, hematuria, mesangiolysis, mononuclear infiltration, and endothelial damage (Westerweel et al. 2012). An autoimmune reaction is one of the major cause of chronic kidney disease, and existing animal models are very useful to explore the molecular aspect of such predisposition.

Immunoglobulin A Nephropathy

Immunoglobulin A (IgA) nephropathy is one of the leading cause of chronic kidney disease and kidney damage worldwide. The condition is characterized by the deposition of IgA1-immune complexes in the mesangium with co-deposition of variable immunoglobulin G/M and complement C3 (Suzuki and Suzuki 2018). Immunoglobulin A nephropathy can be induced in rats by oral and intravenous administration of 0.1% bovine serum albumin for 12 weeks. The rats immunized with bovine serum albumin showed severe proteinuria and a higher urine protein/creatinine ratio. Microscopically, the rat kidneys showed mesangial proliferation and expansion and diffuse deposition of Immunoglobulin A in glomeruli. The kidney tissue showed upregulation of proinflammatory cytokines and activation of kidney NLR family pyrin domain containing-3 inflammasome (He et al. 2015). Similarly, another study used a combination of bovine serum albumin (400 mg/kg, orally), carbon tetrachloride (0.1 ml dissolved in 0.5 ml castor oil, s.c.), and lipopolysaccharide (0.05 mg, i.v.) to induce IgA nephropathy in rats (Wei et al. 2017). Zhang et al. (2010) used a fusion protein MBP-20 consist of maltose-binding protein and a peptide containing 20-amino-acid from *Staphylococcus aureus* to induce IgA nephropathy in Balb/c mice. Mice were immunized with MBP-20 (0.5 mg) for 21 weeks. The mice showed hematuria and increased protein/creatinine ratio. Kidney tissue showed mesangial matrix proliferation and expansion in light microscopy and electron-dense deposits in the mesangium and subendothelium, ultrastructurally. The model was closely found to mimic the clinical and pathological features of IgA nephropathy.

2.6 Alternative *In Vivo* Models for Kidney Injury

2.6.1 Zebrafish Model

There is an increasing interest in Zebrafish (*Danio rerio*) as a model to study kidney diseases. A less maturation time, the large number of offspring, easy handling, low rearing cost, and less ethical constraints make them favourites for the high throughput screening of molecules for various disorders (Katoch and Patial 2020). The transparency of embryos provides an added advantage of studying organs *in situ*. The genetic models of kidney diseases can be easily created in zebrafish. Exposure to nephrotoxic drugs like cisplatin and gentamicin are used to induce acute kidney injury in zebrafish (Hentschel and Bonventre 2005; Sharma et al. 2014). These compounds' toxicity leads to cellular vacuolization, loss of brush border, degeneration of proximal convoluted tubules necrosis, and inflammation. Gentamicin administration to zebrafish at (0–100 mg) different concentrations produced pathological kidney tissue changes. A dose-dependent kidney damage was observed in adult zebrafish, and high doses lead to the degeneration of glomerulus and proximal tubules (Kato et al. 2020). Adriamycin treatment to 3–4 dpf zebrafish (1,2,4 mg/ml) for 24 and 48 h affected the fish survival and caused acute kidney injury. Similarly, puromycin was used to study the podocyte injury in zebrafish. Both the models are used to determine the podocyte injury and glomerular damage in zebrafish larvae and side effects to cardiovascular injury (Rider et al. 2018). Many other agents like acetaminophen, benzo(a)pyrene, sodium benzoate, and mycotoxins are used to induce acute kidney injury in zebrafish (Sharma et al. 2014). Overall, zebrafish has emerged as a promising model to study acute kidney injury, chronic kidney disease, and kidney related drug developmental studies.

2.6.2 *Drosophila Model*

Drosophila is an excellent model for understanding the development of the human kidney system. It is quick, inexpensive, and has extensive functional similarities to mammalian kidney function (Dow et al. 1995; Millet-Boureima et al. 2019). The compact genome of drosophila is wholly sequenced, and around 70% of genes have human homologs. Drosophila has a small body size and fastest filtration rate. They have a separate section for their kidney function, the Malpighian tubules (similar to kidney tubules) and two clusters of nephrocytes within the body cavity (identical to podocytes in the glomerular kidney) (Millet-Boureima et al. 2019). The fly's kidney system is aglomerular, and active transport is responsible for urine formation rather than selective reabsorption (Millet-Boureima et al. 2018; Dow and Romero 2010).

Drosophila nephrocytes were used to study the glomerular filtration barrier disruption and podocyte function in kidney disease. Garland and pericardial are the two types of nephrocytes present in the drosophila. The fenestrated endothelium is absent in nephrocytes, and two barriers include the basement membrane and nephrocyte slit diaphragm. The nephrocytes can be dissected out quickly and used for filtration and uptake studies. Chemical induced kidney injury can be created in the flies by feeding the compound (Marelja and Simons 2019). Malpighian tubules are also used to study kidney physiology and diseases due to their simple anatomy and functional similarity. They contain high cytochrome P450 and glutathione transferase levels and perform the function of detoxification as by mammalian kidney tubules. Malpighian tubules have been used as models to study human calcium oxalate nephrolithiasis and polycystic kidney disease (Millet-Boureima et al. 2018). Research on drosophila resulted in the discovery of sub lethal human V-ATPase mutations leading to kidney tubular acidosis (Dow and Romero 2010). Therefore, drosophila provides a rapid model system to study kidney development, physiology, and disease pathology.

2.7 Chronic Kidney Disease in Animals

There is limited information available on chronic kidney disease in large animals. However, chronic kidney disease was well reported in cats and dogs with more prevalence in older animals. A study reported more than 30% of cats and 10% of dogs over 15 years of age with chronic kidney disease at the University of Minnesota Veterinary Medical Centre. The clinical manifestation of chronic kidney disease in small animals is generally seen as loss or retention of compounds. Moreover, kidneys' inability to regulate water balance results in polydipsia and polyuria (Polzin 2011; Bartges 2012). Potassium, sodium, calcium, and phosphorus are the major minerals affected by chronic kidney disease. However, phosphorus retention is the major effect that occurs during chronic kidney disease, which is exhibited as increased blood phosphorus concentration known as hyperphosphatemia. Alteration in calcium level is the second change that occurs in the case of chronic kidney disease. The calcium level in the body may either decrease or increase, while phosphorus generally increases (Elliott 2006). Animals with chronic kidney disease have increased urine volume, which enhances the excretion of water soluble vitamins. Further, loss of these vitamins results in anorexia which is directly associated with several renal complications. Similarly, humans with renal failure have less ability to excrete certain vitamins. Acid-base abnormalities are most commonly seen in cats and dogs as the kidney's function is to retain bicarbonate ions and excrete hydrogen ions. However, kidney failure results in more retention of hydrogen ions and less reabsorption of bicarbonate ions resulting in metabolic acidosis (Elliott et al. 2003). Proteinuria is also associated with chronic kidney disease in both cats and dogs. An initial urine protein: creatinine ratio greater than 1.0 in dogs is directly associated with a greater risk of developing uremic crisis and death (Vaden and Elliott 2016).

Renal hyperparathyroidism is the secondary consequence of chronic kidney disease. Generally, in human, the bone quality is decreased, and the risk of fracture is increased during this condition. Similar effects are also observed in companion animals, especially in cats and dogs (Segev et al. 2016). Systemic and glomerular hypertension has also been correlated with the development and the progression of kidney diseases in both rodent models and humans. Similarly, Hypoxia has also been reported to contribute in the progression of chronic kidney disease. However, in animals, there relatively less information is available which directly demonstrates the relation between hypoxia and progression of chronic kidney disease (King et al. 2007). Excessive consumption of phosphorous in daily diet leads to phosphorous nephritis in different species like humans, dogs, cats and rats. Feeding of high phosphorous to cats resulted in microalbuminuria and glucosuria, indicating renal damage (Dobenecker et al. 2018). Therefore, diet modification is considered as one of the important interventions in preventing the development of chronic kidney disease in small animals.

2.8 Conclusion and Future Perspective

Kidney diseases are one of the major heath burden on the global economy. Researchers are continuously trying to understand the underlying disease mechanism and to develop highly effective tools for the early detection of disease. Early diagnosis is essential for the management and better prognosis of the disease. In the past few decades, significant progress has been achieved in the identification of clinically relevant kidney markers from body fluids, particularly from plasma/serum and urine. For this, proteomic techniques like electrophoresis, chromatography, protein array, chemical tagging/metabolic labelling coupled with mass spectrometry offer not only for biomarker identification but also increase our understanding of the molecular process underlying kidney disease. In proteomic study, major limitation is the false-positive error, which was rule out using probability based auto validation of peptides by database search. In addition, to achieve high throughput data in the animal experiment, we should consider large sample size, standardized method for sample collection and preparation, advanced proteomic techniques, robust statistical models, proper normalization of data, and extensive validation that often requires strong collaboration with research laboratory across multidiscipline. Furthermore, the selection of body fluids and abundance of biomarkers will be important towards biomarker discovery and diagnosis of kidney disease. The experimental models play a crucial role in the understanding of various disease conditions. This chapter mainly focused on non-transgenic kidney disease models; however, different transgenic models are equally important. None of the experimental models mimics the human disease entirely; however, careful selection of experimental models can provide valuable understandings of disease pathogenesis.

Acknowledgements We acknowledge Director, CSIR- Institute of Himalayan Bioresource Tecnology for his constant support and encouragement. We also thank CSIR, India to provide funding in the form of project grant MLP-0201 and MLP-0204 (IHBT communication no.- 4783).

Declaration of Interest Authors declare no conflict of interest.

References

Abellán CM, Mangold-Gehring S, Micus S, Beddies G, Moritz A, Hartmann E, Lehmann W, Eitner F (2019) A novel model of chronic kidney disease in rats: dietary adenine in combination with unilateral nephrectomy. Kidney Dis 3:135–143. https://doi.org/10.1159/000495750

Aebersold R, Burlingame AL, Bradshaw RA (2013) Western blots versus selected reaction monitoring assays: time to turn the tables? Mol Cell Proteomics 12:2381–2382. https://doi.org/10.1074/mcp.E113.031658

Al-Bayati MA, Giri SN, Raabe OG, Rosenblatt LS, Shifrine MO (1989) Time and dose-response study of the effects of vanadate on rats: morphological and biochemical changes in organs. J Environ Pathol Toxicol Oncol 9:435–455

Alicic RZ, Rooney MT, Tuttle KR (2017) Diabetic kidney disease: challenges, progress, and possibilities. Clin J Am Soc Nephrol 7:2032–2045. https://doi.org/10.2215/CJN.11491116

Bao YW, Yuan Y, Chen JH, Lin WQ (2018). Kidney disease models: tools to identify mechanisms and potential therapeutic targets. Zool Res 39(2):72–86. https://doi.org/10.24272/j.issn.2095-8137.2017.055

Barnett SM, Jackson AH, Rosen BA, Garb JL, Braden GL (2018) Nephrolithiasis and nephrocalcinosis from topiramate therapy in children with epilepsy. Kidney Intl Rep 1:684–690. https://doi.org/10.1016/j.ekir.2018.02.005

Bartges JW (2012) Chronic kidney disease in dogs and cats. Veterinary Clinics Small Anim Pract 42(4):669–692. https://doi.org/10.1016/j.cvsm.2012.04.008

Basile DP, Anderson MD, Sutton TA (2011) Pathophysiology of acute kidney injury. Compr Physiol 2:1303–1353. https://doi.org/10.1002/cphy.c110041

Bellomo R, Kellum JA, Ronco C, Wald R, Martensson J, Maiden M, Bagshaw SM, Glassford NJ, Lankadeva Y, Vaara ST, Schneider A (2017) Acute kidney injury in sepsis. Intensive Care Med 43:816–828. https://doi.org/10.1007/s00134-017-4755-7

Benigni A, Gagliardini E, Remuzzi G (2004) Changes in glomerular perm-selectivity induced by angiotensin II imply podocyte dysfunction and slit diaphragm protein rearrangement. Seminars Nephrol 24:131–140. https://doi.org/10.1016/j.semnephrol.2003.11.005

Bertram JF, Douglas-Denton RN, Diouf B, Hughson MD, Hoy WE (2011) Human nephron number: implications for health and disease. Pediatr Nephrol 26:1529. https://doi.org/10.1007/s00467-011-1843-8

Bikbov B, Purcell CA, Levey AS, Smith M, Abdoli A, Abebe M, Adebayo OM, Afarideh M, Agarwal SK, Agudelo-Botero M, Ahmadian E (2020) Global, regional, and national burden of chronic kidney disease, 1990–2017: a systematic analysis for the global burden of disease study 2017. Lancet 395:709–733. https://doi.org/10.1016/S0140-6736(20)30045-3

Bing P, Maode L, Li F, Sheng H (2006) Expression of kidney transforming growth factor-β and its receptors in a rat model of chronic cyclosporine-induced nephropathy. InTransplantation proceedings 38: 2176-2179. https://doi.org/10.1016/j.transproceed.2006.07.015

Bolignano D, Lacquaniti A, Coppolino G, Donato V, Fazio MR, Nicocia G, Buemi M (2009) Neutrophil gelatinase-associated lipocalin as an early biomarker of nephropathy in diabetic patients. Kidney Blood Press Res 32:91–98. https://doi.org/10.1159/000209379

Bonventre JV, Vaidya VS, Schmouder R, Feig P, Dieterle F (2010) Next-generation biomarkers for detecting kidney toxicity. Nat Biotechnol 28:436–440. https://doi.org/10.1038/nbt0510-436

Bratoeva K, Stoyanov GS, Merdzhanova A, Radanova M (2017) Manifestations of kidney impairment in fructose-induced metabolic syndrome. Cureus 7:9:e1826. https://doi.org/10.7759/cureus.1826

Chawla LS, Kimmel PL (2012) Acute kidney injury and chronic kidney disease: an integrated clinical syndrome. Kidney Int 82:516–524. https://doi.org/10.1038/ki.2012.208

Chen L, Su W, Chen H, Chen DQ, Wang M, Guo Y, Zhao YY (2018) Proteomics for biomarker identification and clinical application in kidney disease. In Advances in clinical chemistry 85:91–113). https://doi.org/10.1016/bs.acc.2018.02.005

Chhimwal J, Sharma S, Kulurkar P, Patial V (2020) Crocin attenuates CCl4-induced liver fibrosis via PPAR-γ mediated modulation of inflammation and fibrogenesis in rats. Hum Exp Toxicol 39:1639–1649. https://doi.org/10.1177/0960327120937048

Dhanapriya J, Dineshkumar T, Sakthirajan R, Shankar P, Gopalakrishnan N, Balasubramaniyan T (2016) Wasp sting-induced acute kidney injury. Clin Kidney J 9:201–204. https://doi.org/10.1093/ckj/sfw004

Diwan V, Mistry A, Gobe G, Brown L (2013) Adenine-induced chronic kidney and cardiovascular damage in rats. J Pharmacol Toxicol Methods 68:197–207. https://doi.org/10.1016/j.vascn.2013.05.006

Dobenecker B, Webel A, Reese S, Kienzle E (2018) Effect of a high phosphorus diet on indicators of renal health in cats. J Feline Med Surg 20(4):339–343. https://doi.org/10.1177/1098612X17710589

Dobrek L, Thor P (2016) Novel biomarkers of acute kidney injury and chronic kidney disease. Polish. Annals of Medicine 24:84–91. https://doi.org/10.1016/j.poamed.2016.10.002

Dow JA, Romero MF (2010) Drosophila provides rapid modeling of kidney development, function, and disease. Am J Physiol Kidney Physiol 299:F1237–F1244. https://doi.org/10.1152/ajprenal.00521.2010

Dow JA, Kelly DC, Davies SA, Maddrell SH, Brown D (1995) A novel member of the major intrinsic protein family in Drosophila-are aquaporins involved in insect malpighian (kidney) tubule fluid secretion. Int J Physiol Lond 489:110–111

du Cheyron D, Daubin C, Poggioli J, Ramakers M, Houillier P, Charbonneau P, Paillard M (2003) Urinary measurement of Na+/H+ exchanger isoform 3 (NHE3) protein as new marker of tubule injury in critically ill patients with ARF. Am J Kidney Dis 42:497–506. https://doi.org/10.1016/s0272-6386(03)00744-3

El Amir YO, Omar W, Khabrani AY, Jahfali AE, Alhakami SM, Dobab NM (2019) Protective effect of avenanthramides against cisplatin induced nephrotoxicity in rats. J Adv Veterinary Anim Res 6:521–527. https://doi.org/10.5455/javar.2019.f377

Elliott DA (2006) Nutritional management of chronic renal disease in dogs and cats. Veterinary Clinics Small Anim Pract 36(6):1377–1384. https://doi.org/10.1016/j.cvsm.2006.08.011

Elliott J, Syme HM, Reubens E, Markwell PJ (2003) Assessment of acid-base status of cats with naturally occurring chronic renal failure. J Small Anim Pract 44(2):65–70. https://doi.org/10.1111/j.1748-5827.2003.tb00122.x

El-Shabrawy M, Mishriki A, Attia H, Emad Aboulhoda B, Emam M, Wanas H (2020) Protective effect of tolvaptan against cyclophosphamide-induced nephrotoxicity in rat models. Pharmacol Res Perspect 8:00659. https://doi.org/10.1002/prp2.659

Faria J, Ahmed S, Gerritsen KG, Mihaila SM, Masereeuw R (2019) Kidney-based in vitro models for drug-induced toxicity testing. Arch Toxicol 93:3397–3418. https://doi.org/10.1007/s00204-019-02598-0

Fenoglio R, Sciascia S, Baldovino S, Roccatello D (2019) Acute kidney injury associated with glomerular diseases. Curr Opin Crit Care 2:573–579. https://doi.org/10.1097/MCC.0000000000000675

Ferlizza E, Campos A, Neagu A, Cuoghi A, Bellei E, Monari E, Dondi F, Almeida AM, Isani G (2015) The effect of chronic kidney disease on the urine proteome in the domestic cat (Felis catus). Vet J 204:73–81. https://doi.org/10.1016/j.tvjl.2015.01.023

Ferlizza E, Isani G, Dondi F, Andreani G, Vasylyeva K, Bellei E, Almeida AM, Matzapetakis M (2020) Urinary proteome and metabolome in dogs (Canis lupus familiaris): the effect of chronic kidney disease. J Proteomic 222:103795. https://doi.org/10.1016/j.jprot.2020.103795

Fine LG, Norman JT (2008) Chronic hypoxia as a mechanism of progression of chronic kidney diseases: from hypothesis to novel therapeutics. Kidney Int 74:867–872. https://doi.org/10.1038/ki.2008.350

Gao D, Zeng LN, Zhang P, Ma ZJ, Li RS, Zhao YL, Zhang YM, Guo YM, Niu M, Bai ZF, Xiao XH (2016) Rhubarb anthraquinones protect rats against mercuric chloride (HgCl2)-induced acute kidney failure. Molecules 21:298. https://doi.org/10.3390/molecules21030298

Gong L, He J, Sun X, Li L, Zhang X, Gan H (2020) Activation of sirtuin1 protects against ischemia/reperfusion-induced acute kidney injury. Biomed Pharmacother 125:110021. https://doi.org/10.1016/j.biopha.2020.110021

Good DM, Zürbig P, Argiles A, Bauer HW, Behrens G, Coon JJ, Dakna M, Decramer S, Delles C, Dominiczak AF, Ehrich JH (2010) Naturally occurring human urinary peptides for use in diagnosis of chronic kidney disease. Mol Cell Proteomics 9:2424–2437. https://doi.org/10.1074/mcp.M110.001917

Grande MT, Pérez-Barriocanal F, López-Novoa JM (2010) Role of inflammation in tubulo-interstitial damage associated to obstructive nephropathy. J Inflamm 7:19. https://doi.org/10.1186/1476-9255-7-19

Gu L, Wang Y, Yang G, Tilyek A, Zhang C, Li S, Yu B, Chai C, Cao Z (2019) Ribes Diacanthum Pall (RDP) ameliorates UUO-induced kidney fibrosis via both canonical and non-canonical

TGF-β signaling pathways in mice. J Ethnopharmacol 231:302–310. https://doi.org/10.1016/j.jep.2018.10.023
Gu LF, Ge HT, Zhao L, Wang YJ, Zhang F, Tang HT, Cao ZY, Yu BY, Chai CZ (2020) Huangkui capsule ameliorates kidney fibrosis in a unilateral ureteral obstruction mouse model through TRPC6 dependent signaling pathways. Front Pharmacol 11:996. https://doi.org/10.3389/fphar.2020.00996
Haidara MA, Al-Ani B, Eid RA, Mohammed ME, Al-Hashem F, Dallak M (2020) Acetaminophen induces alterations to the kidney tubular ultrastructure in a rat model of acute nephrotoxicity protected by resveratrol and quercetin. Int J Morphol 38:585–591. https://doi.org/10.4067/S0717-95022020000300585
Hamzaoui M, Djerada Z, Brunel V, Mulder P, Richard V, Bellien J, Guerrot D (2020) 5/6 nephrectomy induces different kidney, cardiac and vascular consequences in 129/Sv and C57BL/6JRj mice. Sci Rep 10:1524. https://doi.org/10.1038/s41598-020-58393-w
Han X, Jin M, Breuker K, McLafferty FW (2006) Extending top-down mass spectrometry to proteins with masses greater than 200 kilodaltons. Science 314:109–112. https://doi.org/10.1126/science.1128868
Harrison DJ, Kharbanda R, Cunningham DS, McLellan LI, Hayes JD (1989) Distribution of glutathione S-transferase isoenzymes in human kidney: basis for possible markers of kidney injury. J Clin Pathol 42:624–628. https://doi.org/10.1136/jcp.42.6.624
Hart SG (2005) Assessment of kidney injury in vivo. J Pharmacol Toxicol Methods 52:30–45. https://doi.org/10.1016/j.vascn.2005.04.006
He L, Peng X, Liu G, Tang C, Liu H, Liu F, Zhou H, Peng Y (2015) Anti-inflammatory effects of triptolide on IgA nephropathy in rats. Immunopharmacol Immunotoxicol 37:421–427. https://doi.org/10.3109/08923973.2015.1080265
He J, Dijkstra KL, Bakker K, Bus P, Bruijn JA, Scharpfenecker M, Baelde HJ (2020). Glomerular clusterin expression is increased in diabetic nephropathy and protects against oxidative stress-induced apoptosis in podocytes. Scientific Reports 10:14888. https://doi.org/10.1038/s41598-020-71629-z
Hentschel DM, Bonventre JV (2005) Novel non-rodent models of kidney disease. Curr Mol Med 5:537–546. https://doi.org/10.2174/1566524054553469
Himmelfarb J, McMonagle E, McMenamin E (2000) Plasma protein thiol oxidation and carbonyl formation in chronic kidney failure. Kidney Int 58:2571–2578. https://doi.org/10.1046/j.1523-1755.2000.00443.x
Hosseinian S, Rad AK, Mousa-Al-Reza Hadjzadeh NM, Roshan SH, Shafiee S (2016) The protective effect of Nigella sativa against cisplatin-induced nephrotoxicity in rats. Avicenna J Phytomed 6:44–54. PMCID: PMC4884217
Hsu YH, Chiu IJ, Lin YF, Chen YJ, Lee YH, Chiu HW (2020) Lactoferrin contributes a Renoprotective effect in acute kidney injury and early kidney fibrosis. Pharmaceutics 12:434. https://doi.org/10.3390/pharmaceutics12050434
Hua H, Ge X, Wu M, Zhu C, Chen L, Yang G, Zhang Y, Huang S, Zhang A, Jia Z (2018) Rotenone protects against acetaminophen-induced kidney injury by attenuating oxidative stress and inflammation. Kidney Blood Press Res 43:1297–1309. https://doi.org/10.1159/000492589
Huang J, Yao X, Weng G, Qi H, Ye X (2018) Protective effect of curcumin against cyclosporine A-induced rat nephrotoxicity. Mol Med Rep 17:6038–6044. https://doi.org/10.3892/mmr.2018.8591
Ioannou K (2017) Diabetic nephropathy: is it always there? Assumptions, weaknesses and pitfalls in the diagnosis. Hormones (Athens). 16(4):351–361. https://doi.org/10.14310/horm.2002.1755
Isakova T, Wahl P, Vargas GS, Gutiérrez OM, Scialla J, Xie H, Appleby D, Nessel L, Bellovich K, Chen J, Hamm L (2011) Fibroblast growth factor 23 is elevated before parathyroid hormone and phosphate in chronic kidney disease. Kidney Int 79:1370–1378. https://doi.org/10.1038/ki.2011.47
Jayasekara KB, Dissanayake DM, Sivakanesan R, Ranasinghe A, Karunarathna RH, Kumara GW (2015) Epidemiology of chronic kidney disease, with special emphasis on chronic kidney dis-

ease of uncertain etiology, in the north central region of Sri Lanka. J Epidemiol 25:275–280. https://doi.org/10.2188/jea.JE20140074

Johnson RJ, Sanchez-Lozada LG, Nakagawa T (2010) The effect of fructose on kidney biology and disease. J Am Soc Nephrol 21:2036–2039. https://doi.org/10.1681/ASN.2010050506

Jung YH, Han D, Shin SH, Kim EK, Kim HS (2020) Proteomic identification of early urinary-biomarkers of acute kidney injury in preterm infants. Sci Rep 10:4057. https://doi.org/10.1038/s41598-020-60890-x

Kandemir FM, Kucukler S, Eldutar E, Caglayan C, Gülçin İ (2017) Chrysin protects rat kidney from paracetamol-induced oxidative stress, inflammation, apoptosis, and autophagy: a multi-biomarker approach. Sci Pharm 85:4. https://doi.org/10.3390/scipharm85010004

Kashani K, Al-Khafaji A, Ardiles T, Artigas A, Bagshaw SM, Bell M, Bihorac A, Birkhahn R, Cely CM, Chawla LS, Davison DL (2013) Discovery and validation of cell cycle arrest biomarkers in human acute kidney injury. Critical Care 17:1–2. R25. https://doi.org/10.1186/cc12503

Kato Y, Tonomura Y, Hanafusa H, Nishimura K, Fukushima T, Ueno M (2020) Adult zebrafish model for screening drug-induced kidney injury. Toxicol Sci 174:241–253. https://doi.org/10.1093/toxsci/kfaa009

Katoch S, Patial V (2020) Zebrafish: an emerging model system to study liver diseases and related drug discovery. J Appl Toxicol 41:33–51. https://doi.org/10.1002/jat.4031

Khera T, Martin J, Riley S, Steadman R, Phillips AO (2006) Glucose enhances mesangial cell apoptosis. Lab Investig 86:566–577. https://doi.org/10.1038/labinvest.3700418

Kim SS, Song SH, Kim IJ, Jeon YK, Kim BH, Kwak IS, Lee EK, Kim YK (2013) Urinary cystatin C and tubular proteinuria predict progression of diabetic nephropathy. Diabetes Care 36:656–661. https://doi.org/10.2337/dc12-0849

King JN, Tasker S, Gunn-Moore DA, Strehlau G (2007). Prognostic factors in cats with chronic kidney disease. J Vet Intern Med 21(5): 906–916. PMID: 17939542

Kohl K, Herzog E, Dickneite G, Pestel S (2020) Evaluation of urinary biomarkers for early detection of acute kidney injury in a rat nephropathy model. J Pharmacol Toxicol Methods 1:105:106901. https://doi.org/10.1016/j.vascn.2020.106901

L'Azou B, Dubus I, Ohayon-Courtès C, Cambar J (2007) Human glomerular mesangial IP15 cell line as a suitable model for in vitro cadmium cytotoxicity studies. Cell Biol Toxicol 23:267–278. https://doi.org/10.1007/s10565-006-0888-0

L'Imperio V, Smith A, Chinello C, Pagni F, Magni F (2016) Proteomics and glomerulonephritis: a complementary approach in kidney pathology for the identification of chronic kidney disease related markers. PROTEOMICS–Clin Appl 10:371–383. https://doi.org/10.1002/prca.201500075

La Russa D, Giordano F, Marrone A, Parafati M, Janda E, Pellegrino D (2019) Oxidative imbalance and kidney damage in cafeteria diet-induced rat model of metabolic syndrome: effect of bergamot polyphenolic fraction. Antioxidants 8:66. https://doi.org/10.3390/antiox8030066

Laurentius T, Raffetseder U, Fellner C, Kob R, Nourbakhsh M, Floege J, Bertsch T, Bollheimer C, Ostendorf T (2019) Obesity causes inflammatory milieu in kidneys of aging Long-Evans-rats. Zeitschrift Fur Gerontologie Und Geriatrie 52:S145

Lescuyer P, Hochstrasser D, Rabilloud T (2007) How shall we use the proteomics toolbox for biomarker discovery? J Proteome Res 6:3371–3376. https://doi.org/10.1021/pr0702060

Levin AS, Bilous RW, Coresh J (2013) Chapter 1: definition and classification of CKD. Kidney Int Suppl 3:19–62. https://doi.org/10.1038/kisup.2012.64

Li X, Zou Y, Xing J, Fu YY, Wang KY, Wan PZ, Zhai XY (2020) Pretreatment with roxadustat (FG-4592) attenuates folic acid-induced kidney injury through antiferroptosis via Akt/GSK-3β/Nrf2 pathway. Oxidative Med Cell Longev 20:6286984. https://doi.org/10.1155/2020/6286984

Lim AK (2014) Diabetic nephropathy–complications and treatment. Int J Nephrol Renov Dis 7:361–381. https://doi.org/10.2147/IJNRD.S40172

Limbu MH, Lei L, Zhengyuan C, Jing L, Xiaoyi Z, Pingsheng C (2017). An attempt to establish a common animal model for Hepatokidney fibrosis in rats. Pathol Res Intl 2017:8260508. https://doi.org/10.1155/2017/8260508

Lopes JA, Jorge S (2013) The RIFLE and AKIN classifications for acute kidney injury: a critical and comprehensive review. Clin Kidney J 6:8–14. https://doi.org/10.1093/ckj/sfs160

Lopez-Giacoman S, Madero M (2015) Biomarkers in chronic kidney disease, from kidney function to kidney damage. World J Nephrol 4:57–73. https://doi.org/10.5527/wjn.v4.i1.57

Ma JQ, Sun YZ, Ming QL, Tian ZK, Zhang YJ, Liu CM (2020) Effects of gastrodin against carbon tetrachloride induced kidney inflammation and fibrosis in mice associated with the AMPK/Nrf2/HMGB1 pathway. Food Funct 11:4615–4624. https://doi.org/10.1039/d0fo00711k

Machowska A, Sun J, Qureshi AR, Isoyama N, Leurs P, Anderstam B, Heimburger O, Barany P, Stenvinkel P, Lindholm B (2016) Plasma pentosidine and its association with mortality in patients with chronic kidney disease. PLoS One 4:11:e0163826. https://doi.org/10.1371/journal.pone.0163826

MacIsaac RJ, Ekinci EI, Jerums G (2014) Progressive diabetic nephropathy. how useful is microalbuminuria? Contra. Kidney Int 86:50–57. https://doi.org/10.1038/ki.2014.98

Magalhães P, Mischak H, Zürbig P (2016) Urinary proteomics using capillary electrophoresis coupled to mass spectrometry for diagnosis and prognosis in kidney diseases. Current Opin Nephrol Hypertension 25:494–501. https://doi.org/10.1097/MNH.0000000000000278

Mahmoud AM (2014) Hesperidin protects against cyclophosphamide-induced hepatotoxicity by upregulation of PPARγ and abrogation of oxidative stress and inflammation. Can J Physiol Pharmacol 92:717–724. https://doi.org/10.1139/cjpp-2014-0204

Maier TV, Schmitt-Kopplin P (2016) Capillary electrophoresis in metabolomics. In Capillary electrophoresis 2016. Humana Press, New York, pp 437–470. https://doi.org/10.1007/978-1-4939-6403-1_21

Malhotra R, Siew ED (2017) Biomarkers for the early detection and prognosis of acute kidney injury. Clin J Am Soc Nephrol 12:149–173

Marelja Z, Simons M (2019) Filling the gap: drosophila nephrocytes as model system in kidney research. https://doi.org/10.1681/ASN.2019020181

Matovinović MS (2009) 1. Pathophysiology and classification of kidney diseases. Ejifcc 20:2. PMCID: PMC4975264

Matthiesen R, Bunkenborg J, Stensballe A, Jensen ON, Welinder KG, Bauw G (2004) Database-independent, database-dependent, and extended interpretation of peptide mass spectra in VEMS V2. 0. Proteomics 4:2583–2593. https://doi.org/10.1002/pmic.200300792

McGaha TL, Madaio MP (2014) Lupus nephritis: animal modeling of a complex disease syndrome pathology. Drug Discov Today Dis Model 11:13–18. https://doi.org/10.1016/j.ddmod.2014.08.002

Millet-Boureima C, Porras Marroquin J, Gamberi C (2018) Modeling kidney disease "on the fly". BioMed research international. 2018:5697436. https://doi.org/10.1155/2018/5697436

Millet-Boureima C, Chingle R, Lubell WD, Gamberi C (2019) Cyst reduction in a polycystic kidney disease drosophila model using Smac mimics. Biomedicine 7:82. https://doi.org/10.3390/biomedicines7040082

Mischak H, Apweiler R, Banks RE, Conaway M, Coon J, Dominiczak A, Ehrich JH, Fliser D, Girolami M, Hermjakob H, Hochstrasser D (2007) Clinical proteomics: a need to define the field and to begin to set adequate standards. PROTEOMICS–Clin Appl 1:148–156. https://doi.org/10.1002/prca.200600771

Mischak H, Delles C, Vlahou A, Vanholder R (2015) Proteomic biomarkers in kidney disease: issues in development and implementation. Nat Rev Nephrol 11:221–232. https://doi.org/10.1038/nrneph.2014.247

Mitrogianni Z, Barbouti A, Galaris D, Siamopoulos KC (2009) Oxidative modification of albumin in predialysis, hemodialysis, and peritoneal dialysis patients. Nephron Clin Pract 113:c2342–c2340. https://doi.org/10.1159/000235244

Mitsnefes MM, Kathman TS, Mishra J, Kartal J, Khoury PR, Nickolas TL, Barasch J, Devarajan P (2007) Serum neutrophil gelatinase-associated lipocalin as a marker of kidney function in children with chronic kidney disease. Pediatr Nephrol 22:101–118. https://doi.org/10.1007/s00467-006-0244-x

Miyauchi K, Takiyama Y, Honjyo J, Tateno M, Haneda M (2009) Upregulated IL-18 expression in type 2 diabetic subjects with nephropathy: TGF-β1 enhanced IL-18 expression in human kidney proximal tubular epithelial cells. Diabetes Res Clin Pract 83:190–199. https://doi.org/10.1016/j.diabres.2008.11.018

Murphy RA, Stafford RM, Petrasovits BA, Boone MA, Valentovic MA (2017) Establishment of HK-2 cells as a relevant model to study tenofovir-induced cytotoxicity. Int J Mol Sci 18:531. https://doi.org/10.3390/ijms18030531

Nakhoul F, Abassi Z, Morgan M, Sussan S, Mirsky N (2006) Inhibition of diabetic nephropathy in rats by an oral antidiabetic material extracted from yeast. J Am Soc Nephrol 17:127–131. https://doi.org/10.1681/ASN.2005121333

Nashar K, Egan BM (2014) Relationship between chronic kidney disease and metabolic syndrome: current perspectives. Diabetes, Metab Syndrome Obesity Targets Therapy 7:421. https://doi.org/10.2147/DMSO.S45183

Nguyen MT, Devarajan P (2008) Biomarkers for the early detection of acute kidney injury. Pediatr Nephrol 23:2151. https://doi.org/10.1007/s00467-007-0470-x

Ni LA, Saleem M, Mathieson PW (2012) Podocyte culture: tricks of the trade. Nephrology 17:525–531. https://doi.org/10.1111/j.1440-1797.2012.01619.x

Nilsson T, Mann M, Aebersold R, Yates JR, Bairoch A, Bergeron JJ (2010) Mass spectrometry in high-throughput proteomics: ready for the big time. Nat Methods 7:681–685. https://doi.org/10.1038/nmeth0910-681

Nogueira A, Pires MJ, Oliveira PA (2017). Pathophysiological mechanisms of kidney fibrosis: a review of animal models and therapeutic strategies. In Vivo 31:1–22. https://doi.org/10.21873/invivo.11019

Okuda S, Oh Y, Tsuruda H, Onoyama K, Fujimi S, Fujishima M (1986) Adriamycin-induced nephropathy as a model of chronic progressive glomerular disease. Kidney Int 29:502–510. https://doi.org/10.1038/ki.1986.28

Orasanu G, Plutzky J (2009) The pathologic continuum of diabetic vascular disease. J Am Coll Cardiol 53:S35–S42. https://doi.org/10.1016/j.jacc.2008.09.055

Outtandy P, Russell C, Kleta R, Bockenhauer D (2019) Zebrafish as a model for kidney function and disease. Pediatr Nephrol 34:751–762. https://doi.org/10.1007/s00467-018-3921-7

Øvrehus MA, Zürbig P, Vikse BE, Hallan SI (2015) Urinary proteomics in chronic kidney disease: diagnosis and risk of progression beyond albuminuria. Clin Proteomics 12:21. https://doi.org/10.1186/s12014-015-9092-7

Park MR, Wang EH, Jin DC, Cha JH, Lee KH, Yang CW, Kang CS, Choi YJ (2006) Establishment of a 2-D human urinary proteomic map in IgA nephropathy. Proteomics 2006(6):1066–1076. https://doi.org/10.1002/pmic.200500023

Peerapornratana S, Manrique-Caballero CL, Gomez H, Kellum JA (2019) Acute kidney injury from sepsis: current concepts, epidemiology, pathophysiology, prevention and treatment. Kidney Int 96:1083–1099. https://doi.org/10.1016/j.kint.2019.05.026

Pennington SR, Wilkins MR, Hochstrasser DF, Dunn MJ (1997) Proteome analysis: from protein characterization to biological function. Trends Cell Biol 7:1681–1673. https://doi.org/10.1016/S0962-8924(97)01033-7

Picotti P, Aebersold R (2012) Selected reaction monitoring–based proteomics: workflows, potential, pitfalls and future directions. Nat Methods 9:555–566. https://doi.org/10.1038/nmeth.2015

Polichnowski AJ, Griffin K, Sethupathi P, Patel V, Patel M, Picken M, Williamson G, Bidani A (2017) Susceptibility to pressure-induced kidney injury in DOCA-salt vs. angiotensin II hypertensive rats. The FASEB J 31 (1 Suppl):701–702. https://doi.org/10.1096/fasebj.31.1_supplement.701.2

Polzin DJ (2011) Chronic kidney disease in small animals. Veterinary Clinics Small Anim Pract 41(1):15–30. https://doi.org/10.1016/j.cvsm.2010.09.004

Pontillo C, Jacobs L, Staessen JA, Schanstra JP, Rossing P, Heerspink HJ, Siwy J, Mullen W, Vlahou A, Mischak H, Vanholder R (2017) A urinary proteome-based classifier for the early

detection of decline in glomerular filtration. Nephrol Dialysis Transplant 32:1510–1506. https://doi.org/10.1093/ndt/gfw239
Provenzano M, Rotundo S, Chiodini P, Gagliardi I, Michael A, Angotti E, Borrelli S, Serra R, Foti D, De Sarro G, Andreucci M (2020) Contribution of predictive and prognostic biomarkers to clinical research on chronic kidney disease. Int J Mol Sci 21:5846. https://doi.org/10.3390/ijms21165846
Qi MY, Wang XT, Xu HL, Yang ZL, Cheng Y, Zhou B (2020) Protective effect of ferulic acid on STZ-induced diabetic nephropathy in rats. Food Funct 11:3706–3718. https://doi.org/10.1039/c9fo02398d
Rangel Silvares R Nunes Goulart da Silva Pereira E, Eduardo Ilaquita Flores E, Lino Rodrigues K, Ribeiro Silva A, Goncalves-de-Albuquerque CF, Daliry A (2019) High-fat diet-induced kidney alterations in rats with metabolic syndrome: endothelial dysfunction and decreased antioxidant defense. Diabetes Metab. Syndr. Obes 12:1773–1781. https://doi.org/10.2147/DMSO.S211253
Raptis V, Kapoulas S, Grekas D (2013). Role of asymmetrical dimethylarginine in the progression of kidney disease. Nephrology 18:11–21. https://doi.org/10.2217/bmm-2019-0008
Rider SA, Bruton FA, Collins RG, Conway BR, Mullins JJ (2018) The efficacy of Puromycin and Adriamycin for induction of glomerular failure in larval zebrafish validated by an assay of glomerular permeability dynamics. Zebrafish 15(3):234–242. https://doi.org/10.1089/zeb.2017.1527
Rodríguez-Romo R, Berman N, Gómez A, Bobadilla NA (2015) Epigenetic regulation in the acute kidney injury to chronic kidney disease transition. Nephrology 20:736–743. https://doi.org/10.1111/nep.12521
Romagnani P, Remuzzi G, Glassock R, Levin A, Jager KJ, Tonelli M, Massy Z, Wanner C, Anders HJ (2017) Chronic kidney disease. Nat Rev Dis Primers 3:1–24. https://doi.org/10.1038/nrdp.2017.88
Romanova Y, Laikov A, Markelova M, Khadiullina R, Makseev A, Hasanova M, Rizvanov A, Khaiboullina S, Salafutdinov I (2020) Proteomic analysis of human serum from patients with chronic kidney disease. Biomol Ther 10:257. https://doi.org/10.3390/biom10020257
Roscioni SS, De Zeeuw D, Hellemons ME, Mischak H, Zürbig P, Bakker SJ, Gansevoort RT, Reinhard H, Persson F, Lajer M, Rossing P (2013) A urinary peptide biomarker set predicts worsening of albuminuria in type 2 diabetes mellitus. Diabetologia 52:259–267. https://doi.org/10.1007/s00125-012-2755-2
Safhi MM (2018) Nephroprotective effect of Zingerone against CCl4-induced kidney toxicity in Swiss albino mice: molecular mechanism. Oxidative Med Cell Longev. https://doi.org/10.1155/2018/2474831
Satoh M, Kumar A, Kanwar YS, Reeves WH (1995) Anti-nuclear antibody production and immune-complex glomerulonephritis in BALB/c mice treated with pristane. Proc Natl Acad Sci 92:10934–10938. https://doi.org/10.1073/pnas.92.24.10934
Ścibior A, Gołębiowska D, Adamczyk A, Niedźwiecka I, Fornal E (2014) The kidney effects of vanadate exposure: potential biomarkers and oxidative stress as a mechanism of functional kidney disorders—preliminary studies. BioMed Res Intl:740105. https://doi.org/10.1155/2014/740105
Segev G, Meltzer H, Shipov A (2016) Does secondary renal osteopathy exist in companion animals? Veterinary Clinics Small Anim Pract 46(6):1151–1162. https://doi.org/10.1016/j.cvsm.2016.06.006
Sgambat K, Amatya K, Moudgil A (2019) Nutritional challenges across the spectrum of chronic kidney disease. Asian J Pediat Nephrol 2:2. https://doi.org/10.4103/AJPN.AJPN/2/19
Sharma P, Sharma S, Patial V, Singh D, Padwad YS (2014) Zebrafish (Danio rerio): a potential model for nephroprotective drug screening. Clin Queries Nephrol 3:97–105. https://doi.org/10.1016/j.cqn.2014.11.002
Sharma S, Sharma P, Kulurkar P, Singh D, Kumar D, Patial V (2017) Iridoid glycosides fraction from Picrorhiza kurroa attenuates cyclophosphamide-induced kidney toxicity and peripheral neuropathy via PPAR-γ mediated inhibition of inflammation and apoptosis. Phytomedicine 36:108–117. https://doi.org/10.1016/j.phymed.2017.09.018

Sharma D, Gondaliya P, Tiwari V, Kalia K (2019) Kaempferol attenuates diabetic nephropathy by inhibiting RhoA/rho-kinase mediated inflammatory signalling. Biomed Pharmacother 109:1610–1619. https://doi.org/10.1016/j.biopha.2018.10.195
Shaw I, Rider S, Mullins J, Hughes J, Péault B (2018) Pericytes in the kidney vasculature: roles in health and disease. Nat Rev Nephrol 14:521–534. https://doi.org/10.1038/s41581-018-0032-4
Sieber J, Weins A, Kampe K, Gruber S, Lindenmeyer MT, Cohen CD, Orellana JM, Mundel P, Jehle AW (2013) Susceptibility of podocytes to palmitic acid is regulated by stearoyl-CoA desaturases 1 and 2. Am J Pathol 183:735–744. https://doi.org/10.1016/j.ajpath.2013.05.023
Simon-Friedt BR, Wilson MJ, Blake DA, Yu H, Eriksson Y, Wickliffe JK (2015) The RPTEC/TERT1 cell line as an improved tool for in vitro nephrotoxicity assessments. Biol Trace Elem Res 166:66–71. https://doi.org/10.1007/s12011-015-0339-y
Stacchiotti A, Borsani E, Rodella L, Rezzani R, Bianchi R, Lavazza A (2003) Dose-dependent mercuric chloride tubular injury in rat kidney. Ultrastruct Pathol 27:253–259. https://doi.org/10.1080/01913120309921
Sugiura T, Wada A (2009) Resistive index predicts kidney prognosis in chronic kidney disease. Nephrol Dialysis Transplant 24:2780–2785. https://doi.org/10.1093/ndt/gfp121
Sui W, Dai Y, Zhang Y, Chen J, Liu H (2010). Proteomic profiling of immunoglobulin a nephropathy in serum using magnetic bead based sample fractionation and matrix-assisted laser desorption/ionization-time of flight mass spectrometry. Saudi Med J 31:1378–1379. PMID: 21136007
Sun Y, Ge X, Li X, He J, Wei X, Du J, Sun J, Li X, Xun Z, Liu W, Zhang H (2020) High-fat diet promotes kidney injury by inducing oxidative stress and mitochondrial dysfunction. Cell Death Dis 11:1–14. https://doi.org/10.1038/s41419-020-03122-4
Suzuki H, Suzuki Y (2018) Murine models of human IgA nephropathy. InSeminars Nephrol 38:513–520. https://doi.org/10.1016/j.semnephrol.2018.05.021
Tan RJ, Zhou L, Zhou D, Lin L, Liu Y (2013) Endothelin receptor a blockade is an ineffective treatment for adriamycin nephropathy. PLoS One 8:e79963. https://doi.org/10.1371/journal.pone.0079963
Tan RZ, Zhong X, Li JC, Zhang YW, Yan Y, Liao Y, Wen D, Diao H, Wang L, Shen HC (2019) An optimized 5/6 nephrectomy mouse model based on unilateral kidney ligation and its application in kidney fibrosis research. Kidney Failure 41:555–566. https://doi.org/10.1080/0886022X.2019.1627220
Tannu NS, Hemby SE (2006) Two-dimensional fluorescence difference gel electrophoresis for comparative proteomics profiling. Nat Protoc 1:1732. https://doi.org/10.1038/nprot.2006.256
Tsalamandris C, Allen TJ, Gilbert RE, Sinha A, Panagiotopoulos S, Cooper ME, Jerums G (1994) Progressive decline in kidney function in diabetic patients with and without albuminuria. Diabetes 43:649–655. https://doi.org/10.2337/diab.43.5.649
Uchino S (2010) Creatinine. Curr Opin Crit Care 16:562–567. https://doi.org/10.1097/MCC.0b013e32833ea7f3
Uwaezuoke SN (2017) The role of novel biomarkers in predicting diabetic nephropathy: a review. Intl J Nephrol Renovasc Dis 10:221. https://doi.org/10.2147/IJNRD.S143186
Vaden SL, Elliott J (2016) Management of proteinuria in dogs and cats with chronic kidney disease. Veterinary Clinics Small Anim Pract 46(6):1115–1130. https://doi.org/10.1016/j.cvsm.2016.06.009
Vaidya VS, Ramirez V, Bobadilla NA, Bonventre JV (2005) A microfluidics based assay to measure kidney injury molecule-1 (Kim-1) in the urine as a biomarker for early diagnosis of acute kidney injury. J Am Soc Nephrol 16:192A. (Abstract)
Vaidya VS, Ferguson MA, Bonventre JV (2008) Biomarkers of acute kidney injury. Annu Rev Pharmacol Toxicol 48:463–493. https://doi.org/10.1146/48.113006.094615
Vallon V (2011) The proximal tubule in the pathophysiology of the diabetic kidney. Am J Phys Regul Integr Comp Phys 300:R1009–R1022. https://doi.org/10.1152/ajpregu.00809.2010
Van JA, Scholey JW, Konvalinka A (2017) Insights into diabetic kidney disease using urinary proteomics and bioinformatics. J Am Soc Nephrol 28:1050–1061. https://doi.org/10.1681/ASN.2016091018

Wada T, Furuichi K, Sakai N, Iwata Y, Yoshimoto K, Shimizu M, Takeda SI, Takasawa K, Yoshimura M, Kida H, Kobayashi KI (2000) Up-regulation of monocyte chemoattractant protein-1 in tubulointerstitial lesions of human diabetic nephropathy. Kidney Int 58:1492–1499. https://doi.org/10.1046/j.1523-1755.2000.00311.x

Wang H, Naghavi M, Allen C, Barber RM, Bhutta ZA, Carter A (2016) Global, regional, and national life expectancy, all-cause mortality, and cause-specific mortality for 249 causes of death, 1980–2015: a systematic analysis for the Global Burden of Disease Study 2015. Lancet 388:1459–1544. https://doi.org/10.1016/S0140-6736(16)31012-1

Wang D, Luo Y, Myakala K, Orlicky DJ, Dobrinskikh E, Wang X, Levi M (2017a) Serelaxin improves cardiac and kidney function in DOCA-salt hypertensive rats 7:1–0. https://doi.org/10.1038/s41598-017-09470-0

Wang J, Masehi-Lano JJ, Chung EJ (2017b) Peptide and antibody ligands for kidney targeting: nanomedicine strategies for kidney disease. Biomater Sci 5:1450–1459. https://doi.org/10.1039/c7bm00271h

Wei MG, Sun W, He WM, Ni L, Yang YY (2015) Ferulic acid attenuates TGF-β1-induced kidney cellular fibrosis in NRK-52E cells by inhibiting Smad/ILK/snail pathway. Evid Based Complement Alternat Med. https://doi.org/10.1155/2015/619720

Wei L, Du Y, Jia L, Ma X, Chen Z, Lu J, Tian L, Duan Z, Dong F, Lv Z, Yao G (2017) Therapeutic effects of FK506 on IgA nephropathy rat. Kidney Blood Press Res 42:983–998. https://doi.org/10.1159/000485346

Westerweel PE, Joles JA, den Ouden K, Goldschmeding R, Rookmaaker MB, Verhaar MC (2012) ACE inhibition in anti-Thy1 glomerulonephritis limits proteinuria but does not improve kidney function and structural remodeling. Nephron Extra 2:9–16. https://doi.org/10.1159/000335750

Wu Q, Fenton RA (2018) Proteomic approaches in kidney disease biomarker discovery. Am J Physiol Kidney Physiol 315:F1817–F1821. https://doi.org/10.1152/ajprenal.00421.2018

Wu J, Zheng C, Wan X, Shi M, McMillan K, Maique J, Cao C (2020) Retinoic acid alleviates cisplatin-induced acute kidney injury through activation of autophagy. Front Pharmacol 11:987. https://doi.org/10.3389/fphar.2020.00987

Yang HC, Zuo Y, Fogo AB (2010) Models of chronic kidney disease. Drug Discov Today Dis Model 7:13–19. https://doi.org/10.1016/j.ddmod.2010.08.002

Yates JR III (1998) Mass spectrometry and the age of the proteome. J Mass Spectrom 33:1–19. https://doi.org/10.1002/(SICI)1096-9888

Yates JR III (2004) Mass spectral analysis in proteomics. Annu Rev Biophys Biomol Struct 33:297–316. https://doi.org/10.1146/annurev.biophys.33.111502.082538

Yuan HT, Li XZ, Pitera JE, Long DA, Woolf AS (2003) Peritubular capillary loss after mouse acute nephrotoxicity correlates with down-regulation of vascular endothelial growth factor-A and hypoxia-inducible factor-1α. Am J Pathol 163:2289–2301. https://doi.org/10.1016/S0002-9440(10)63586-9

Zalups RK (2000). Molecular interactions with mercury in the kidney. Pharmacol Rev 52:113–144. PMID: 10699157

Zhang L, Ye F, He Y, Kong D, Han C, Zhao Z, Zhu J, Meng H, Liu X, Jin X (2010) Establishment of a mouse IgA nephropathy model with the MBP-20-peptide fusion protein. Anat Rec Adv Integr Anat Evol Biol 293:1729–1737. https://doi.org/10.1002/ar.21225

Zhang Y, Ou M, Lin H, Lai L, Chen H, Chen J, Sui W, Xue W, Zhang R, Gan Q, Tang D (2020) Proteomic analysis of differentially expressed proteins in the serum of patients with acute kidney allograft rejection using iTRAQ labelling technology. Mol Med Rep 22:2329–2341. https://doi.org/10.3892/mmr.2020.11299

Zhao YY, Vaziri ND, Lin RC (2015) Lipidomics: new insight into kidney disease. InAdvances in clinical chemistry (Vol. 68, pp. 153-175). Elsevier. https://doi.org/10.1016/bs.acc.2014.11.002

Zhao J, Qi X, Dai Q, He X, Dweep H, Guo M, Luo Y, Gretz N, Luo H, Huang K, Xu W (2016) Toxicity study of ochratoxin A using HEK293 and HepG2 cell lines based on microRNA profiling. Hum Experim Toxicol 36:8–22. https://doi.org/10.1177/0960327116632048

Zheng Z, Deng G, Qi C, Xu Y, Liu X, Zhao Z, Zhang Z, Chu Y, Wu H, Liu J (2019). Porous Se@ SiO_2 nanospheres attenuate ischemia/reperfusion (I/R)-induced acute kidney injury (AKI) and inflammation by antioxidative stress. Intl J Nanomed 14:215. https://doi.org/10.2147/IJN.S184804

Zhong J, Yang HC, Fogo AB (2017) A perspective on chronic kidney disease progression. Am J Physiol Kidney Physiol 312:F375–F384. https://doi.org/10.1152/ajprenal.00266.2016

Zhou H, Pisitkun T, Aponte A, Yuen PS, Hoffert JD, Yasuda H, Hu X, Chawla L, Shen RF, Knepper MA, Star RA (2006) Exosomal Fetuin-a identified by proteomics: a novel urinary biomarker for detecting acute kidney injury. Kidney Intl 1847–1857. https://doi.org/10.1038/sj.ki.5001874

Zürbig P, Jerums G, Hovind P, MacIsaac RJ, Mischak H, Nielsen SE, Panagiotopoulos S, Persson F, Rossing P (2012) Urinary proteomics for early diagnosis in diabetic nephropathy. Diabetes 61:3304–3313. https://doi.org/10.2337/db12-0348

Chapter 3
Generation of Gene Edited Pigs

S. Navarro-Serna, C. Piñeiro-Silva, R. Romar, J. Parrington, and J. Gadea

Abstract

Background:

The porcine species (*Sus scrofa domesticus*) has had a great impact as a source of food worldwide, but also it is a very useful animal for biomedical applications. Genetic engineering involves modifying the sequence of DNA. Nowadays, genetic engineering is a common procedure in many laboratories, in part due to the development of simple, accessible, cheap, and effective programmable endonucleases such as those in the CRISPR/Cas9 approach. A key advantage of gene editing is the possibility of producing pigs with desired characteristics

Mayor Advances:

Here we review the latest advances in the production of genetically modified pigs with a particular focus on the use of CRISPR/Cas9 gene editing in the porcine species, as a way to produce genetically modified pigs, and with a consideration of advantages and limitations, as well as new approaches, with regard to this technology.

Keywords Endonucleases · Embryo microinjection · Oocyte electroporation · Somatic cell nuclear transfer · Disease resistance · Animal production · Human disease bio-models · Bioproducts

Authors Sergio Navarro-Serna and Celia Piñeiro-Silva have contributed equally to this work.

S. Navarro-Serna · R. Romar · J. Gadea (✉)
Physiology of Reproduction Group, Department of Physiology, Universidad de Murcia, Murcia, Spain

Instituto Murciano de Investigación Biosanitaria, IMIB-Arrixaca-UMU, Murcia, Spain
e-mail: jgadea@um.es

C. Piñeiro-Silva
Physiology of Reproduction Group, Department of Physiology, Universidad de Murcia, Murcia, Spain

J. Parrington
Department of Pharmacology, University of Oxford, Oxford, UK

V. K. Yata et al. (eds.), *Sustainable Agriculture Reviews 57*, Sustainable Agriculture Reviews 57, https://doi.org/10.1007/978-3-031-07496-7_3

3.1 Introduction

The porcine species (*Sus scrofa domesticus*) has had a great impact in the meat industry worldwide, being the second most consumed meat source in the world in 2020 after poultry according to the Food and Agriculture Organization of the United Nations (FAO 2020). Since its domestication 9000 years ago from the wild boar (*Sus scrofa*) (Giuffra et al. 2000), humans have been selecting for breeding those porcine variants with desired characteristics for food production and animal husbandry. This genetic selection has most recently allowed a significant increase of pig production in comparison with that obtained a few decades ago (Rocadembosch et al. 2016; Koketsu et al. 2017). However, classical breeding and genetic selection techniques are slow processes that carry an economic and time cost which can be avoided by the use of genetic engineering (Yang and Wu 2018). In recent decades, techniques have been developed to accelerate genetic selection by gene editing. Genetic engineering involves modifying the sequence of DNA, which makes it possible to introduce new genes or change the pattern of gene expression in edited cells, including modifications in sequences that encode proteins, regulatory sequences or sequences that produce non-coding RNAs (Yang and Wu 2018).

Nowadays, genetic engineering is a common procedure in many laboratories, in part due to the development of simple, accessible, cheap, and effective programmable endonucleases such as those in the CRISPR/Cas9 approach. The contribution of Jennifer Doudna and Emmanuelle Charpentier to the development of this approach, discovered in microorganisms but now used to edit the genome of practically any organism (Jinek et al. 2012) was acknowledged by the award of the Nobel Prize in Chemistry to them in 2020 and has opened the door to easier genetic modification of mammals, and specifically in our case, of pigs.

Since the first generation of gene edited pigs using CRISPR/Cas9 in 2014 (Hai et al. 2014), new findings have been published relating to strategies for the efficient production of gene edited pigs. A key advantage of gene editing is the possibility of producing pigs with desired characteristics, for example, pigs with better meat production (Wang et al. 2015a), less fat deposition (Zheng et al. 2017) or resistance to viral diseases (Whitworth et al. 2016), far more rapidly than would be possible using traditional breeding methods. Furthermore, genetically modified pigs can be produced for use in biomedical research due to the similarity of pigs and humans in relation to physiology, anatomy, size, and metabolic profile (Zettler et al. 2020). In addition, pigs have advantages over other animal models like non-human primates because they are cheaper to produce, are associated with less ethical issues, and because of better development of porcine embryo manipulation techniques (Yang and Wu 2018). Reflecting this, porcine models of human diseases like diabetes mellitus (Tanihara et al. 2020b; Zettler et al. 2020), neuromuscular diseases (Crociara et al. 2019) and ones for xenotransplantation (Hein et al., 2020), have recently been produced.

In recent reported studies, differences in procedure have been found with respect to the type of methodology and the strategies used to obtain modified embryos, the molecular nature of Cas9, the method and conditions of CRISPR/Cas9 delivery into embryos, and the embryo transfer protocol. Up till now, there has not been a standard procedure for generation of gene edited pigs, so it is very important to study all the factors that have had an influence on the efficiency of this technique, to determine what are the best options in the design of protocols to optimize the CRISPR/Cas9 approach. For that reason, the objective of this chapter is to summarize the knowledge about the advances of production of genetically modified pigs with the use of CRISPR/Cas9 system in the porcine species, focusing on the production of genetically modified pigs, its advantages and limitations, as well as the new approaches for this system.

3.2 History of Gene Editing in Pigs

3.2.1 Gene Editing Before Programmable Endonucleases

The first strategy to produce genetically modified mammals was pronuclear injection which involves the direct introduction of a DNA construct into the pronuclei of the fertilized egg that can integrate randomly in the genome of the zygote. Using this approach, transgenic mice were produced in the early 1980s (Gordon et al. 1980; Gordon and Ruddle 1981; Palmiter et al. 1982) followed by other mammals such as pigs, sheep or rabbits in 1985 (Brem et al. 1985; Hammer et al. 1985) (Fig. 3.1). Other approaches used to introduce foreign DNA into pig embryos were developed, such as the use of retroviruses (Petters et al. 1987), lentiviruses (Hofmann et al. 2003) or sperm-mediated DNA transfer (Sperandio et al. 1996; Lavitrano et al. 1997; Garcia-Vazquez et al. 2010). These approaches were limited by the fact that the foreign DNA randomly inserted into the host genome and did not allow the specific editing of endogenous genes, so other approaches were needed. In mice, homologous recombination was used to genetically modify embryonic stem cells (ESCs) in culture and these were then introduced into early embryos to produce chimeras. Despite the low efficiency of this approach, the generation of genetically modified animals by homologous recombination was a significant improvement in terms of the on-target nature of the introduced mutation and the increase in mutation rates (Robertson et al. 1986; Hooper et al. 1987).

The lack of success in obtaining ESCs from other mammalian species, apart from rats and humans, made it impossible to apply this technique to mammals in general (Yang and Wu 2018). However, the development of cloning by somatic cell nuclear transfer, which involves the introduction of somatic cells into enucleated oocytes, solved the problem and made it possible to generate fully genetically modified animals in one generation by first performing homologous recombination in these somatic cells (McCreath et al. 2000). Although the first attempt to generate

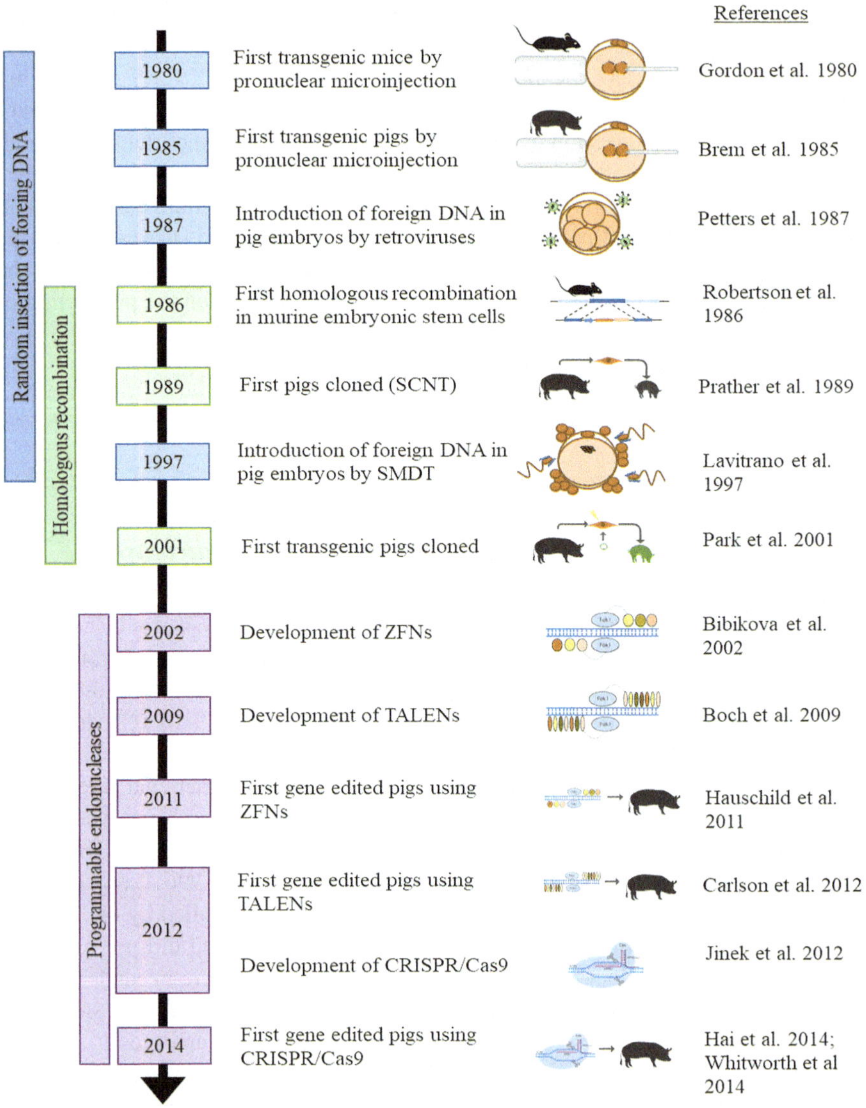

Fig. 3.1 History of pig gene editing: Timeline of milestones in the history of gene editing in swine species

genetically modified animals using homologous recombination was made in the 1990s, it was not until 2001 that the first application of this technique in pigs was reported (Park et al. 2001; Dai et al. 2002; Lai et al. 2002). While being a great advance in terms of gene editing in livestock, a limitation was that the efficiency was still low.

3.2.2 Programable Endonucleases

The production of genetically modified mammals has been greatly helped by the development of programmable endonucleases. Three such types of endonucleases have been developed so far: ZFNs (zinc finger nucleases), TALENs (transcription activator- like effector nucleases) and the CRISPR/Cas system (CRISPR-associated protein). All of them are composed principally of two domains: a DNA binding domain that recognises the target DNA sequence and a cleavage domain that produces a double-strand break (DSB) in the target DNA sequence (Petersen 2017).

Genetic modifications mediated by programmable endonucleases are due to mistakes produced by the cell machinery in the process of repairing the DSB. After the DSB occurs, the cells can repair these breakages by two cellular mechanisms: non-homologous end joining (NHEJ), or homology-directed repair (HDR) (Fig. 3.2). In most cases (over 90%), this repair happens via the NHEJ mechanism. During this process in which the two ends of the break are brought together and ligated, small insertions or deletions (indels) often occur, which can modify the reading frame of the gene (if the indel is not a multiple of three) causing a premature appearance of a stop codon and producing a knock-out. However, if a donor DNA is introduced the repair can occur via the HDR pathway. In this case, part of an endogenous gene can be subtly changed or alternatively one or more complete genes can be introduced, producing a knock-in (Whitelaw et al. 2016; Petersen 2017).

The first programmable endonuclease used to produce genetically modified animals was ZFNs, as first reported in a Drosophila model in 2002 (Bibikova et al. 2002).However, it was not until 2011 that this type of endonuclease was used to

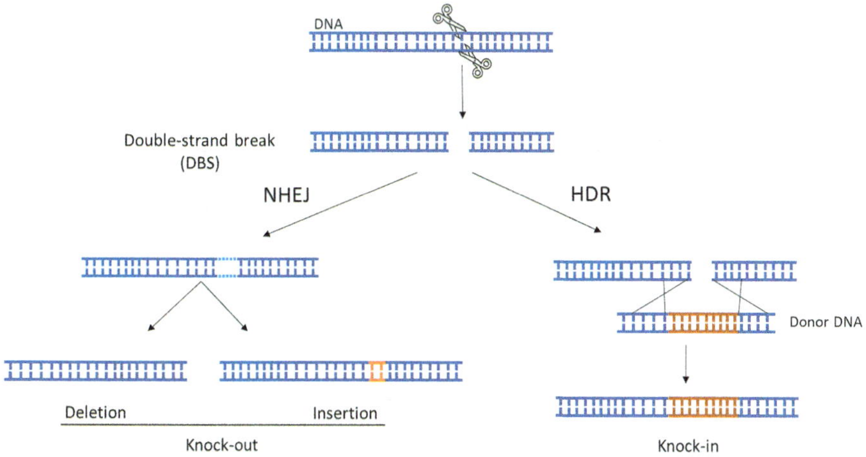

Fig. 3.2 DNA repair mechanism: Schematic showing the DNA repair process after cutting by an endonuclease. Two types of repair can occur after the double-strand break: non-homologous end joining (NHEJ) in the absence of a donor DNA sequence or homology-directed repair (HDR) in the presence of a donor DNA sequence. NHEJ produces insertions or deletions that can cause knock-out and HDR can generate knock-ins through the integration of exogenous DNA

generate knock-out pigs (Hauschild et al. 2011). ZFNs consist of the nuclease domain of the restriction enzyme FokI and multiple zinc finger protein sequences combined to recognize a specific DNA sequence. Each zinc finger protein is formed by 30 amino acids which recognizes a sequence of 3 bp (Wolfe et al. 2000; Bibikova et al. 2001). FokI must dimerize to produce a DSB, so ZFNs needs to be designed and used as pairs (Vanamee et al. 2001; Miller et al. 2007).

The second type of programmable endonuclease developed (Boch et al. 2009; Christian et al. 2010) and used to produced genetically modified pigs was TALENs (Carlson et al. 2012). This type of nuclease consist of the FokI restriction enzyme domain as is the case with ZFNs; in contrast however, the DNA binding domain is formed by peptides that can recognize a single base pair, and which are called transcription activator-like effectors (Boch et al. 2009; Christian et al. 2010). The fact that TALEN domains can recognize a single base allows greater design possibilities with TALEN nucleases, these being more specific, easy to use, and efficient than ZFNs, which recognize groups of three bases (Hockemeyer et al. 2011; Moore et al. 2012). In addition, the use of TALENs seems to be less cytotoxic than ZFNs (Yao et al. 2016).

After ZFNs and TALENs, the development of a third type of programmable endonuclease called CRISPR/Cas9 was reported (Jinek et al. 2012; Cong et al. 2013; Mali et al. 2013). The CRISPR/Cas9 system was identified in archaea and eubacteria as an adaptative immune mechanism to protect the cell from viral infections (Mojica et al. 2000; Mojica et al. 2005). CRISPR/Cas systems can be classified into two classes, and six types, but the one most used for gene editing involves the Cas9 from *Streptococcus pyogenes*, and is classified as class 2, type II.

In contrast to the ZFNs and TALENs, which are composed of proteins, the CRISPR/Cas9 system consists of Cas9 protein and a single guide RNA (sgRNA). Cas9 protein has two nuclease domains, RuvC and HNH, which cut the target and complementary sequences respectively to generate DSBs. The sgRNA allows the endonuclease to cut at the target sequence and contains a sequence of 20 nucleotides complementary to the target sequence, cis-repressed RNA (crRNA), and a Cas9 binding sequence (trans-activating crRNA). To allow cleavage of the DNA sequence, the sgRNA should contain a DNA sequence that is complementary to the sgRNA and an adjacent protospacer-adjacent motif (PAM) (Yuk et al. 2014). The PAM sequence differs depending on the species, and the system from *S. pyogenes* needs a 5′-NGG-3′ PAM sequence (Petersen 2017; Yang and Wu 2018).

Compared with ZFNs and TALENs, the CRISPR/Cas9 system is highly efficient and adaptable as a way to produce mutations in developing embryos in part due to the difficult of designing and generating novel ZFN and TALEN endonucleases (Hai et al. 2014; Zhou et al. 2016). The first time that the use of the CRISPR/Cas9 system was described in pigs was in 2014 (Hai et al. 2014), when it was used to produce knock-out pigs for the v-WF gene, whose deficiency in humans causes severe von Willebrand disease, by microinjection of Cas9 mRNA and sgRNA into the cytoplasm of one-cell zygote obtained by *in vitro* fertilization. Since that time, a large number of gene edited pigs have been reported with a number of different applications, as will be explained further in Sect. 3.4.

3.3 How to Generate a Gene Edited Pig?

3.3.1 Methodology

3.3.1.1 Editing of Somatic Cells and Somatic Cell Nuclear Transfer (SCNT)

Briefly, the SCNT technique consists of the introduction of a genetically modified somatic cell into an enucleated oocyte to produce an embryo with the desired genetic characteristics. The first cloned piglets (non-genetically modified) were produced in 1989 by Prather *et al.* using as donor cells porcine embryo blastomeres (Prather et al. 1989).. Subsequently, the first genetically modified piglets were produced in 2001, by introducing the GFP gene contained in a retroviral vector into foetal fibroblasts (Park et al. 2001), but it was not until 2014 that the CRISPR/Cas9 system was first used to produced genetically modified pigs by SCNT (Sato et al. 2014; Whitworth et al. 2014). This technique is widely used, but it has low efficiency (10-30% of blastocyst rate and 0.5-5% piglets/transferred embryos) due to the difficulty of the procedure and epigenetic dysregulations in cloned embryos, which can cause development defects in the offspring such as weakness or low birth weight (Beaujean et al. 2015).

On the other hand, the mayor advantage of this procedure over the direct modification of embryos is that the genotype of the offspring is known as the donor cells with the desired genotype are chosen. This is a great advantage when multiple genes are targeted or when the purpose is a knock-in or a conditional transgenesis (Li et al. 2015; Whitelaw et al. 2016). In general, the SCNT procedure can be divided into three steps: (a) selection, culture and genetic modification of donor somatic cells, (b) oocyte enucleation and nuclear transfer and (c) embryo culture (Fig. 3.3).

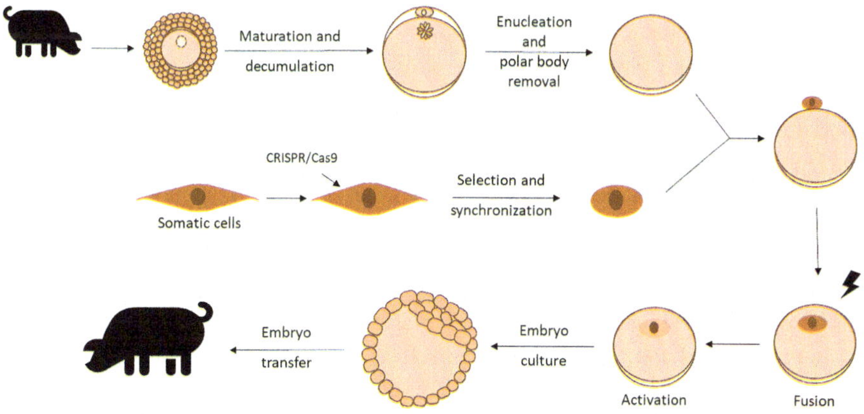

Fig. 3.3 Schematic of the process of somatic cell nuclear transfer (SCNT). Somatic cells are edited by the CRISPR/Cas9 system. After being selected and synchronized, cells are fused with *in vitro* matured enucleated oocytes. Subsequently, cloned zygotes are cultured *in vitro* until the time of embryo transfer

Selection, Culture, and Genetic Modification of Donor Cells

This step involves the selection and culture of the donor somatic cells. The first thing to consider is the donor cell type. In general, these cells need to have a correct karyotype and long lifespan, exhibit good proliferation, and they should be transfectable (Beaujean et al. 2015).More than 200 different cell types have been used as donor cells to produce cloned pigs, each one presenting advantages and disadvantages (Liu et al. 2015). Theoretically, the less differentiated cells are the ones easiest to reprogram. This reprogramming involves essential epigenetic changes that occur during embryo development (Lee and Prather 2014; Liu et al. 2015).

Although there are few studies that demonstrate that one cellular type is better than another, foetal fibroblasts are the most common cell type used for SCNT to produce genetically modified pigs (Whitworth et al. 2014; Chen et al. 2020; Li et al. 2020a); they are highly dedifferentiated cells (Liu et al. 2015). In addition, adult fibroblast cells can be obtained from valuable animals like genetically modified pigs, so they can be used to bypass one step of the process (Sheets et al. 2018; Xu et al. 2020).

Liu et al. achieved a greater overall efficiency in terms of live birth rate using foetal fibroblasts as donor cells in comparison with adult fibroblasts, observing less abnormalities in the piglets produced (deformities, mummified and stillbirth piglets) (Liu et al. 2015). In addition, Zhai et al. observed a higher blastocyst rate using bone marrow- derived mesenchymal stem cells in comparison with foetal fibroblasts. Furthermore, cloned embryos using as donor cells bone marrow-derived mesenchymal stem cells were more similar to those produced by conventional *in vitro* fertilization (IVF) in terms of expression of pluripotency genes, degree of apoptosis, and number of cells per blastocyst (Zhai et al. 2018). One thing to consider is that both studies were performed without transfected somatic cells.

Once a suitable cell type is chosen, the cells are cultured to purify them (when obtained from an animal) and to achieve the necessary cell concentration. Next, these cells need to be transfected (Kurome et al. 2015). There are many transfection methods available: viral vectors (adenovirus, retrovirus…), chemical transfection methods (nanoparticles, lipofection…) and physical transfection methods (electroporation, magnetotransfection…) (Fajrial et al. 2020). Regarding the production of genetically modified pigs, the authors usually prefer the electroporation method, being the one most used to generate knock-out and knock-in models (Lai et al. 2016; Cho et al. 2018; Han et al. 2020; Xu et al. 2020; Zhang et al. 2020) and because it is simpler, more efficient and reproducible (Fajrial et al., 2020). This method involves subjecting the cells to an electric field that induces the formation of pores in the cell membrane through which the components of the CRISPR/Cas9 system can enter the cell (Fajrial et al. 2020). However, some authors use other methods like lipofection (Luo et al. 2019; Fischer et al. 2020; Zhu et al. 2020a). This method is based on the use of lipid molecules that encapsulate the CRISPR/Cas9 system components forming vesicles. These vesicles are introduced into the cell by endocytosis, releasing the components. One thing to consider when using this method is that the components in the vesicle are only going to reach the cytoplasm so it is not the best option when using plasmid DNA because that needs to reach the nucleus to function (Fajrial et al. 2020).

Something that needs to be taken into account is the format in which the CRISPR/Cas9 system is delivered. This can be plasmid DNA (plasmid encoding Cas9 and sgRNA), mRNA (mRNA encoding for Cas9 and separate sgRNA) and ribonucleoprotein (Cas9 protein and sgRNA) (Fajrial et al. 2020). When using DNA vectors, these require previous transcription and translation to produce functional ribonucleoprotein in the cell. This increases the lag time between the transfection and the expression of the system, which increases the risk of off-target events in addition to the high persistence of the system. Furthermore, plasmid DNA can be inserted into the genome of the cell, so this possibility should be assessed when selecting the most suitable donor cell to perform SCNT (Wang et al. 2017a; Fajrial et al. 2020).

In contrast, mRNA does not have risk of insertional mutagenesis and, as it does not require the entry into the nucleus to function, its expression is faster. Furthermore, the use of mRNA results in transient expression, reducing the persistence time of the system in the cell, which reduces the risk of off-target events (Wang et al. 2017a; Fajrial et al. 2020). In general, when talking about SCNT the most common delivery format is DNA, with only one group performing the transfection using mRNA to produce piglets by SCNT (Guo et al. 2019a). In addition, some authors have used ribonucleoprotein to transfect different porcine somatic cells, so this delivery format could be used as well to produce cloned pigs (Park et al. 2016; Elkhadragy et al. 2021).

Once the cells are edited, the next step is select a colony with the desired mutation. The cells are cultured to produce a single-cell colony (monoclonal colony) and once they reach the appropriate cell density, part of the colony is used to detect (usually by PCR and sequencing) the presence of mutations while the other part continues in culture. The colonies with the desired mutation are used as donor cells (Beaujean et al. 2015; Xie et al. 2020b; Zhang et al. 2020). Sometimes a selection procedure can be performed when the plasmid used has a selection marker. Some authors use antibiotics like geneticin (Han et al. 2020; Huang et al. 2020a) and puromycin (Fischer et al. 2020; Zhu et al. 2020b), others a fluorescent marker like the enhanced green fluorescent protein (EGFP), to select the transfected cells (Li et al. 2020b; Shi et al. 2020).

Oocyte Enucleation and Nuclear Transfer

In this phase mature oocytes are enucleated. The oocytes are usually collected from ovaries from prepuberal gilts obtained from local slaughterhouses, and then *in vitro* matured (Beaujean et al. 2015). Some authors have shown that a selection of the best oocytes can be helpful to achieve a better cloning efficiency. For example, treatment with a hyperosmotic medium with sucrose assists the selection of high-quality oocytes and improves enucleation efficiency due to the formation of a swelling around the metaphase plate (Dang-Nguyen et al. 2018).

Once mature, the meiotic spindle and the polar body of the oocytes are removed. There are two enucleation techniques. The first one is called blind enucleation and is performed with micromanipulation equipment (Polejaeva et al. 2000; Liu et al. 2017) (Fig. 3.4). The second one is called handmade cloning and consists of the bisection of 1/3 of the oocyte (where the polar body is located) with a splitting blade (Du et al. 2007; Liu et al. 2017; Li et al. 2019b). Although Liu et al. achieved a greater blastocyst rate when performing handmade cloning and it is cheaper and easier (Liu et al. 2017), blind enucleation is the most common enucleation method used to produce genetically modified pigs.

With both methods, the manipulation medium needs to be supplemented with cytochalasin B in order to avoid the extrusion of the pseudo-second polar body and to maintain the diploidy of the resulting embryos (Beaujean et al. 2015). In addition, in order to increase the enucleation rate, a treatment with demecolcine can be used. With this treatment a swelling around the meiotic spindle occurs and it becomes easier to remove (Gao et al. 2019). The next phase is the nuclear transfer itself. First, the somatic cells need to be synchronised. Usually, cells in the G0/G1 phase of the cell cycle are used to perform SCTN because that way the diploidy of the resulting embryo is maintained (if another phase is used, the re-replication of the DNA can occur before the first cell cycle of the embryo). This synchronisation is usually performed by serum starvation (Hatada 2017). Next, one donor cell is placed in the perivitelline space of the enucleated oocyte with micromanipulator equipment (if blind enucleation is performed) (Kurome et al. 2015; Cho et al. 2018). If handmade cloning is performed, two cytoplasts are attached to a single donor cell (Du et al. 2007; Liu et al. 2017).

These cytoplast-cell complexes are then fused and activated. The fusion allows the nucleus of the donor cell to enter the oocyte cytoplasm and the activation releases the oocyte from its state of meiotic arrest state (Hatada 2017). The fusion is performed by electric pulses (Kurome et al. 2015). The activation can be electrical or chemical, but the first method is more common (Whitworth et al. 2009). Furthermore, it can be performed at the same time as the fusion (simultaneous activation) or a few

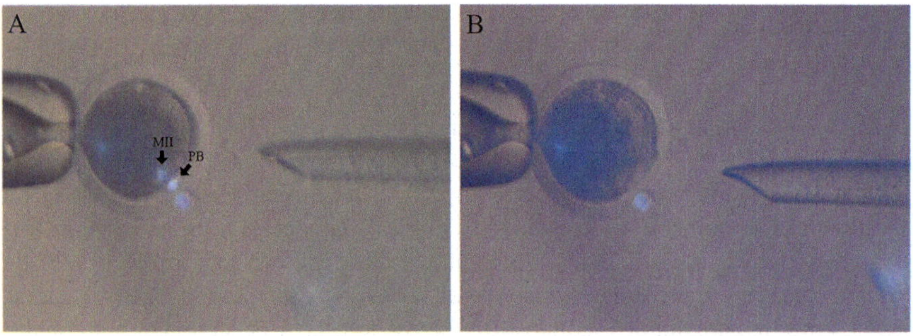

Fig. 3.4 Mature oocyte enucleation. Enucleation process for somatic cell nuclear transfer in porcine oocytes. (**a**) Mature oocyte stained with Hoechst 33342. (**b**) Mature oocyte enucleated without a metaphase plate and polar body. MII: Metaphase II; PB: polar body

hours later (delayed activation) (Hatada 2017; Guo et al. 2019b). Guo et al. in their meta-analysis observed that the delayed activation increases blastocyst rate (without improving cell number per blastocyst) to a greater degree than the simultaneous activation, probably because the cytoplasm of the oocyte needs time to mix with the cytoplasm and the nucleus of the donor cell (Guo et al. 2019b), but simultaneous activation is more commonly used when producing genetically modified pigs.

Embryo Culture

After fusion and activation, the reconstructed embryos are cultured until embryo transfer. For the first 3 hours the medium needs to contain cytochalasin B to avoid the extrusion of the pseudo-second polar body (Hatada 2017). Next, embryos are cultured in *in vitro* culture mediums such as porcine zygote medium (PZM) or North Carolina State University 23 medium (NCSU-23) until ready for embryo transfer (Im et al. 2004; Lee and Prather 2014).

During embryo culture, the remodelling of the chromatin of the somatic cell occurs. This process is required to achieve the totipotent state that the embryo needs to reach to develop correctly, but it is dysregulated in cloned embryos which leads to the low efficiency of the process (Vajta et al. 2007; Lee and Prather 2014). DNA methylation and histone acetylation/deacetylation play important roles in embryo development and cellular function and errors in these processes contribute to the low efficiency of the SCNT procedure (Diao et al. 2013; Kumar et al. 2013). For that reason, donor cells and embryos have been treated with different compounds like DNA methyltransferases inhibitors and histone deacetylase inhibitors (Zhao et al. 2009; Diao et al. 2013; Kumar et al. 2013).

The additive most commonly used in the production of genetically modified pigs is Scriptaid, a histone deacetylase inhibitor (Whitworth et al. 2014; Cho et al. 2018; Boettcher et al. 2020; Zhu et al. 2020b). This compound induces histone acetylation, which changes the structure of the chromatin (more relaxed form), allowing transcription. Furthermore, it promotes DNA demethylation, a key process in nuclear reprogramming and remodelling (Zhao et al. 2009; Park et al. 2012).

The use of other treatments has led to an improvement of the efficiency of the SCNT procedure, but they have not yet been used to produce genetically modified pigs. These compounds include other histone deacetylase inhibitors such as oxamflatin, which led to an improvement in blastocyst rate, embryo quality (total cell number per blastocyst and gene expression), and overall efficiency (offspring) in comparison with control cloned embryos and cloned embryos treated with Scriptaid (Park et al. 2012); sodium butyrate, with which a greater blastocyst rate and a positive effect on embryo gene expression were achieved (Liu et al. 2012); valproic acid, which led to an increase in blastocyst rate and embryo quality (total cell number per blastocyst and gene expression) (Miyoshi et al. 2016; Lv et al. 2020) and suberoylanilide hydroxamic acid, which led to an improvement in blastocyst rate and embryo quality (gene expression) (Sun et al. 2017).

Regarding DNA methyltransferase inhibitors, the treatment with 5-aza-2′-deoxycytidine and zebularine led to improved blastocyst rate and embryo quality (gene expression) in comparison with a control group (Huan et al. 2013; Taweechaipaisankul et al. 2019). Furthermore, donor cells can also be treated with these compounds to improve embryo development. For example, sodium butyrate, 5-aza-2′-deoxycytidine, zebularine and RG108 (DNA methyltransferase inhibitor) were used (Diao et al. 2013), leading to improved blastocyst rate and less DNA fragmentation in comparison with the control cloned embryos.

3.3.1.2 Editing of Gametes and Embryos

The delivery of CRISPR/Cas9 components into *in vivo* or *in vitro* derived zygotes either as mRNAs or ribonucleoprotein makes it possible to achieve gene editing in embryos without the need to perform a cloning step, thereby making this an easier process compared to SCNT due to no need for enucleation and cell fusion. Therefore, this methodology makes it possible to obtain a higher yield in terms of quality and quantity of embryos compared to SCNT (Tanihara et al. 2016).

Gene editing in gametes and embryos has disadvantages such as not knowing the resulting sequence of the target gene after CRISPR/Cas9 cutting and subsequent random DNA strand repair. As a consequence, after the process of embryo production, gene editing and embryo transfer, animals may have no mutation, or be heterozygous and a high percentage of animals might be mosaics, having cells and tissues with different combinations of wild type and knock-out alleles. Also, if the objective of gene editing is to generate a knock- in pig, there is the problem that homologous recombination is less frequent than non- homologous end joining, and the percentage of embryo knock-ins will therefore be low compared to embryos obtained by SCNT. Since the first generation of genetically modified pigs with CRISPR/Cas9 in 2014 (Hai et al. 2014; Whitworth et al. 2014), intracytoplasmic injection has been the most used method for releasing CRISPR/Cas9 into the embryo. Recently, other methods of CRISPR/Cas9 delivery into zygotes such as electroporation have been developed and are in the process of being optimized (Tanihara et al. 2016; Nishio et al. 2018).

CRISPR/Cas9 Delivery by Intracytoplasmic Microinjection

Intracytoplasmic microinjection is a process similar to intracytoplasmic sperm injection (ICSI) (Fig. 3.5). While a holding pipette stabilizes the zygote, a thin micropipette is used to pierce the zona pellucida and the cytoplasmic membrane, and Cas9 and sgRNA is delivered into the cytoplasm. Although this method is less aggressive towards the integrity of the zygote than SCNT, intracytoplasmic micro-injection can cause stress which affects embryo quality and therefore the percentage of embryos able to give rise to a piglet (Hai et al. 2014; Wang et al. 2015c).

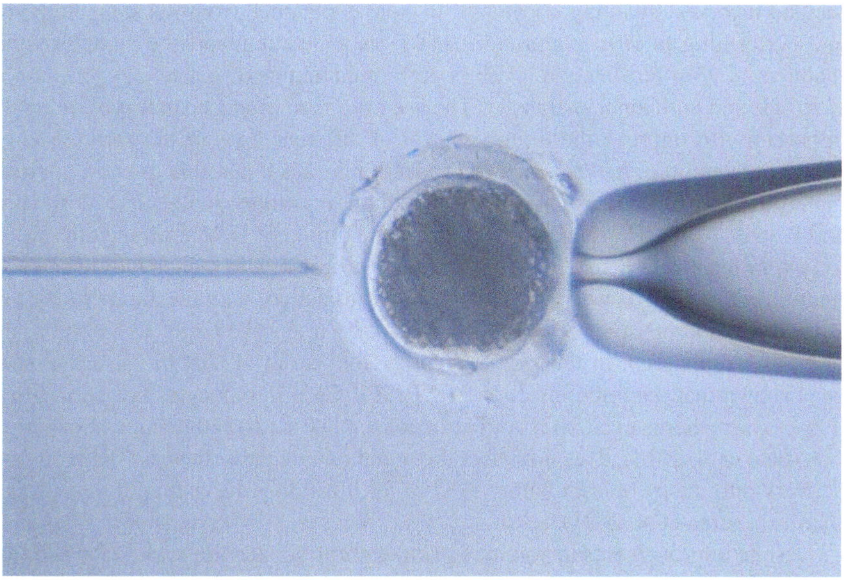

Fig. 3.5 Mature oocyte microinjection: *In vitro* matured oocyte being subjected to intracytoplasmic microinjection with CRISPR/Cas9 gene-editing components

The importance of producing embryos with high quality to be transferred is one of the critical steps. In the literature, we find that both *in vivo* (Hai et al. 2014; Peng et al. 2015; Wang et al. 2015b; Petersen et al. 2016; Wang et al. 2016; Burkard et al. 2017; Chuang et al. 2017; Park et al. 2017) and *in vitro* zygotes (Whitworth et al. 2014; Park et al. 2017; Whitworth and Prather 2017; Hinrichs et al. 2018) have been used for intracytoplasmic microinjection.

In vivo embryo production consists of collecting zygotes by natural mating or artificial insemination after oestrus detection. Around 14-24 hours after insemination, the zygotes have been flushed from oviducts and collected. These embryos can be obtained from sows or gilts sacrificed at the slaughterhouse (Petersen et al. 2016; Chuang et al. 2017) or also by surgical collection (Peng et al. 2015; Wang et al. 2015c; Burkard et al. 2017; Park et al. 2017). Both *in vivo* zygote production using slaughtered sows or via surgery present some difficulties in terms of accurately and reliably obtaining newly formed zygotes because the ovulation takes place over a 4 hour windows of time in synchronized sows (Park et al. 2017). Timing is an important problem in *in vivo* production because is very probable that DNA replication takes place before injection or when zygote cleavage happens, in which case there are only two options: discard the embryo or inject both cells knowing that you would thereby produce a mosaic (Peng et al. 2015).

In vivo production of embryos is an intensive labour and numerous sows are required to obtain a sufficient number of *in vivo* derived zygotes (Wang et al. 2015b). Consequently, other authors have used *in vitro* produced embryos (Whitworth et al. 2014; Park et al. 2017; Whitworth and Prather 2017; Hinrichs et al. 2018). This method involves obtaining oocytes from ovaries obtained from the slaughterhouse and performing *in vitro* maturation (IVM). After cumulus-oocyte complexes are matured, *in vitro* fertilization (IVF) is performed and next, putative zygotes are *in vitro* cultured until embryo transfer. The use of *in vitro* produced embryos offers the opportunity to improve the timing control of different parts of the early development of the embryo. *In vitro* embryo production makes it possible to know approximately the time of fertilization better than with *in vivo* embryo collection (Whitworth and Prather 2017). This can help to regulate the timing of DNA editing with respect to *in vitro* fertilization and DNA replication. One strategy to reduce mosaicism is to microinject as soon as possible, even before fertilization, with the aim of being that the ribonucleoprotein is present in the cytoplasm to cut the target DNA before the first DNA replication in the zygote (Lamas-Toranzo et al. 2019b). It has already been shown that the microinjection of CRISPR/Cas9 into oocytes before fertilization (Navarro-Serna et al. 2021) or parthenote embryos just at the moment of activation (Sato et al. 2018), does not affect the mutation rate, nevertheless CRISPR/Cas9 delivery into the cytoplasm before *in vitro* fertilization reduces the mosaicism rate (Navarro-Serna et al. 2021).

Despite improvements in porcine *in vitro* embryo production such as the addition of reproductive fluids (Canovas et al. 2017; Paris-Oller et al. 2021), cytokines (Yuan et al. 2017), or culture media (Redel et al. 2016), embryo quality has not yet been assessed similar to embryos produced *in vivo*. One problem with *in vitro* produced embryos is the high rate of polyspermy that takes place in *in vitro* fertilization process in swine species (Romar et al. 2019). Due to this problem, some authors have used other IVF options, for instance the use of *in vivo* matured oocytes, but this approach is very laborious and difficult and requires a large number of animals similar to that for *in vivo* zygote production (Park et al. 2017).

Regarding the molecular nature of CRISPR/Cas9, this has been microinjected in porcine embryos as DNA, mRNA, or ribonucleoprotein. There is no evidence that there are differences in mutation rate using such different strategies (Navarro-Serna et al. 2021) but due to the possibility of genome integrity, microinjected plasmids have been the least used. Despite this risk, DNA integrations were not found in studies in which CRISPR/Cas9 plasmids were used in pigs (Petersen et al. 2016; Chuang et al. 2017). The most used forms for microinjection have been Cas9 mRNA and sgRNA (Hai et al. 2014; Kwon et al. 2015; Peng et al. 2015; Burkard et al. 2017; Hinrichs et al. 2018; Chen et al. 2019a; Ostedgaard et al. 2020; Xie et al. 2020a). The combination of Cas9 protein and sgRNA as an injected ribonucleoprotein is also used (Sheets et al. 2016; Park et al. 2017; Sheets et al. 2018) and has the advantage that the activity is instantaneous after ribonucleoprotein formation without the need for Cas9 protein to be synthesized in the zygote.

In addition to differences in the molecular nature of the CRISPR/Cas9, variations have also been found in the concentration and/or proportion of Cas9 and sgRNA microinjected. There is not a standard concentration because the amount of CRISPR/Cas9 microinjected varies depending on the efficiency of the designed sgRNA and the molecular environment in which the target sequence is found. In the literature, the potential toxic effect of CRISPR/Cas9 microinjection in embryo development has been tested by performing injection with water or Cas9 mRNA and sgRNA in parthenogenetic activated oocytes. The blastocyst rate was similar in both groups, so CRISPR/Cas9 does not seem to have toxic effects (Hai et al. 2014; Whitworth et al. 2014; Wang et al. 2015b; Yu et al. 2016). However intracytoplasmic microinjection can produce some defects in embryo development (Yu et al. 2016). Thus, CRISPR/Cas9 has been microinjected at a concentration of between 25-1000ng/μl in porcine embryos.

Another difference besides the method of embryo injection is the site in which zygotes were injected. In almost all studies, CRISPR/Cas9 components were injected into the cytoplasm (Hai et al. 2014; Whitworth et al. 2014; Zhou et al. 2016; Burkard et al. 2017; Whitworth and Prather 2017) but in Chuang *et al.* 2017, the zygotes were injected in the pronucleus (Chuang et al. 2017). Pronuclear microinjection is difficult in embryos of domestic animals due to the cytoplasm being opaque due to lipid droplets present in the cytoplasm and the pronucleus microinjection requires centrifugation of the zygote to expose the pronucleus (Wall et al. 1985).

In addition, other strategies have been described such as combining SCNT and intracytoplasmic microinjection (Sheets et al. 2016; Sheets et al. 2018). In Sheets *et al.* 2016, SCNT was performed with foetal fibroblasts and matured oocytes with the aim being to microinject CRISPR/Cas9 components as with *in vivo* or *in vitro* derived embryos (Sheets et al. 2016) and in Sheets *et al.* 2018, genetically modified zygotes were produced by *in vivo* zygote microinjection and these were used to obtain genetically modified foetal fibroblast cells which were used to obtain piglets by SCNT (Sheets et al. 2018). Despite the limitation of *in vitro* embryo production, the quality and quantity of *in vitro* produced embryos is sufficient for authors to report the generation of healthy piglets.

CRISPR/Cas9 Delivery by Electroporation

In addition to microinjection, the increasing use of electroporation as a method of introducing CRISPR/Cas9 into the cytoplasm of porcine zygotes has been of significant importance in the last few years (Tanihara et al. 2016; Nishio et al. 2018). This technique involving placing the zygotes in media with Cas9 (protein or mRNA) and sgRNA and subjecting the zygotes to an electric current that allows the formation of membrane pores through which CRISPR/Cas9 components can enter (Fig. 3.6).

Transfection by electroporation has been in development for many types of mammalian cells, for example foetal fibroblasts for SCNT but to penetrate the zygote membrane is more difficult because the zona pellucida must be penetrated before this membrane. To reduce this barrier, the zona pellucida can be partially digested by acid Tyrode's solution in mouse zygotes, improving CRISPR/Cas9 entry (Chen et al. 2016) but in porcine studies, CRISPR/Cas9 editing by electroporation has been reported without zona pellucida digestion. Another option to improve the penetration of the zona pellucida is to use molecules with a smaller size

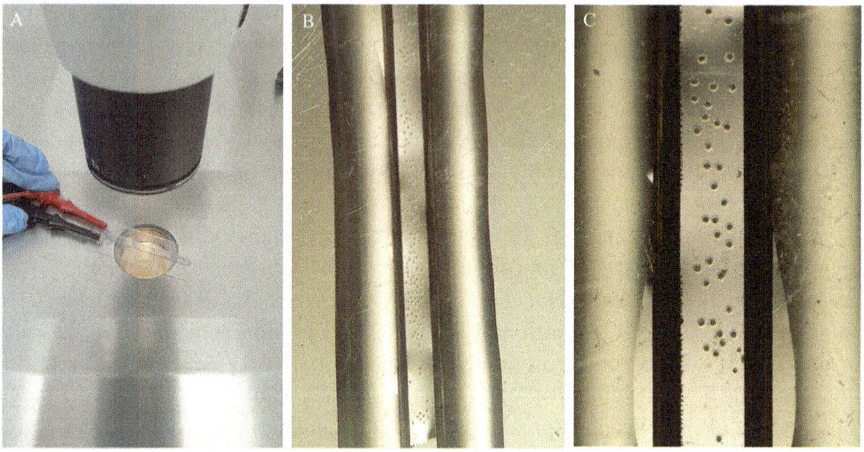

Fig. 3.6 Mature oocyte electroporation: *In vitro* matured oocyte during electroporation process. (**a**) plate electroporation device. (**b**) oocytes in electroporation device at low magnification. (**c**) oocytes in electroporation device at higher magnification

than Cas9 protein (160 kDa) with respect to Cas9 mRNA (~1500 kDa), this can improve mutation rate (Tanihara et al. 2016). Another limitation is that the need to put the zygotes into a medium with a mixture of Cas9 and sgRNA requires using more sgRNA and Cas9 than for CRISPR/Cas9 microinjection.

Since 2016, the successful generation of genetically modified pigs using this technique has been reported (Tanihara et al. 2016; Nishio et al. 2018; Tanihara et al. 2018; Hirata et al. 2019; Tanihara et al. 2019a; Hirata et al. 2020; Le et al. 2020; Tanihara et al. 2020a; Yamashita et al. 2020). Some protocols using different electroporation devices, and variations in the number and polarity of pulses, pulse length, Cas9 and single guide RNA proportion and concentration, electroporation medium and voltage strength have been described.

Two main devices for electroporation have been used: electroporation plates and electroporation cuvettes. The use of the last one has not been reported for porcine zygotes, so there seems to be a preference for the use of electroporation plates. Some electroporation devices offer the opportunity to perform bipolar pulses (Nishio et al. 2018; Yamashita et al. 2020). However, it has been reported that the use of bipolar pulses does not increase the mutation rate but produces a decrease of blastocyst rate compared to that with unipolar pulses (Nishio et al. 2018). In addition to polarity, machines such as the NEPA21 Super Electroporator (NEPAGENE, Ichikawa, Japan) offer the option of making two types of pulses: first an electroporation pulse, which involves performing a high voltage pulse to destabilize the membrane and later transfer pulses to create an input current of CRISPR/Cas9 present in the electroporation media (Yamashita et al. 2020).

Regarding pulses, a balance must be found between the damage caused to embryo quality and quantity and the proportion of mutant embryos obtained. A greater number of pulses (Tanihara et al. 2016), voltage strength (Nishio et al. 2018) and pulse length (Tanihara et al. 2016) can negatively affect embryo development.

Currently, the electroporation parameters used are between 3-5 pulses, 20-30 V and 1ms of pulse length.

Similar to microinjection, electroporation also makes it possible to deliver CRISPR/Cas9 into *in vitro* matured oocytes rather than zygotes (Hirata et al. 2019). As with microinjection, in addition to reducing the amount of mosaicism, this strategy offers other benefits. The zona pellucida of the mature oocytes has a higher permeability than the zona pellucida of zygotes, allowing molecules of up to 170 kDa to cross it, compared to those of 110 kDa that can cross the zona pellucida of zygotes (Legge 1995). This situation makes the membrane of mature oocytes more permeable and, as a consequence, a high concentration of ribonucleoprotein can enter the cytoplasm and therefore a higher mutation rate can be obtained (Hirata et al. 2019).

Compared with electroporation, CRISPR/Cas9 microinjection and SCNT requires greater and more laborious work because micromanipulation needs more qualified personnel to microinject cell after cell in the shortest possible time, and this also requires a lot of time (Navarro-Serna et al. 2019). In contrast, groups of between 50-100 zygotes can be electroporated at the same time, allowing more rapid work and a shorter time of exposure of the embryos to a hostile environment outside the incubator.

3.3.1.3 Embryo Transfer

Once produced, gene-edited embryos can be transferred to recipient animals. Commercial application of embryo transfer (ET) in swine is not as well developed as other assisted reproductive techniques such as artificial insemination or oestrous synchronization. ET was considered unfeasible for decades mainly because of the requirements of surgical techniques for embryo collection and embryo deposition into recipients, alongside challenges in the preservation of embryos (Martinez et al. 2019). This situation has drastically changed in recent decades with the current technology allowing non-surgical ET (nsET) in a non-sedated recipient (Li et al. 1996). Despite piglets having been born from *in vitro*-produced embryos following nsET (Yoshioka et al. 2003; Yoshioka et al. 2012), when working with gene-edited *in vitro*-derived embryos, the conventional surgical ET (sET) by mid laparotomy is the most common methodology followed by researchers due to the compromised quality and limited number of embryos obtained. As such, the low quality of *in vitro*-produced porcine embryos is compounded by the difficulties arising from a surgical intervention. Therefore, in order to maximize sET efficiency with gene edited embryos it is crucial to investigate the most relevant ET- related factors impacting the final pregnancy rate and fertility such as the selection of recipients, embryo handling, age and number of embryo transferred, ET position, and the use of aiding parallel strategies such as the use of co-transferred helper embryos or the infusion of seminal plasma.

Whenever possible, females with an excellent reproductive history should be used as recipients since reproductive performance is one of the well-known factors influencing ET success in the pig. In fact, when ET is performed in first-oestrus gilts, the survival of embryos is approximately 20% lower compared with that from older gilts (reviewed by (Youngs 2001)). The precise control and detection of oestrous in the recipient animal and the timing of *in vitro* fertilization (IVF) to produce embryos is another key factor for a successful ET. It has been demonstrated that after sET the efficiency of piglet production (percentage number of piglet(s) born based on the number of embryos transferred) is greater in recipients whose oestrous cycle is asynchronous to that of donors with a 1-day delay (8.3%) compared to those recipients with a 2-day (1.5%) or 3-day (0.9%) delay (Yoshioka et al. 2012). The same has been described when transferring *in vivo*-derived embryo by nsET (Angel et al. 2014). From the practical point of view, it means that oocytes must be inseminated in the lab 24-48 hours before oestrous beginning in the recipient animal. This leads to an environment where uterine histotrophic secretions from a "younger" uterus are compatible with metabolic needs of the "older" embryos. Following this schedule, our group has achieved around 35% of transferred recipients reaching pregnancy to term after sET of wild type *in vitro*-derived embryos (Paris-Oller et al. 2021) and CRISPR/Cas9 gene-edited embryos (Navarro-Serna et al. 2021). The final yield of ET also depends on the protocols followed to generate the embryo. Thus, the current efficiency of sET with SCNT-embryo is very low with results indicating that transfer of cloned embryos cultured for a longer time after reconstruction (22-24 hours vs. 4-6 hours) decreases the recipient's pregnancy rate and farrowing rate suggesting that long *in vitro* culture time negatively affects the development of transferred cloned porcine embryos (Shi et al. 2015).

Another factor to consider for ET is the embryo's age, which is linked to the ET site, and the number of embryos to transfer. In an ideal situation, once gene-edited zygotes are produced, it is advisable to reduce the culture time to minimize the stress derived from *in vitro* culture conditions (Garcia-Vazquez et al. 2010). That would mean transferring 2-4 cell stage embryos into the recipient's oviduct. Despite there not being extended studies in pigs relating to unilateral or bilateral transfer, it has been shown after sET of cloned embryos into double oviducts that there is an increase in pregnancy rate, farrowing rate of recipients, and the developmental rate of transferred embryos, compared to the situation with unilateral oviduct transfer (Shi et al. 2015). Based on these findings, splitting the total number of embryos to transfer, and performing a bilateral deposition should result in a higher efficiency with sET compared to a unilateral approach. As for morulae or blastocysts, the site for sET is the uterus, preferably within 30 cm from the tip of the uterine horn. Although there are not extensive studies relating to the most suitable sET site for embryos, it seems clear that the uterine body should be avoided since under *in vivo* conditions morula/blastocysts remain near the tip of the uterine horn until day 6 to 7 of the oestral cycle, progressing subsequently toward the uterine body (Hunter et al. 1967; Dziuk 1985). Information in the literature about the optimum total number of transferred embryo varies with numbers ranging from 14 to 50 wild type *in vitro*-derived blastocysts and litter sizes of 8-10 piglets (Marchal et al. 2001;

Kikuchi et al. 2002; Yoshioka et al. 2002; Yoshioka et al. 2003; Somfai et al. 2014; Paris-Oller et al. 2021) although some groups report 100% pregnancy rates after sET with only 20-25 blastocysts per recipient (Yoshioka et al. 2002). However, when transferring gene-edited *in vitro*- derived embryos, it is necessary to increase these numbers up 80-200 per recipient (Onishi et al. 2000; Shi et al. 2015; Navarro-Serna et al. 2021).

Although porcine embryos can develop well *in vitro* to the blastocyst stage, their subsequent development *in utero* after ET in the uterine horns of surrogates can be improved. One strategy to achieve this is to use helper embryos since typically pigs require at least four foetuses for a successful pregnancy and the presence of helper embryos might assist the full development of gene-edited embryos (Onishi et al. 2000). This strategy consists in transferring gene-edited embryo concomitantly with other embryos with the objective being to help establish and/or maintain pregnancy in recipients. Recently, the use of co-transferred helper embryos has been used in an attempt of increase efficiency of sET using cloned embryos (Shi et al. 2015). These authors employed helper embryos (50 per recipient) with different origins such as parthenogenetic, *in vitro*- and *in vivo*-derived (by inseminating the recipient before sET) embryos but either type of helper embryos could aid establishment and/or maintenance of pregnancy with SCNT embryos. On the other hand, another strategy would be the use of seminal plasma. It has been reported that infusion of heterologous seminal plasma prior to AI of recipients upregulates the expression of embryo development related genes in day 6 wild type pig embryos (Martinez et al. 2020; Tajima et al. 2020). However, the likely beneficial effect of this strategy to increase sET efficiency with gene-edited embryos remains to be determined.

3.3.2 How to Detect Mutations?

In the literature, we can find different methods to detect DNA modifications, which change depending on the information that we want to obtain: detection of mutations, identification of the mutant sequence, or evaluation of mosaicism.

3.3.2.1 Electrophoresis-Based Techniques

The simplest method is to perform PCR and then agarose gel electrophoresis. This method is used to evaluate large insertions or deletions for example when the strategy is to generate large deletions and two sgRNAs are designed to remove part of an exon or even complete exons (Whitworth and Prather 2017; Wu et al. 2017; Hirata et al. 2020; Koppes et al. 2020). PCR product digestion with mismatch-sense endonucleases, such as T7 endonuclease I are also used to detect mutations produced by CRISPR/Cas9 and other programmable endonucleases (Wang et al. 2015a; Kang et al. 2016; Bloom et al. 2017; Li et al. 2017a; Xie et al. 2017). This method involves performing a PCR of the target region of the embryo/animal sample. Next, the PCR product is mixed with a wild type amplicon, denatured, and reannealed. As a result, a heteroduplex will be formed with one wild type chain and one sample chain. When a sample of DNA has mutations, a heteroduplex with mismatches are formed and these mismatches are detected and cut by T7 endonuclease I after incubation with the endonuclease. Finally, agarose gel electrophoresis is performed with the digestion product. When the band has the size of the PCR product, this signifies that the sample is wild type and when a smaller band appears, T7 endonuclease I has cut it, so the sample is mutant (Bloom et al. 2017). This method makes it possible to obtain results quickly and easily, however the T7 endonuclease I assay only gives information about the presence of mutations in the sample.

Other methods make it possible not only to detect mutations in the sample but also to detect the number of bases that are inserted or deleted in the alleles and the number of alleles in a sample. The fluorescent PCR-capillary gel electrophoresis technique is accurate enough to differentiate one base-pair differences between alleles, so this technique can indicate the presence or absence of a frameshift in the coding sequence of the gene (Ramlee et al. 2017). This technique involves performing a conventional PCR with one primer labelled with a fluorochrome and later the sample is run in a capillary gel electrophoresis device linked to a genetic analyser (Ramlee et al. 2015). In order to discriminate the wild type allele in comparison to the other alleles in a sample, a PCR product of the sample is mixed with another wild type amplicon labelled with a different fluorochrome (Ramlee et al. 2017).

The fluorescent PCR-capillary gel electrophoresis technique allows detection of not only the presence of mutations, but also the number of alleles (Ramlee et al. 2017). The detection of alleles, based on their size, makes it possible to differentiate between wild type samples, heterozygous samples with one allele being wild type and another mutant, heterozygous samples with two different alleles mutated (Ramlee et al. 2017) or also mosaicism with the presence of more than two alleles of different sizes (Navarro-Serna et al. 2021). In addition, the knowledge of the size of alleles makes it possible to know if the difference in the number of nucleotides causes a frameshift and gives rise to a knock-out phenotype. In an interesting example, wild type (WT) sample and edited samples (Fig. 3.7a) as well as wild type (WT) sample and homozygotic edited samples (Fig. 3.7b) were shown difference in the migration time and sequence.

Fig. 3.7 (continued) bottom of each figure. (**a**) sequence and capillary electrophoresis graph of WT sample and edited sample with mosaicism. In the capillary electrophoresis graph three alleles can be seen with the edited sample in blue and WT control sample in green. (**b**) sequence of wild type sample and homozygotic edited sample. In capillary electrophoresis graph one allele can be seen with the edited sample in blue and the WT control sample in green

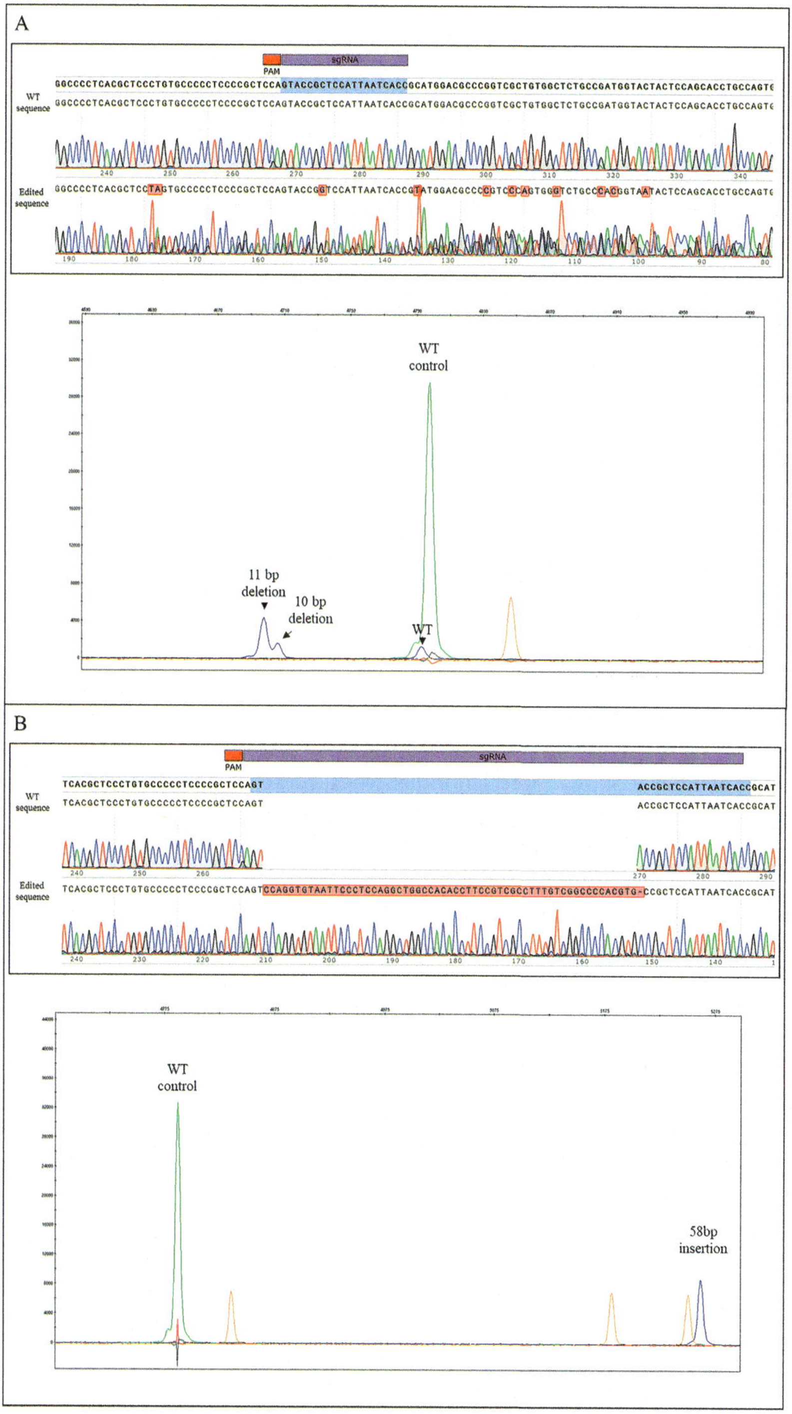

Fig. 3.7 Evaluation of mutation by sanger sequencing and capillary electrophoresis: Comparison of WT and edited sequences at the top and capillary electrophoresis graph at the

3.3.2.2 Sequencing Techniques

Although electrophoresis-based techniques offer a lot of useful information, they do not make it possible to know the sequences of alleles detected. For this, it is necessary to perform sequencing. Sanger sequencing is the method most reported in the literature for detection of mutations (Burkard et al. 2017; Whyte et al. 2018; Chen et al. 2019a; Whitworth et al. 2019; Hirata et al. 2020). In this method, the target sequence is amplified by PCR and then sequenced directly (Sakurai et al. 2014). Sequencing the PCR product directly produces a convoluted spectrum with overlapping peaks in the Cas9 cut region, so, this makes it possible to detect mutations in the sample, but the sequence cannot be known if the sample is not from a homozygous organism (Dehairs et al. 2016).

To solve this, the PCR product can be purified and amplicons can be cloned into vectors for bacterial propagation and single colony cloning sequencing (Lamas-Toranzo et al. 2019a). Clonal sequencing makes it possible to detect the different alleles present in a sample, and discern between wild type samples, heterozygous samples with one allele wild type and another mutant, heterozygous samples with two different alleles mutated, and mosaic samples. The difference with respect to the fluorescent PCR-capillary gel electrophoresis technique is that clonal sequencing makes it possible to know the sequence, also giving information about base substitutions and differences between alleles with the same number of inserted or deleted base pairs but with different sequences.

This method requires more laborious work; however the development of software to analyse sequences with overlapping peaks produced by DNA repair after Cas9 cleavage has allowed the analysis *in silico* of mutation without the need for clonal sequencing. Software such as Tracking of Indels by Decomposition (TIDE) () makes it possible to determinate the spectrum and frequency of targeted mutations designed to analyse a pool of cells (Brinkman et al. 2018). The ability of the software to decompose sequences into as many alleles as detected makes it possible not only to evaluate mutations (Le et al. 2020; Tanihara et al. 2020b) but also to detect mosaicism (Yamashita et al. 2020).

Finally, the most complete and accurate method is next-generation sequencing. This method makes it possible to obtain information about all the alleles present in a sample (Tanihara et al. 2019a); nevertheless, it is more expensive than other methods mentioned above unless a large number of samples are analysed. This technique consists of performing PCR on samples adding a barcode to identify each sample (Hirata et al. 2019; Tanihara et al. 2019a; Le et al. 2020). Next, all samples to be analysed are mixed together in a pool and sequenced. Finally, the data are analysed and interpreted. This method is a good option when a large number of samples have to be analysed, such as with candidate regions for off-target mutations (Le et al. 2020).

In conclusion, we currently have a large number of options for evaluating mutations after gene editing using CRISPR/Cas9, which can be chosen based on the information required for the research.

3.3.3 Negative Aspects of Gene Editing

3.3.3.1 Mosaicism

Mosaicism is the most important problem for gene editing in embryos (Hai et al. 2014). This involves the presence of two or more cell lines with different alleles of the target gene in one organism due to the activity of Cas9 in the zygote after the first DNA replication or even after the subsequent cell divisions. A problem with the generation of mosaic organisms is that they cannot be used to provide samples to study because the distribution and proportion of the different alleles with different functionalities around the organism is unknown (Navarro-Serna et al. 2020). Furthermore, there is a possibility that the edited alleles are not found in the germ line and therefore the animal cannot be used as a founder of a colony. On the other hand, there may be mosaic organism that have all the alleles edited and therefore these organisms could present the desired phenotype (Yamashita et al. 2020).

In species that reach reproductive age quickly, such as mice, mosaicism is not an important problem because a second generation can be obtained after crossing the animals obtained in the first generation. Therefore, a second generation can be obtained with the guarantee that these are not mosaics and that the allele detected in one tissue is the same as that found in all tissues of the organism. However, in a livestock species such as the pig, the time to reach reproductive age and gestation time is longer, so mosaicism leads to a greater time being required to obtain piglets with the required mutation (Mehravar et al. 2019).

One way to solve this problem is to perform embryo production by SCNT because these embryos have the required mutation, which is previously analysed in cell culture. However, due to the difficulties associated with cloning, gene editing directly in zygotes is of great interest despite disadvantages such as mosaicism. In the literature, several strategies have been described to try to reduce the mosaicism such as, performing the delivery of CRISPR/Cas9 as early as possible with respect to the first DNA replication in the zygote (Lamas-Toranzo et al. 2019b; Navarro-Serna et al. 2021), the enhancement of repair mechanisms after the Cas9 cut to decrease mosaicism (Yamashita et al. 2020), modifications of Cas9 protein (Tu et al. 2015), or editing strategies using multiple sgRNAs for the same target gene.

The timing of CRISPR/Cas9 delivery into the zygote cytoplasm with respect to the first DNA replication in an important factor that affects the degree of mosaicism (Navarro- Serna et al. 2021). In *in vivo* collected embryos, the control of timing is so low that even unfertilized oocytes or two-cell embryos may be obtained (Peng et al. 2015). Due to this lack of control, it is unknown whether at the time of micro-injection the first DNA replication of the zygote has already taken place or not. Due to this lack of control, the *in vivo* collected zygotes are injected just after being collected (Hai et al. 2014; Petersen et al. 2016; Burkard et al. 2017; Gadea et al. 2018) and therefore, this is the reason that high rates of mosaicism in live-born piglets are reported by many authors (Peng et al. 2015; Wang et al. 2015c; Yu et al. 2016; Zhou et al. 2016). On the other hand, DNA replication in porcine zygotes produced *in vitro* has been described to begin between 8-9 hours post- insemination (Jeong et al. 2007; Navarro-Serna et al. 2021). Despite this, in most studies, *in vitro* produced zygotes have been microinjected between 5 (Park et al. 2017) and 14 hours post-insemination (Whitworth et al. 2014; Whitworth and Prather 2017).

The effect of CRISPR/Cas9 microinjection time has been studied in both parthenogenetic pig embryos (Tao et al. 2016) and *in vitro* produced embryos (Navarro-Serna et al. 2021). When an oocyte is activated by parthenogenetic activation, DNA synthesis occurs between 5-6 hours post-activation (Liu et al. 1996). In a study which used parthenogenetic pig embryos, activated oocytes were injected at 3, 8 or 18 hours after activation. In this study, the time of microinjection did not affect cleavage and mutation rate but the percentage of embryo development was less in the 3 hours group (Tao et al. 2016). Despite the mutation rate not being affected by the timing, the biallelic mutation rate was influenced by the microinjection time after parthenogenetic activation in pigs and the rate was higher when microinjection was performed earlier, that is at 3 and 8 hours post- activation (Tao et al. 2016). The mosaicism rate was also affected, and 1/3 of mosaic embryos were obtained when the microinjection occurred before DNA replication and 100% after DNA replication (Tao et al. 2016). The same results were obtained when *in vitro* produced embryos were microinjected at three different times: in oocytes before *in vitro* insemination, 5-6 hours after insemination and 10-11 hours after insemination (Navarro-Serna et al. 2021). *In vitro* embryo production and mutation rate was not affected by the time of microinjection, but mosaicism was lower in embryos microinjected as soon as possible after the first zygotic DNA replication (Navarro-Serna et al. 2021). These results suggest that performing the microinjection time earlier may be a good strategy to reduce mosaicism, such that even the microinjection of matured oocytes before insemination is a good option.

When zygotes are obtained by SCNT and then microinjected, timing is also an important factor as following somatic cell injection an oocyte at metaphase-like stage after 2 hours and pronuclear structure is formed within 6 hours (Bui et al. 2006). In Sheets *et al.* 2016 the time of microinjection into cloned zygotes was 4 hours after fusion (Sheets et al. 2016). In addition to delivering CRISPR/Cas9 components as soon as possible, other strategies have been described to reduce mosaicism, such as reducing the half-life of Cas9 protein by tagging Cas9 with ubiquitin-proteasomal degradation signals (Tu et al. 2017). The reduction of the half-life of Cas9 leads to the concentration of the protein deceasing earlier than wild-type Cas9, so that the protein concentration is reduced when the successive DNA replications take place. As a consequence, this study reports a decrease in mosaicism after using the modified Cas9 in primate embryos (Tu et al. 2017).

Another approach described to reduce mosaicism is to shorten the repair time of the cuts produced by CRISPR/Cas9 (Yamashita et al. 2020). The authors that used this strategy concluded that not only should Cas9 cut the target DNA before the first DNA replication, but the repair and consequent creation of the mutation should also take place before the DNA replication. If the repair occurs after DNA replication, each replicated chain will have a different degree of repair, resulting in more than two alleles and in the production of a mosaic organism. To solve this problem, the authors decided to introduce murine three-prime repair exonuclease 2 (mTrex2) into porcine zygotes at the same time as CRISPR/Cas9 by electroporation. The results showed that co-delivery of CRISPR/Cas9 and the repair protein mTrex2 can reduce mosaicism (Yamashita et al. 2020).

Despite the improvements described, currently there is not any methodology to completely remove the mosaicism in gene editing directly in embryos so, another option described is to use multiple guides for the same target gene in order to generate knock- out piglets in which at least all cells of the organism have a loss of objective protein function and a knock-out phenotype, even it is due to different alleles with different mutations (Whitworth and Prather 2017; Wu et al. 2017; Zuo et al. 2017; Hirata et al. 2020).

In conclusion, currently, the only method that allows the generation of one hundred percent homozygous knock-out pigs edited by CRISPR/Cas9 is through SCNT. However, if the goal is to generate a KO colony, one-step gene editing in embryos is enough to achieve this.

3.3.3.2 Off-Target Mutations

The main advantage of programmable endonucleases is their high specificity for the target sequence to be modified. However, there is the possibility that DNA sequences similar to the sgRNA have sufficient consensus. This could mean that the Cas9 protein cuts in other regions of the genome different than the target sequence. These modifications outside the target region are called off-target mutations. The editing of other sequences in the genome of the organism could affect the expression or integrity of other genes that could lead to phenotypic variations (Wang et al. 2015b). This could complicate the phenotypic study of the animal models generated because the phenotype that is observed might not be due to the modification of the target gene but instead be due to alterations in other similar sequences in other genes.

Strategies to increase the specificity of editing have been designed, such as the development of Cas9 nickases. Unlike the wild type Cas9, which generates a double- strand break in the target site that is then repaired by NHEJ or HDR, Cas9 nickase produces a single-strand nick which is repaired without mistakes. This system needs sgRNA pairs to produce two nicks in close proximity and generate a double-strand break. The requirement to design two sgRNAs that recognize the target gene reduces the possibility of off-target editing occurring but the likelihood of HDR taking place is less than for double-strand breaks generated with wild-type Cas9 editing (Shen et al. 2014). Previous studies showed that the CRISPR/Cas9 system might cut sequences with divergences from the PAM sequence (Wang et al. 2016) and these mistakes could cause off-target editing, nevertheless the presence of off-target mutations in genetically modified piglets produced by CRISPR/Cas9 has not been reported (Navarro-Serna et al. 2020). Studies of off-target mutations in pigs detected *de novo* mutations in genetically modified pigs with a frequency close to that previously estimated in humans, which indicates that in this study, CRISPR/Cas9 does not significantly increase the rate of off- target mutations (Wang et al.

2016). Furthermore, another study in porcine embryos found that the frequency of off-target mutations was very low, even when using a high concentration of CRISPR/Cas9 complex (Le et al. 2020).

In conclusion, the low incidence of off-target mutations reported is due to the use of bioinformatic guide design tools and the knowledge available about whole genome sequences of *Sus scrofa* in data bases. This has made it possible to design sgRNAs associated with a very low possibility of producing off-target mutations(Le et al. 2020).

3.3.3.3 Chromosome Aberrations

The generation of genetically modified pigs using CRISPR/Cas9 involves a complicated gamete manipulation process. In addition to gene editing, other processes are involved such as oocyte *in vitro* maturation, collection and *in vitro* culture of embryos produced *in vivo*, cloning, microinjection, electroporation, *in vitro* fertilization and culture, and embryo transfer.

In the literature, it has been reported that assisted reproductive techniques in pigs can produce chromosome aberrations such as aneuploidies (Hornak et al. 2009) and chromosome translocations (Garcia-Vazquez et al. 2011). Studies in porcine blastocysts showed that around forty percent of *in vitro* produced embryos had chromosomal aberrations (Ulloa Ullo et al. 2008; Hornak et al. 2009). Although it is possible that many of these abnormalities are not compatible with generating a liveborn piglet, some of them are compatible. Indeed, in our studies we have detected the presence of reciprocal chromosomal translocation in piglets produced by ICSI (Garcia-Vazquez et al. 2011) and also when we generated gene edited pigs, a double chromosomal translocation was detected in one homozygous knock-out pig produced *in vitro* (Navarro-Serna et al. 2021). Despite the risk of chromosomal aberrations, the karyotype has not been analyzed in other studies in which CRISPR/Cas9 gene editing pigs were produced. This means that there may be chromosomal alterations that cause health problems in animal models which are falsely attributed to the desired gene modification.

3.4 Applications of Gene Edited Pigs

The ability to edit genes allows insertion of exogenous sequences (knock-in) or elimination of gene function (knock-out), which could have unlimited applications for use in basic science, agriculture, and biomedicine. In these areas, the production of genetically modified pigs makes it possible to investigate areas of interest such as gene expression, protein structure, intracellular mechanisms and gene functions (Whitelaw et al. 2016; Yao et al. 2016; Burkard et al. 2017; Wells and Prather 2017). The use of gene edited pigs is higher in biomedicine than in agriculture. This is due to the severe restriction of consumption of food products derived from genetic modified organisms (GMO's). This topic will be revised in Sect. 3.6 of this chapter.

3.4.1 Basic Science

Knock-out and knock-in pigs are used to study the function of different genes or proteins as complementary studies with murine models in different areas of knowledge (Table 3.1). For example in the developmental biology area, a knock-out model for OCT4 was developed to study the function of this transcription factor in the early development of the pig embryo (Kwon et al. 2015) and Lai *et al.* produced a knock-in model introducing the coding gene of a fluorescent protein under the

Table 3.1 Gene edited pigs with applications in basic science

Research field	Gene	Study	Methodology	Model	References
Development biology	OCT4	Preimplantation embryo development	Injection	KO	Kwon et al. (2015)
Development biology	OCT4	Pluripotency cells reporter	SCNT	KI	Lai et al. (2016)
Immunology	PBD2	Protection against infection	SCNT	KI	Huang et al. (2020a)
Immunology	IgM	B-cell deficiency	SCNT	KO	Chen et al. (2015)
Metabolism	GRB10	Insulin resistance and obesity	Injection	KO	Sheets et al. (2016)
Metabolism	MC3R	Fat metabolism, energy homeostasis	SCNT	KO	Yin et al. (2019)
Metabolism	IRX3	Body mass, fat metabolism and obesity	SCNT	KO	Zhu et al. (2020a)
Organogenesis	NGN3	Pancreas development	Injection and SCNT	KO	Sheets et al. 2018)
Organogenesis	SIX1 and SIX4	Kidney development	SCNT	KO	Wang et al. (2019b)
Organogenesis	EDA	Submucosal glands development	Injection	KO	Ostedgaard et al. (2020)
Organogenesis	ETV2	Hematoendotelial linages	SCNT	KO	Das et al. (2020)
Reproduction	PTGS2	Embryonic development and luteal function	SCNT	KO	Pfeiffer et al. (2020)
Reproduction	BMP15	Female fertility and follicular development	SCNT	KO	Shi et al. (2020))
Reproduction	SRY	Sex determination	Injection	KO	Kurtz et al. (2021)
Cellular biology	TPC2	Calcium signalling	Injection	KO	Navarro-Serna et al. (2021)

SCNT Somatic cell nuclear transfer. Injection: Intracytoplasmic microinjection of oocytes/embryos

control of the promoter of OCT4 in order to localize and monitor pluripotency (Lai et al. 2016).

Different models have been developed to study the role of different genes in organogenesis. For example, pigs with a knock-out in NGN3 (Sheets et al. 2018), SIX1 and SIX4 (Wang et al. 2019b), ETV2 (Das et al. 2020) or EDA (Ostedgaard et al. 2020) were used to study pancreas, kidney, hematoendotelial and submucosal gland development, respectively. Other genetically modified pigs were produced to investigate the functions of different genes in fat metabolism, insulin resistance and obesity, such as GRB10 (Sheets et al. 2016), MC3R (Yin et al. 2019) and IRX03 (Zhu et al. 2020a), respectively.

Other examples of models produced in the area of basic science are pigs with knock-out in PTGS2 to study its function in embryonic development and luteal function (Pfeiffer et al. 2020), in BMP15 to investigate its function on follicular development (Shi et al. 2020) and in PBD2, a gene with antimicrobial activity (Huang et al. 2020a). We have developed a TPC2 KO pig model to explore the functional role of TPC2 in calcium signalling pathways (Navarro-Serna et al. 2021). The information generated with the use of gene edited pig models will be complementary to knowledge derived from cell and rodent models (Hryhorowicz et al. 2020). Most of the pig models generated are KO (Table 3.1), probably because the first steps in the study of the function of a gene is the knock out and on the other hand the efficiency for KO models is higher than for KI models.

3.4.2 *Agricultural Production*

In terms of agricultural applications, the CRISPR/Cas9 system can be used with the objective of creating animals with an improved carcass composition, decreasing input requirements, animals with an improved milk production, or animals that are resistant to diseases (Gadea and Garcia Vazquez 2010; Murray and Maga 2016; Wells and Prather 2017; Yang and Wu 2018).

3.4.2.1 Animal Health

Porcine infectious diseases are a great problem for pig production, with huge economic costs and a blockade of the international market and interchange. One of the main applications of genome modification in animals for agriculture is the generation of animals resistant to diseases (Whitelaw and Sang 2005; Lassnig and Müller 2015). (Table 3.2). In order to improve the general immune defence of the animals, Han *et al.* generated knock-in porcine models for lactoferrin, an immune-active protein with antimicrobial and antiviral activity, to improve milk quality and subsequently piglet development (Han et al. 2020).

In relation to viral diseases, two main strategies could be applied using CRISPR/Cas9 technology to inhibit or block viral infection, one of them involves targeting

Table 3.2 Gene edited pigs with applications in animal health

Gene	Study	Methodology	Model	References
CD163 and CD1D	Resistance to PRRS	SCNT and Injection	KO	Whitworth et al. (2014)
CD163	Resistance to PRRS	Injection	KO	Burkard et al. (2017)
TMPRRSS2	Resistance to influenza virus	Injection	KO	Whitworth et al. (2017)
"shRNA"	Resistance to classical swine fever virus	SCNT	KI	Xie et al. (2018)
CD163	Resistance to PRRS	SCNT	KO	Yang et al. (2018)
CD163	Resistance to PRRS	SCNT	KO	Wang et al. (2019a)
CD163	Resistance to PRRS	SCNT	KO	Guo et al. (2019a)
CD163	Resistance to PRRS	Electroporation	KO	Tanihara et al. (2019b)
CD163/ hCD163	Resistance to PRRS	SCNT	KO/ KI	Chen et al. (2019b)
pAPN	Resistance to coronavirus	Injection	KO	Whitworth et al. (2019)
pAPN	Resistance to transmissible gastroenteritis virus (TGEV)	SCNT	KO	Luo et al. (2019)
CD163 & pAPN	Resistance to PRRS and TGE	SCNT	KO*2	Xu et al. (2020)
pRSAD2	Resistance to classical swine fever and pseudorabies	SCNT	KI	Xie et al. (2020b)
Lactoferrin	antibacterial activities in milk	SCNT	KI	Han et al. (2020)

SCNT: Somatic cell nuclear transfer. Injection: Intracytoplasmic microinjection of oocytes/embryos

host genes that are essential for viral infection (cell receptors, etc) and the other directly targeting viral DNA as a potential antiviral strategy (Soppe and Lebbink 2017). Porcine reproductive and respiratory syndrome (PRRS) is one of the most important panzootic infectious viral diseases in pigs and due to the generation of late-term abortions and stillbirths, this disease causes important economic losses to the farming industry, (Wells and Prather 2017). For that reason, the most common model for disease resistance produced using CRISPR/Cas9 is the one resistant to PRRS.

The first attempt to modify the receptors of the virus in porcine alveolar macrophages involved generating a KO pig for CD169 (Prather et al. 2013). However, the KO pigs suffered the same course of the PRRS disease. The same group later generated other KO models for CD163 and CD1D as a possible way to block the interaction between the virus and the macrophages (Whitworth et al. 2014). CD163 has been identified as a putative fusion receptor for the PRRS virus, so different authors from different labs produced knock-out pigs for CD163 that are resistant to PRRS

(Burkard et al. 2017; Yang et al. 2018; Guo et al. 2019a; Tanihara et al. 2019b; Wang et al. 2019a). Furthermore, Whitworth *et al.* and Xu *et al.* disrupted CD1D (a major histocompatibility complex protein) and pAPN (receptor of transmissible gastroenteritis (TGE) virus), respectively, producing double knock-out animals resistant to PRRS and the ones in the second study resistant to TGE, as well (Xu et al. 2020). Another strategy to achieve this resistance is replacing the SRCR5 domain of the CD163 gene with the corresponding domain of human CD163 (Chen et al. 2019b).

In addition to PRRS, other diseases were targeted such as those linked to infection by different viruses including coronaviruses. With this aim, Whitworth *et al.* and Luo *et al.* produced porcine models lacking aminopeptidase-N, a protein present on the surface of epithelial cells that has been suggested as a receptor for different coronaviruses (Luo et al. 2019; Whitworth et al. 2019). Going further, Xie *et al.* produced a knock-in model, piglets that produced pRSAD2, an enzyme with antiviral activities against a wide range of viruses such as classic swine fever virus or pseudorabies virus (Xie et al. 2020b).

Another indirect application of the CRISPR/Cas9 system in the control of viral infections that is not directly via generation of resistant pigs, involves developing mutations in the virus that could facilitate the design of valuables vaccines or by the use of CRISPR/Cas9 for the diagnosis of the disease. A clear example of these alternatives is in African swine fever (ASF), by use of this methodology for vaccines (Borca et al. 2018; Hubner et al. 2018) and diagnosis (Bai et al. 2019; He et al. 2020).

One strategy to control viral diseases is the use of specific small hairpin RNAs (shRNA) to reduce the susceptibility to infection by very contagious viral diseases that leads to important economic losses in the pig industry, such as foot and mouth disease virus (Hu et al. 2015). With the application of CRISPR/Cas9 technology, shRNA and SCNT the knock-in (KI) animals are protected against classical swine fever virus (Xie et al. 2018).

3.4.2.2 Animal Production Improvement

In addition, animal production can be improved by targeting different genes that do not involve disease resistance. The most common one is the gene encoding for myostatin, a negative regulator of muscle growth. Different authors have developed knock-out models for myostatin which exhibit greater muscle mass, enhancing in this way the quality of the product (Wang et al. 2015a; Tanihara et al. 2016; Li et al. 2020b; Zhu et al. 2020b). Another approach for improving muscle development is the one proposed by Liu *et al.*, who disrupted the insulin growth factor 2 (IGF2) gene (Liu et al. 2019) (Table 3.3).

Other authors have developed models involving the introduction of one or more foreign genes. Knock-in pigs for Fat-1 have been produced to improve meat quality. This gene encodes a fatty acid desaturase which converts n-6 polyunsaturated fatty acids to n-3 poly- unsaturated fatty acids, that provide more health benefits (Li et al. 2018). Another example is the knock-in for uncoupling protein 1 (UCP1). This protein is located in the inner mitochondrial membrane and regulates heat production,

Table 3.3 Gene edited pigs with applications in animal production improvement

Research field	Gene	Study	Methodology	Model	References
Growth	MSTN	Muscle development	SCNT and Talen	KO	Kang et al. (2017)
Growth	MSTN	Muscle development	SCNT	KO	Wang et al. (2015a)
Growth	MSTN	Muscle development	SCNT	KO	Wang et al. 2017b)
Growth	MSTN	Muscle development	SCNT	KO	Li et al. (2020b)
Growth	MSTN	Muscle development	SCNT	KO	Zhu et al. (2020b)
Growth	MSTN	Muscle development	Electroporation	KO	Tanihara et al. (2016)
Growth	IGF2	Muscle development	SCNT	KO	(Liu et al. 2019)
Metabolism	UCP1	Thermoregulation	SCNT	KI	Zheng et al. (2017)
Meat quality	Fat-1	Fatty acids n-3PUFAs	SCNT	KI	Li et al. (2018)
Pollution reduction	β-glucanase, xylanase, and phytase	Production of digestive enzymes	SCNT	KI*3	Li et al. (2020a)

SCNT: Somatic cell nuclear transfer

but it is absent in pigs. These knock-in animals can better maintain their body temperature and showed decreased fat deposition, improving in this way production efficiency (Zheng et al. 2017).

3.4.2.3 Pollution Reduction

Pigs lack several enzymes in their digestive tract that hydrolyse plant cell walls to release the nutrients that could be absorbed during digestion. An interesting approach is the integration of bacterial enzymes in the salivary glands of the pig to hydrolyse the complex carbohydrates such as the phytate that contains phosphorus in the diet. The objective of this methodology is to increase the intestinal absorption of these nutrients and reduce the presence of this compound in the manure, this being an ecological contamination problem in some areas.

These models have been previously produced by other technologies different than endonucleases like pronuclear injection (Golovan et al. 2001). Using the PSP/APPA transgene (parotid secretary protein promoter linked to the *Escherichia coli* appA phytase gene) one study produced pigs that expressed the functional enzyme in the saliva with a resulting increase in phosphorus digestibility and reduction of phosphorus in the manure (Golovan et al. 2001; Forsberg et al. 2014a). This model was known as the Enviropig and supported by the Guelph University in Canada. The Enviropig's characteristics have been described (Murray et al. 2007; Golovan et al. 2008; Forsberg et al. 2013; Forsberg et al. 2014a; Forsberg et al. 2014b), as has the manure that they produce (Mao et al. 2008). The animals were terminated in 2012 (Clark 2015).

Later using electroporation and SCNT another study generated pigs with three microbial enzymes, β-glucanase, xylanase, and phytase in the salivary glands (Zhang et al. 2018b). The expression of these enzymes led to a reduction in nitrogen and phosphorous in manure and an increase in growth rates (Zhang et al. 2018b). In another study PiggyBac Transposons and SNCT were used to generate pigs that expressed in their saliva four enzymes; pectinase, xylanase, phytase, and TeEGI (cellulase and β-glucanase) using somatic cell cloning (Wang et al. 2020).

Recently, the application of CRISPR/Cas9 and SCNT led to the generation of a triple knock-in (Li et al. 2020a). The authors integrated into the porcine genome three genes encoding three microbial enzymes (β-glucanase, xylanase, and phytase), which are produced in the salivary glands of the knock-in pigs. This model improves feed efficiency and reduces environmental impact because these enzymes degrade non-starch polysaccharides and phytate in plants, which can significantly promote the digestion of nitrogen and phosphorus in formula feed (Li et al. 2020a).

3.4.3 *Biomedicine*

The third group of applications is related to biomedicine. This group includes models produced to improve xenotransplantation, to produce different bioproducts and to mimic and study several human diseases.

3.4.3.1 Xenotransplantation

Organ transplantation is the only option for patients with severe organ failure, but there are not sufficient donors to cover the large number of patients that need one (Niemann and Petersen 2016). Xenotransplantation is a potential approach to solve this problem. The pig is considered the most suitable species for this purpose due to its ease of breeding and the similarities with humans regarding physiology and organ size and function. The major problem for clinical application of xenotransplants is the adverse immune reaction of the host (Niemann and Petersen 2016; Naeimi Kararoudi et al. 2018; Niu et al. 2021). The first immune reaction of the host body is the hyperacute rejection (HAR), which is induced by pre-existing antibodies that principally target α-Gal antigens that exist on the surface of porcine cells (Fu et al. 2020). Different knock-out pigs were produced to eliminate these antigens in order to decrease the HAR. With this purpose, the GGTA1 gene, encoding for a galactosyltransferase that catalyses the formation of the α-Gal antigen, was disrupted producing GGTA1 knock-out pigs (Petersen et al. 2016; Chuang et al. 2017; Tanihara et al. 2020a). In addition to this modification, other genes were targeted at the same time to produce triple and quadruple knock-out pigs (Zhang et al. 2018a; Fischer et al. 2020; Hein et al. 2020; Tanihara et al. 2021).; in addition, there have been other targets (Li et al. 2015; Sake et al. 2019; Fu et al. 2020) (Table 3.4).

Other examples of genetically modified pigs with xenotransplantation applications are the ones produced by SCNT by Hinrichs *et al.* who disrupted the growth hormone receptor (GHR) in GGTA1-deficient cells expressing the human cluster of differentiation (hCD46) and human thrombomodulin (hTHBD) to reduce the size of organ donor pigs for preclinical studies (Hinrichs et al. 2020) or the pigs with severe combined immunodeficiency produced by Boettcher et al., who depleted the IL2RG gene in pigs within a naturally occurring disruption of DCLRE1C (ARTEMIS) background (Boettcher et al. 2020).

In addition to problems related to immune system responses against xenotransplants, the transplantation of organs from one species to another may be associated with other difficulties such as the transmission of endogenous retroviruses. Porcine endogenous retroviruses (PERV) are gamma retroviruses which can infect human cells and integrate into the human genome in cell culture (Yang et al. 2015). Even though no study has observed PERV transmission to humans, they could potentially integrate into the host genome and lead to immunodeficiency and tumorigenesis, so these retroviruses need to be annulled in order to generate a pig that can be used as an organ donor. For this reason, some studies have focused on generating pigs free of PERV (Niu et al. 2017; Li et al. 2019a; Niu et al. 2021).

In the case of liver transplantation, coagulation and blood factors are crucial for the success of the transplant so Li *et al.* proposed a knock-in model to solve this problem. They produced by SCNT, pigs expressing the humanized liver proteins

Table 3.4 Gene edited pigs with applications in xenotransplantation

Gene	Methodology	Model	Year	References
GGTA1, CMAH, iGb3S	SCNT	KO*3	2015	Li et al. (2015)
GGTA1	Injection	KO	2016	Petersen et al. (2016)
Porcine endogenous retroviruses	SCNT	KO	2017	Niu et al. (2017)
GGTA1	Injection	KO	2017	Chuang et al. (2017)
GGTA1, β4GalNT2 and CMAH	SCNT	KO*3	2018	Zhang et al. (2018a)
hF7 and hAlbumin	SCNT	KI*2	2019	Li et al. (2019a)
B-2-microglobulin	SCNT	KO	2019	Sake et al. (2019)
GHR	SCNT	KO	2020	Hinrichs et al. (2020)
GGTA1, CMAH, β4GalNT2 and β2M	SCNT	KO * 4	2020	Fischer et al. (2020)
GGTA, CIITA and β2M	SCNT	KO*3	2020	Fu et al. (2020)
GGTA1, CMAH and β2M	SCNT	KO*3	2020	Hein et al. (2020)
IL2RG	SCNT	KO	2020	Boettcher et al. (2020)
GGTA1	Electroporation	KO	2020	Tanihara et al. (2020a)

SCNT: Somatic cell nuclear transfer. Injection: Intracytoplasmic microinjection of oocytes/embryos

blood-coagulation factor VII (hF7) and albumin (hALB), replacing the previous background of pig F7, with a negative PERV background (Li et al. 2019a).

3.4.3.2 Bioproducts

Pigs can be used as bioreactors to produce different bioproducts (reviewed by (Gadea and Garcia Vazquez 2010; Bertolini et al. 2016)), but this application is the least developed so far. For now, only two genetically modified pigs were developed using CRISPR/Cas9 to synthetise bioproducts. The first one is the one proposed by Peng *et al.*, a knock-in model for human serum albumin, the most abundant plasma protein which is needed for essential processes such as maintenance of plasma oncotic pressure, or transportation of small molecules. This protein is needed in cases of severe diseases such as liver failure or traumatic shock (Peng et al. 2015).

The second example is a knock-out model for IgM. These animals do not produce B-cells, which is the first step for developing pigs that produce humanized polyclonal antibodies that can be used in clinical medicine (Chen et al. 2015).

3.4.3.3 Models of Human Diseases

Pigs are an excellent animal model for understanding the pathological processes of human diseases and developing therapeutic strategies because of their similarity to humans in terms of anatomy, physiology, and genetics (Perleberg et al. 2018). A lot of different models of human diseases have been developed using CRISPR/Cas9 since 2014, covering diverse areas from oncology to hearing loss (Table 3.5).

The area in which most disease models have been developed is for neuroscience. Parkinson's disease is the most common neurodegenerative movement disorder in the elderly, so Parkinson models have been produced using different strategies. Zhou *et al.* disrupted two genes, PARK2 and PINK1 producing a double knock-out model (Zhou et al. 2015). The PARK2 gene encodes a protein called parkin, a component of multiprotein E3 ubiquitin ligase complex and PINK1 gene encodes PTEN-induced putative kinase 1, a mitochondrial serine/threonine-protein kinase. The depletion of either of these two genes produces autosomal recessive early-onset Parkinson's disease in humans (Zhou et al. 2015). In addition to this combination, Wang *et al.* produced a triple knock-out targeting PARK2, PINK1 and DJ1, to model early-onset Parkinson's disease (Wang et al. 2016).

Duchenne's muscular dystrophy is an X-linked hereditary muscular dystrophy and people who have this disease suffer a severe and progressive clinical course of muscle weakness, loss of ability to move, and finally death, but no treatment has yet been developed. The disease is caused by a mutation in the dystrophin gene. For that reason, knock-out pigs for dystrophin were produced (Yu et al. 2016; Wu et al. 2018). Limb-girdle muscular dystrophy also has no treatment and causes muscle wasting. Expression levels of the FBXO40 gene decrease in limb-girdle muscular dystrophy patients, so a knock-out model of FBXO40 was developed to study this

Table 3.5 Gene edited pigs as model of human diseases

Gene	Human disease	Area	Methodology	Model	Year	References
LMNA	Hutchinson-Gilford progeria syndrome	Ageing	SCNT	KI	2019	Dorado et al. (2019)
ApoE and LDLR	Atherosclerosis	Cardiovascular	SCNT	KO *2	2017	Huang et al. (2017)
OSBPL2	Deafness	Deafness	SCNT	KO	2019	Yao et al. (2019)
Mutant GJB2 CDS	Hearing loss	Dermatology	Injection	KI	2020	Xie et al. (2020a)
HR	Atrichia	Dermatology	SCNT	KO	2019	Gao et al. (2019)
vWF	Von Willebrand disease (vWD)	Haematology	Injection	KO	2014	Hai et al. (2014)
F9/hF9	Haemophilia B	Haematology	SCNT	KO/KI	2020	Chen et al. (2020)
TYR, IL2RG, and RAG1	Albinism and immunodeficiency	Immunology	Injection	KO*3	2019	Chen et al. (2019a)
NLRP3	Cryopyrin-associated periodic syndrome	Immunology	SCNT	KO	2020	Li et al. (2020c)
TYR	Albinism	Melanin biosynthesis	SCNT	KO	2015	Zhou et al. (2015)
MITF	Hypopigmentation, deafness, Waardenburg and Tietz syndromes	Melanin biosynthesis	Injection	KO	2015	Wang et al. (2015b)
INS	Diabetes mellitus	Metabolism	SCNT	KO	2018	Cho et al. (2018)
GHR	Laron Syndrome	Metabolism	Injection	KO	2018	Hinrichs et al. (2018)
hIAPP	Diabetes mellitus	Metabolism	SCNT	KI	2019	Zou et al. (2019)
PDX1	Diabetes mellitus	Metabolism	Electroporation	KO	2020	Tanihara et al. (2020b)
PARK2 and PINK1	Parkinson	Neurology	SCNT	KO*2	2015	Zhou et al. (2015)
parkin, DJ-1, PINK1	Parkinson	Neurology	Injection	KO *3	2016	Wang et al. (2016)
DMD	Muscle dystrophy	Neurology	Injection	KO	2016	Yu et al. (2016)
TPH2	Neuropsychiatric disorders	Neurology	SCNT	KO	2017	Li et al. (2017b)

(continued)

Table 3.5 (continued)

Gene	Human disease	Area	Methodology	Model	Year	References
SCNA	Parkinson	Neurology	SCNT	KO	2018	Zhu et al. (2018)
DMD	Muscle dystrophy	Neurology	SCNT	KO	2018	Wu et al. (2018)
FBXO40	Muscle dystrophy	Neurology	SCNT	KO	2018	Zou et al. (2018)
Phenylalanine hydroxylase (PHA)	Phenylketonuria (PKU)	Neurotoxicity	Injection	KO	2020	Koppes et al. (2020)
RUNX3	Cancer	Oncology	SCNT	KO	2016	Kang et al. (2016)
MITF	Hypopigmentation, deafness, Waardenburg and Tietz syndromes	Oncology	Injection	KO	2017	Hai et al. (2017)
TP53	Cancer	Oncology	Electroporation	KO	2018	Tanihara et al. (2018)
COL2A1	Type II collagenopathy	Skeletal development	SCNT	KO	2020	Zhang et al. (2020)

SCNT: Somatic cell nuclear transfer. Injection: Intracytoplasmic microinjection of oocytes/embryos

disease and the function of FBXO40 in skeletal muscle development (Zou et al. 2018).

Another group of models are the one related to metabolic diseases such as diabetes mellitus. Diabetes mellitus is a chronic disease, characterized by high blood glucose levels, polyuria, polydipsia, and weight loss. It is one of the most common public health problems worldwide and large animal models for the evaluation of different treatments are required. The major cause of diabetes is the deficiency in functional insulin because of abnormal insulin secretion and/or decreased physiological responses to insulin. For that reason, different models of genetically modified pigs were produced (Zettler et al. 2020). Cho *et al.* produced insulin deficient pigs by disrupting the INS gene, so the piglets were not able to produce insulin (Cho et al. 2018). Tanihara *et al.* targeted the pancreatic duodenal homeobox 1 (PDX1) gene, producing a monoallelic disruption. The biallelic

mutation of this gene causes abnormal development of the pancreas and death during infanthood, but a monoallelic mutation of the PDX1 gene impairs insulin secretion from pancreatic β-cells, causing diabetes (Tanihara et al. 2020b). Islet amyloid polypeptide (IAPP) is a polypeptide hormone that has a toxic effect on β-cells when it aggregates which leads to the progressive failure of insulin secretion. Human IAPP is one of the most highly aggregated polypeptides and some studies considered the amyloidosis of human IAPP as a potentially important cause of diabetes mellitus type II. Therefore, Zou *et al.* developed a knock-in model that express human IAPP to study the pathogenesis of diabetes mellitus type II (Zou et al. 2019).

Cancer is one of the most common cause of death worldwide so suitable animal models are needed to study this group of diseases and its treatment. Runt-related transcription factor 3 (RUNX3) is known as a tumour suppressor gene which, when absent, contributes to gastrointestinal cancer development, so a knock-out model for RUNX3 was developed to study this type of cancer (Kang et al. 2016). In addition, the TP53 gene encodes a transcription factor that acts as a tumour suppressor by promoting senescence or apoptosis following DNA damage induced by cell stress. Mutations in this gene are associated with cancer in humans, in particular with Li Fraumeni multiple cancer syndrome. To study this disease, Tanihara *et al.* produced knock-out piglets by disrupting TP53 (Tanihara et al. 2018).

Haemophilia B is an inherited X-linked bleeding disorder caused by a dysfunction in the F9 gene which encodes the coagulation factor IX, a vitamin K-dependent plasma protein that participates in the intrinsic blood coagulation pathway. In patients with haemophilia B, recurrent spontaneous bleeding mainly occurs in the synovial joints causing chronic pain, immobility, and an important reduction in quality of life. Chen *et al.* developed a pig model to study haemophilia B by depleting the F9 gene. Furthermore, they used the CRISPR/Cas9 system to introduce the human F9 gene into knock-out fibroblasts for F9, to determine if this gene therapy procedure could ameliorate the bleeding phenotype (Chen et al. 2020).

Other examples of production of models for human diseases are summarised in Table 3.5 and include models for deafness (Yao et al. 2019; Xie et al. 2020a), atherosclerosis (Huang et al. 2017), albinism (Zhou et al. 2015; Chen et al. 2019a) and Hutchinson- Gilford progeria syndrome (Dorado et al. 2019), among others.

3.5 Future Directions

3.5.1 Base Editors

In addition to the conventional use of Cas9 to generate INDELs, modifications of this Cas protein have been developed. Among the modified forms of Cas9, base editors stand out. These proteins are synthetic enzymes derived from modifications of *Streptococcus pyogenes* Cas9 which induces single-nucleotide changes in the DNA sequence without cutting the DNA double strand of DNA (Komor et al. 2016; Kim et al. 2017).

Base editors are characterized by the presence of Cas9 with a defective catalytical domain, called dead Cas9 or with an impaired catalytical domine, called Cas9 nickase and fused with a deaminase (Kim et al. 2017). The main advantage of base editor is the ability to generate desired mutations without double-stranded DNA breaks and DNA donor template to produce a knock-in, allowing therefore the possibility of producing the desired stop codons or precise modifications to make personalized disease models caused by pair base substitution (Kim et al. 2017; Eid et al. 2018).

Depending on the presence of a cytidine deaminase or an adenine deaminase, the base editors system can be classified in two groups: cytosine base editors that convert C:G pairs to T:A pairs and adenine base editors that convert A:T pairs to G:C pairs.

The enzymatic activity of a cytosine base editor involves the conversion of cytosine into uracil by deamination. Thus, a cytosine that pairs with guanine becomes an uracil that has base-paring proprieties of thymine and it pairs with adenine (Kim et al. 2017). A cytosine base editor requires the presence of the target cytosine within a 5-nucleotide window near the PAM sequence, within the position 4 to 8 (Komor et al. 2016). Despite the precision, this 5-nucleotide window can be a problem because the enzyme can modify all cytosine in that range, inducing undesired changes to the target locus (Komor et al. 2016).

In 2016, the first cytosine base editor (BE1) was developed, using rat APOBEC1 as deaminase and a dead Cas9 plus XTEN domain (Komor et al. 2016). Subsequentely, a second generation of base editor (BE2) was designed by the addition of uracil DNA glycosylase inhibitor (UGI) to the complex APOBEC-XTEN-dCas9-UGI (Komor et al. 2016). UGI inhibits uracil DNA glycosylase, which removes uracil from DNA and initiates base-excision repair with the reversion of the U:G pair to C:G pair, decreasing the efficiency of base editors (Komor et al. 2016). Up to this moment, the maximum efficiency that could be achieved was 50% because only one chain of the double stranded DNA was modified.

With the objective of going beyond the limit of inducing changes in the non-edited chain, a third generation was developed (BE3). In this generation, the HNH domain of Cas9 was restored (APOBEC-XTEN-dCas9(A840H)-UGI) (Komor et al. 2016). Therefore, dCas9 was substituted by nickase Cas9 to cut the DNA strand containing the unedited guanine to stimulate the repair of this chain. Therefore, cutting in the non-edited change increases the possibilities of solving the U:G mismatch in U:A respect C:G (Komor et al. 2016).

Subsequently, new versions have been developed. A fourth generation (BE4) of cytidine base editor was designed with the addition of a second copy of UGI to the C terminus of the construct (Komor et al. 2017). Also, to increase the stability of the double strand, the bacteriophage Mu-originated Gam protein was added (BE4-Gam) (Yuan et al. 2020). Gram protein of bacteriophage Mu binds to the end of DSBs and protects them from degradation, as this would reduce the indel formation during the process of base editing (Komor et al. 2017).

The addition of nuclear localization signals (NLSs) to the cytosine base editor was also described (Koblan et al. 2018). This modification increased the mutation rate but simultaneously produced other unwanted mutations (Yuan et al. 2020).

In porcine embryos, the use of cytosine base editors has been described (Xie et al. 2019; Su et al. 2020; Yuan et al. 2020), but not yet adenine base editors. In these studies, cytosine base editors of different generations were used. In all studies, the strategy used was to produce premature stop codons. In this way, knock-out embryos were generated to model diseases such as Duchenne muscular dystrophy (Xie et al. 2019; Su et al. 2020), albinism (Xie et al. 2019), Hutchinson-Gilford Progeria syndrome (Xie et al. 2019), and the absence of cell of immune system (Xie et al. 2019). In the field of xenotransplantation; cytosine base editors have also been used for also for knock-out porcine endogenous retroviruses (Xie et al. 2019) and simultaneously knock-out three genes (GGTA1, B4GalNT2 and CMAH) to remove the expression of alpha-1,2-galactose in pigs, the major hyperacute rejection xeno-antigen (Yuan et al. 2020).

3.5.2 Conditional Models

The Cre–loxP system is a powerful tool for conditional models, that is successfully used for murine models. This system makes it possible to investigate genes of interest in a specific organ/tissue in a specific moment or time (Smedley et al. 2011). The Cre-loxP system needs two elements. First, Cre-driver animals are generated in which Cre recombinase is expressed by a promoter that specifically targets the cell or tissue of interest. Second, specific genes are engineered to be flanked by loxP in specific animals (floxed animals). Conditional knockout pig are generated by breeding the Cre-driver animals with floxed ones. The specificity and timing of recombination are controlled by use of a promoter and/or enhancer.

Although many Cre–loxP mouse models have been established, there are few pig models available. Some Cre pigs have been generated for specific tissues as germ cells by using the VASA promoter (Song et al. 2016), for astrocytes using the promoter of the pig glial fibrillary acidic protein (*pGFAP*) gene (Hwang et al. 2018), alveolar epithelial cells (Luo et al. 2014b) or kidney collecting duct cells (Luo et al. 2014a). Also, different authors have developed models with reporters for monitoring Cre activity in vivo (Li et al. 2009; Li et al. 2014). Additionally, different strategies have been developed for the efficient deletion of the IoxP flanked selectable marker like use of neomycin to avoid possible side effects (Whitworth et al. 2018; Huang et al. 2020b). In pigs, this conditional gene expression strategy has been used to promote oncogenic expression. The Oncopig is a transgenic pig with Cre-inducible TP53R167H and KRASG12D mutations (Schook et al. 2015; Schook et al. 2016). This commercially available model have been used to study liver and pancreatic cancer (Schachtschneider et al. 2017; Boas et al. 2020).

3.6 Legal and Ethical Regulations

The use of gene editing in organisms with the aim of achieving genetic advantages in a short time has been highly controversial. After the application of gene editing technologies, and the development of programmable endonucleases that are as easy and cheap to use as CRISPR/Cas9, it was necessary to create legislation to regulate the use of this technology.

Currently, genetically modified organisms are covered by the same regulations as transgenic organisms (Lamas-Toranzo et al. 2017; Wasmer 2019). This legislation also includes organisms with simple genetic modifications generated by genetic engineering, such as mutations that affect a single base. However, these mutations cannot be distinguished from organisms bred by conventional techniques, such as those that arise from random mutagenesis (Wasmer 2019). The use of genetically modified organisms is strictly regulated around the world. Next, we will present the legislative situation of genetically modified organisms in Europe, United States of America (USA) and China.

European Union

In the European Union (EU) the regulations covering genetically modified animals for human and animal consumption are somewhat restrictive such that nowadays only genetically modified plants are authorised for that purpose (European Commission Register for Genetically Modified Organisms, https://webgate.ec.europa.eu/dyna/gm_register) . So that a genetically modified organism can be commercialized or released to the environment it needs the approval of the European Food Safety Agency (EFSA). The EFSA assesses the risks that the genetically modified organism may present to the environment, and human health and animal safety, in the EU, and decides if this organism can be approved or not according European regulations on genetically modified food and feed (No 1829/2003 and No 503/2013).

If the genetically modified organism or its bioproducts have medical purposes, it needs to be authorised by the European Medicines Agency (EMA). When the medicinal product contains genetically modified organisms an environmental impact study must be performed and sent to the EMA in addition to the typical reports (administrative, quality, non-clinical and clinical data). However, regarding genetically modified animals, they usually produce a bioproduct that will be part of a medicine. This way, the medical product does not contain a genetically modified organism and does not need an environmental impact study (EMA 2006).

In the case of genetically modified animals that produce bioproducts the EMA provides a guideline on the approaches that should be employed in order to achieve satisfactory quality for biological active substances (EMA 2013). The medicinal product containing components derived from transgenic animals must follow the Regulation (EC) No 726/2004. The effect of the transgene on the health and longevity of the animals must be supervised. Furthermore, a monitoring protocol should be followed in order to assess the health and wellbeing of the animals and check specific infections. It is important to confirm that the bioproduct is free from microorganisms such bacteria, fungi, mycoplasma or virus (EMA 2013). To date, three bioproducts produced by genetically modified animals have been approved in the EU: Antithrombin (ATryn®) from goat milk approved in 2006 but withdrawn in 2018 (Adiguzel et al. 2009; EMA 2019). Human C1- inhibitor (Ruconest) from rabbit milk approved in 2010 (EMA 2020). Sebelipase α (Kanuma) from egg hen approved in 2015 (EMA 2015).

Tissues and organs from genetically modified animals for the objective of xenotransplantation must follow two guidelines: (a) the guideline on xenogeneic cell-based medicinal products and (b) the guideline on the quality, preclinical and clinical aspects of gene transfer medicinal products (EMA 2001, 2009). The therapy must follow the common testing and development procedures, but genetically modified animals should be fully characterised and confirmation of the nature of the inserted, deleted or modified gene must be provided (EMA 2009)

USA

In the USA the regulations covering genetically modified organisms are less strict. To approve the production of an animal with medical and consumption purposes, a new "animal drug application" must be proposed to the Food and Drug Administration describing the characteristics of the animals and its environmental impact and food safety. Furthermore, a compositional and nutritional analysis must be performed if the aim of the product is animal or human consumption in other to compare the composition of the genetically modified animal and the wild-type animal and study any possible toxicological or nutritional hazard to consumers (FDA 2017).

There are two genetically modified animals that have been approved for human consumption in the USA by the Food and Drug Administration the AquAdvantage Salmon (Clifford 2014), which was approved in 2015 and the GalSafe pig, approved in 2020. The last one was also approved for human therapeutics (FDA 2020).

The GalSafe pigs have a disruption of the GGTA1 gene, so they do not produce α-1,3- galactosyltransferase and there are no α-Gal antigens in the surface of their cells. This is an advantage for xenotransplantation, as mentioned above, but this model also has other commercial potential because people that suffer from α-gal syndrome, an allergy to red meat, are able to consume meat from these animals (FDA 2020).

Regarding drugs and biological products derived from genetically modified animals, they should follow the same approval procedure as the ones derived from other sources, providing data of pre-clinical and clinical studies, manufacture and safety (Federal Regulations CFR § 601.2. Applications for biologics licenses; procedures for filing. And CFR §314.5. Applications for FDA approval to market a new drug). There are three products produced by genetically modified animals approved for commercialization and these are the same as those approved in the EU.

China

China has a strict law concerning genetically modified animals. To date, no genetically modified animal has been approved in China for consumption or for medical purposes. The Ministry of Agriculture and Rural Affairs of the State Council of China is the organization that provides the license to a genetically modified organism to be produced and commercialised. To achieve the approval, documents describing the genetically modified animal and its safety must be provided in addition to the usual information (2019).

The economic impact of the application of these new technologies in the pig industry has been evaluated and quantified (Novoselova et al. 2013; Van Eenennaam et al. 2020). According to some authors the delay in the regulation of these animals will have an economic impact and also a reduction in global food security (Van Eenennaam et al. 2020; Feng and Yang 2019).

3.7 Conclusions

The generation of gene edited pigs with new endonucleases has important applications in the field of agriculture and livestock production and in the biomedicine sector. The develop of more efficient protocols will facilitate the extension and applications to different approaches. These improvements will have worldwide impact in the economy and in the health of the population in terms of food security and control and treatment of human diseases.

Acknowledgements This work was supported by MINECO-FEDER (AGL 2015-66341-R), Fundación Séneca 20040/GERM/16 and 21105/PDC/19; and FPU fellowship to SNS (FPU16/04480) from Spanish Ministry of Education, Culture and Sport.

References

Adiguzel C, Iqbal O, Demir M, Fareed J (2009) European community and US-FDA approval of recombinant human antithrombin produced in genetically altered goats. Clin Appl Thromb Hemost 15(6):645–651. https://doi.org/10.1177/1076029609339748

Angel MA, Gil MA, Cuello C, Sanchez-Osorio J, Gomis J, Parrilla I, Vila J, Colina I, Diaz M, Reixach J, Vazquez JL, Vazquez JM, Roca J, Martinez EA (2014) An earlier uterine environment favors the in vivo development of fresh pig morulae and blastocysts transferred by a nonsurgical deep-uterine method. J Reprod Dev 60(5):371–376

Bai J, Lin H, Li H, Zhou Y, Liu J, Zhong G, Wu L, Jiang W, Du H, Yang J, Xie Q, Huang L (2019) Cas12a-Based On-Site and Rapid Nucleic Acid Detection of African Swine Fever. Front Microbiol 10:2830. https://doi.org/10.3389/fmicb.2019.02830

Beaujean N, Jammes H, Jouneau A (2015) Nuclear reprogramming: methods and protocols. methods in molecular biology, Methods and protocols,1222, 2nd edn. Springer New York, New York

Bertolini LR, Meade H, Lazzarotto CR, Martins LT, Tavares KC, Bertolini M, Murray JD (2016) The transgenic animal platform for biopharmaceutical production. Transgenic Res 25(3):329–343. https://doi.org/10.1007/s11248-016-9933-9

Bibikova M, Carroll D, Segal DJ, Trautman JK, Smith J, Kim YG, Chandrasegaran S (2001) Stimulation of homologous recombination through targeted cleavage by chimeric nucleases. Mol Cell Biol 21(1):289–297. https://doi.org/10.1128/MCB.21.1.289-297.2001

Bibikova M, Golic M, Golic KG, Carroll D (2002) Targeted chromosomal cleavage and mutagenesis in Drosophila using zinc-finger nucleases. Genetics 161(3):1169–1175

Bloom K, Ely A, Arbuthnot P (2017) A T7 endonuclease I assay to detect talen-mediated targeted mutation of HBV cccDNA. In: Methods in molecular biology, vol 1540. Humana Press Inc., pp 85–95. https://doi.org/10.1007/978-1-4939-6700-1_8

Boas FE, Nurili F, Bendet A, Cheleuitte-Nieves C, Basturk O, Askan G, Michel AO, Monette S, Ziv E, Sofocleous CT, Maxwell AWP, Schook LB, Solomon SB, Kelsen DP, Scherz A, Yarmohammadi H (2020) Induction and characterization of pancreatic cancer in a transgenic pig model. PLoS One 15(9):e0239391. https://doi.org/10.1371/journal.pone.0239391

Boch J, Scholze H, Schornack S, Landgraf A, Hahn S, Kay S, Lahaye T, Nickstadt A, Bonas U (2009) Breaking the code of DNA binding specificity of TAL-type III effectors. Science 326(5959):1509–1512. https://doi.org/10.1126/science.1178811

Boettcher AN, Li Y, Ahrens AP, Kiupel M, Byrne KA, Loving CL, Cino-Ozuna AG, Wiarda JE, Adur M, Schultz B, Swanson JJ, Snella EM, Ho CS, Charley SE, Kiefer ZE, Cunnick JE, Putz EJ, Dell'Anna G, Jens J, Sathe S, Goldman F, Westin ER, Dekkers JCM, Ross JW, Tuggle CK (2020) Novel Engraftment and T Cell Differentiation of Human Hematopoietic Cells in ART (-/-) IL2RG (-/Y) SCID Pigs. Front Immunol 11:100. https://doi.org/10.3389/fimmu.2020.00100

Borca MV, Holinka LG, Berggren KA, Gladue DP (2018) CRISPR-Cas9, a tool to efficiently increase the development of recombinant African swine fever viruses. Scientific Rep 8(1):3154. https://doi.org/10.1038/s41598-018-21575-8

Brem G, Brenig B, Goodman HM, Selden RC, Graf F, Kruff B, Springman K, Hondele J, Meyer J, Winnacker EL, Kräußlich H (1985) Production of transgenic mice, rabbits and pigs by microinjection into pronuclei. Reprod Domest Anim 20(4):251–252. https://doi.org/10.1111/j.1439-0531.1985.tb00423.x

Brinkman EK, Kousholt AN, Harmsen T, Leemans C, Chen T, Jonkers J, van Steensel B (2018) Easy quantification of template-directed CRISPR/Cas9 editing. Nucleic Acids Res 46(10):e58. https://doi.org/10.1093/nar/gky164

Bui HT, Van Thuan N, Wakayama T, Miyano T (2006) Chromatin remodeling in somatic cells injected into mature pig oocytes. Reproduction 131(6):1037–1049. https://doi.org/10.1530/rep.1.00897

Burkard C, Lillico SG, Reid E, Jackson B, Mileham AJ, Ait-Ali T, Whitelaw CB, Archibald AL (2017) Precision engineering for PRRSV resistance in pigs: Macrophages from genome edited pigs lacking CD163 SRCR5 domain are fully resistant to both PRRSV genotypes while maintaining biological function. PLoS pathogens 13(2):e1006206. https://doi.org/10.1371/journal.ppat.1006206

Canovas S, Ivanova E, Romar R, Garcia-Martinez S, Soriano-Ubeda C, Garcia-Vazquez FA, Saadeh H, Andrews S, Kelsey G, Coy P (2017) DNA methylation and gene expression changes derived from assisted reproductive technologies can be decreased by reproductive fluids. eLife 6:e23670. doi:https://doi.org/10.7554/eLife.23670

Carlson DF, Tan W, Lillico SG, Stverakova D, Proudfoot C, Christian M, Voytas DF, Long CR, Whitelaw CB, Fahrenkrug SC (2012) Efficient TALEN-mediated gene knockout in livestock. Proc Natl Acad Sci U S A 109(43):17382–17387. https://doi.org/10.1073/pnas.1211446109

Chen F, Wang Y, Yuan Y, Zhang W, Ren Z, Jin Y, Liu X, Xiong Q, Chen Q, Zhang M, Li X, Zhao L, Li Z, Wu Z, Zhang Y, Hu F, Huang J, Li R, Dai Y (2015) Generation of B cell- deficient pigs by highly efficient CRISPR/Cas9-mediated gene targeting. J Genet Genom Yi chuan xue bao 42(8):437–444. https://doi.org/10.1016/j.jgg.2015.05.002

Chen S, Lee B, Lee AY, Modzelewski AJ, He L (2016) Highly Efficient Mouse Genome Editing by CRISPR Ribonucleoprotein Electroporation of Zygotes. J Biol Chem 291 (28):14457- 14467. https://doi.org/10.1074/jbc.M116.733154

Chen B, Gu P, Jia J, Liu W, Liu Y, Liu W, Xu T, Lin X, Lin T, Liu Y, Chen H, Xu M, Yuan J, Zhang J, Zhang Y, Xiao D, Gu W (2019a) Optimization strategy for generating gene- edited tibet minipigs by synchronized oestrus and cytoplasmic microinjection. Int J Biol Sci 15(12):2719–2732. https://doi.org/10.7150/ijbs.35930

Chen J, Wang H, Bai J, Liu W, Liu X, Yu D, Feng T, Sun Z, Zhang L, Ma L, Hu Y, Zou Y, Tan T, Zhong J, Hu M, Bai X, Pan D, Xing Y, Zhao Y, Tian K, Hu X, Li N (2019b) Generation of Pigs Resistant to Highly Pathogenic-Porcine Reproductive and Respiratory Syndrome Virus through Gene Editing of CD163. Int J Biol Sci 15(2):481–492. https://doi.org/10.7150/ijbs.25862

Chen J, An B, Yu B, Peng X, Yuan H, Yang Q, Chen X, Yu T, Wang L, Zhang X, Wang H, Zou X, Pang D, Ouyang H, Tang X (2020) CRISPR/Cas9-mediated knockin of human factor IX into swine factor IX locus effectively alleviates bleeding in hemophilia B pigs. Haematologica. https://doi.org/10.3324/haematol.2019.224063

Cho B, Kim SJ, Lee EJ, Ahn SM, Lee JS, Ji DY, Lee K, Kang JT (2018) Generation of insulin- deficient piglets by disrupting INS gene using CRISPR/Cas9 system. Transgenic Res 27(3):289–300. https://doi.org/10.1007/s11248-018-0074-1

Christian M, Cermak T, Doyle EL, Schmidt C, Zhang F, Hummel A, Bogdanove AJ, Voytas DF (2010) Targeting DNA double-strand breaks with TAL effector nucleases. Genetics 186(2):757–761. https://doi.org/10.1534/genetics.110.120717

Chuang CK, Chen CH, Huang CL, Su YH, Peng SH, Lin TY, Tai HC, Yang TS, Tu CF (2017) Generation of GGTA1 mutant pigs by direct pronuclear microinjection of CRISPR/Cas9 plasmid vectors. Anim Biotechnol 28(3):174–181. https://doi.org/10.1080/10495398.2016.1246453

Clark JL (2015) Killing the enviropigs. J Anim Ethics 5(1):20–30

Clifford H (2014) AquAdvantage® Salmon - a pioneering application of biotechnology in aquaculture. BMC Proc 8(S4):O31. https://doi.org/10.1186/1753-6561-8-s4-o31

Cong L, Ran FA, Cox D, Lin S, Barretto R, Habib N, Hsu PD, Wu X, Jiang W, Marraffini LA, Zhang F (2013) Multiplex genome engineering using CRISPR/Cas systems. Science 339(6121):819–823. https://doi.org/10.1126/science.1231143

Crociara P, Chieppa MN, Vallino Costassa E, Berrone E, Gallo M, Lo Faro M, Pintore MD, Iulini B, D'Angelo A, Perona G, Botter A, Formicola D, Rainoldi A, Paulis M, Vezzoni P, Meli F, Peverali FA, Bendotti C, Trolese MC, Pasetto L, Bonetto V, Lazzari G, Duchi R, Perota A, Lagutina I, Quadalti C, Gennero MS, Dezzutto D, Desiato R, Boido M, Ghibaudi M, Valentini MC, Caramelli M, Galli C, Casalone C, Corona C (2019) Motor neuron degeneration, severe myopathy and TDP-43 increase in a transgenic pig model of SOD1- linked familiar ALS. Neurobiol Dis 124:263–275. https://doi.org/10.1016/j.nbd.2018.11.021

Dai Y, Vaught TD, Boone J, Chen SH, Phelps CJ, Ball S, Monahan JA, Jobst PM, McCreath KJ, Lamborn AE, Cowell-Lucero JL, Wells KD, Colman A, Polejaeva IA, Ayares DL (2002) Targeted disruption of the alpha1,3-galactosyltransferase gene in cloned pigs. Nat Biotechnol 20(3):251–255. https://doi.org/10.1038/nbt0302-251

Dang-Nguyen TQ, Nguyen HT, Somfai T, Wells D, Men NT, Viet-Linh N, Noguchi J, Kaneko H, Kikuchi K, Nagai T (2018) Sucrose assists selection of high-quality oocytes in pigs. Anim Sci J Nihon chikusan Gakkaiho 89(6):880–887. https://doi.org/10.1111/asj.13015

Das S, Koyano-Nakagawa N, Gafni O, Maeng G, Singh BN, Rasmussen T, Pan X, Choi KD, Mickelson D, Gong W, Pota P, Weaver CV, Kren S, Hanna JH, Yannopoulos D, Garry MG, Garry DJ (2020) Generation of human endothelium in pig embryos deficient in ETV2. Nat Biotechnol 38(3):297–302. https://doi.org/10.1038/s41587-019-0373-y

Dehairs J, Talebi A, Cherifi Y, Swinnen JV (2016) CRISP-ID: decoding CRISPR mediated indels by Sanger sequencing. Scientific reports 6(1):28973. https://doi.org/10.1038/srep28973

Diao YF, Naruse KJ, Han RX, Li XX, Oqani RK, Lin T, Jin DI (2013) Treatment of fetal fibroblasts with DNA methylation inhibitors and/or histone deacetylase inhibitors improves the development of porcine nuclear transfer-derived embryos. Anim Reprod Sci 141(3-4):164–171. https://doi.org/10.1016/j.anireprosci.2013.08.008

Dorado B, Ploen GG, Barettino A, Macias A, Gonzalo P, Andres-Manzano MJ, Gonzalez-Gomez C, Galan-Arriola C, Alfonso JM, Lobo M, Lopez-Martin GJ, Molina A, Sanchez-Sanchez R, Gadea J, Sanchez-Gonzalez J, Liu Y, Callesen H, Filgueiras-Rama D, Ibanez B, Sorensen CB, Andres V (2019) Generation and characterization of a novel knockin minipig model of Hutchinson-Gilford progeria syndrome. Cell Discov 5(1):16. https://doi.org/10.1038/s41421-019-0084-z

Du Y, Kragh PM, Zhang Y, Li J, Schmidt M, Bogh IB, Zhang X, Purup S, Jorgensen AL, Pedersen AM, Villemoes K, Yang H, Bolund L, Vajta G (2007) Piglets born from handmade cloning, an innovative cloning method without micromanipulation. Theriogenology 68(8):1104–1110. https://doi.org/10.1016/j.theriogenology.2007.07.021

Dziuk P (1985) Effect of migration, distribution and spacing of pig embryos on pregnancy and fetal survival. J Reprod Fertil Suppl 33:57–63

Eid A, Alshareef S, Mahfouz MM (2018) CRISPR base editors: genome editing without double-stranded breaks. Biochem J 475(11):1955–1964. https://doi.org/10.1042/bcj20170793

Elkhadragy L, Regan MR, W MT, Goli KD, Patel S, Garcia K, Stewart M, Schook LB, Gaba RC, Schachtschneider KM (2021) Generation of genetically tailored porcine liver cancer cells by CRISPR/Cas9 editing. Biotechniques 70(1):37–48. https://doi.org/10.2144/btn-2020-0119

EMA (2001) Note for guidance on the quality, preclinical and clinical aspects of gene transfer medicinal products. European Medicines Agency (EMA), London

EMA (2006) Guideline on environmental risk assessments for medicinal products consisting of, or containing, genetically modified organisms (GMOs). European Medicines Agency (EMA), London

EMA (2009) Guideline on xenogeneic cell-based medicinal products. European Medicines Agency (EMA), London

EMA (2013) Guideline on Quality of biological active substances produced by transgene expression in animals. European Medicines Agency (EMA), London

EMA (2015) Kanuma sebelipase alfa European Medicines Agency (EMA)

EMA (2019) ATryn. Withdrawal of the marketing authorisation in the European Union EMA (2020) Ruconest conestat alfa European Medicines Agency (EMA)

Fajrial AK, He QQ, Wirusanti NI, Slansky JE, Ding X (2020) A review of emerging physical transfection methods for CRISPR/Cas9-mediated gene editing. Theranostics 10(12):5532–5549. https://doi.org/10.7150/thno.43465

FAO (2020) Food outlook biannual report on global food markets – June 2020. Food Outlook, Rome. https://doi.org/10.4060/ca9509en

FDA (2017) Guidance for industry regulation of intentionally altered genomic DNA in animals

FDA (2020) GalSafe pig. NADA 141-542. US Food & Drug Administration.

Feng J, Yang F (2019) The Regulation of Genetically Modified Food in China. Biotechnol Law Report 38(5):289–293. https://doi.org/10.1089/blr.2019.29137.jf

Fischer K, Rieblinger B, Hein R, Sfriso R, Zuber J, Fischer A, Klinger B, Liang W, Flisikowski K, Kurome M, Zakhartchenko V, Kessler B, Wolf E, Rieben R, Schwinzer R, Kind A, Schnieke A (2020) Viable pigs after simultaneous inactivation of porcine MHC class I and three xenoreactive antigen genes GGTA1, CMAH and B4GALNT2. Xenotransplantation 27(1):e12560. https://doi.org/10.1111/xen.12560

Forsberg CW, Meidinger RG, Liu M, Cottrill M, Golovan S, Phillips JP (2013) Integration, stability and expression of the E. coli phytase transgene in the Cassie line of Yorkshire Enviropig. Transgenic Res 22(2):379–389. https://doi.org/10.1007/s11248-012-9646-7

Forsberg CW, Meidinger RG, Ajakaiye A, Murray D, Fan MZ, Mandell IB, Phillips JP (2014a) Comparative carcass and tissue nutrient composition of transgenic Yorkshire pigs expressing phytase in the saliva and conventional Yorkshire pigs. J Anim Sci 92 (10):4417–4439. https://doi.org/10.2527/jas.2014-7780

Forsberg CW, Meidinger RG, Murray D, Keirstead ND, Hayes MA, Fan MZ, Ganeshapillai J, Monteiro MA, Golovan SP, Phillips JP (2014b) Phytase properties and locations in tissues of transgenic pigs secreting phytase in the saliva. J Anim Sci 92(8):3375–3387. https://doi.org/10.2527/jas.2014-7782

Fu R, Fang M, Xu K, Ren J, Zou J, Su L, Chen X, An P, Yu D, Ka M, Hai T, Li Z, Li W, Yang Y, Zhou Q, Hu Z (2020) Generation of GGTA1−/−β2M−/−CIITA−/− Pigs Using CRISPR/Cas9 technology to alleviate xenogeneic immune reactions. Transplantation 104(8):1566–1573. https://doi.org/10.1097/tp.0000000000003205

Gadea J, Garcia Vazquez F (2010) Applications of transgenic pigs in biomedicine and animal production. Itea-Informacion Tecnica Economica Agraria 106(1):30–45

Gadea J, Garcia-Vazquez FA, Hachem A, Bassett A, Romero-Aguirregomezcorta J, Canovas S, Romar R, Parrington J (2018) Generation of TPC2 knock out pig embryos by CRISPR- Cas technology. Reprod Domest Anim 53:87–88

Gao QS, Xuan MF, Luo ZB, Paek HJ, Kang JD, Yin XJ (2019) Hairless-knockout piglets generated using the clustered regularly interspaced short palindromic repeat/CRISPR- associated-9 exhibit abnormalities in the skin and thymus. Exp Anim 68(4):519–529. https://doi.org/10.1538/expanim.19-0018

Garcia-Vazquez FA, Ruiz S, Matas C, Izquierdo-Rico MJ, Grullon LA, De Ondiz A, Vieira L, Aviles-Lopez K, Gutierrez-Adan A, Gadea J (2010) Production of transgenic piglets using ICSI-sperm-mediated gene transfer in combination with recombinase RecA. Reproduction 140(2):259–272. https://doi.org/10.1530/REP-10-0129

Garcia-Vazquez FA, Hernandez-Caravaca I, Martin M, Gomez E, Rodriguez A, Sanchez-Sanchez R, Gadea J (2011) Two cases of reciprocal chromosomal translocation (4; 7)(p+; q-) (2; 8) (q-; q+) in piglets produced by ICSI. Reprod Domes Anim 46(4):728–730. https://doi.org/10.1111/j.1439-0531.2010.01707.x

Giuffra E, Kijas JM, Amarger V, Carlborg O, Jeon JT, Andersson L (2000) The origin of the domestic pig: independent domestication and subsequent introgression. Genetics 154(4):1785–1791

Golovan SP, Meidinger RG, Ajakaiye A, Cottrill M, Wiederkehr MZ, Barney DJ, Plante C, Pollard JW, Fan MZ, Hayes MA, Laursen J, Hjorth JP, Hacker RR, Phillips JP, Forsberg CW (2001) Pigs expressing salivary phytase produce low-phosphorus manure. Nat Biotechnol 19(8):741–745. https://doi.org/10.1038/90788

Golovan SP, Hakimov HA, Verschoor CP, Walters S, Gadish M, Elsik C, Schenkel F, Chiu DK, Forsberg CW (2008) Analysis of Sus scrofa liver proteome and identification of proteins differentially expressed between genders, and conventional and genetically enhanced lines. Comparative biochemistry and physiology Part D, Genomics & proteomics 3(3):234–242. https://doi.org/10.1016/j.cbd.2008.05.001

Gordon JW, Ruddle FH (1981) Integration and stable germ line transmission of genes injected into mouse pronuclei. Science 214(4526):1244–1246. https://doi.org/10.1126/science.6272397

Gordon JW, Scangos GA, Plotkin DJ, Barbosa JA, Ruddle FH (1980) Genetic transformation of mouse embryos by microinjection of purified DNA. Proc Natl Acad Sci U S A 77(12):7380–7384. https://doi.org/10.1073/pnas.77.12.7380

Guo C, Wang M, Zhu Z, He S, Liu H, Liu X, Shi X, Tang T, Yu P, Zeng J, Yang L, Cao Y, Chen Y, Liu X, He Z (2019a) Highly Efficient Generation of Pigs Harboring a Partial Deletion of the CD163 SRCR5 Domain, Which Are Fully Resistant to Porcine Reproductive and Respiratory Syndrome Virus 2 Infection. Front Immunol 10 (1846):1846. doi:https://doi.org/10.3389/fimmu.2019.01846

Guo Z, Lv L, Lui D, Fu B (2019b) Delayed activation methods and development of pig somatic cell nuclear transfer embryos: a meta-analysis. Large Anim Rev 25(2):69–73

Hai T, Teng F, Guo R, Li W, Zhou Q (2014) One-step generation of knockout pigs by zygote injection of CRISPR/Cas system. Cell Res 24(3):372–375. https://doi.org/10.1038/cr.2014.11

Hai T, Guo W, Yao J, Cao C, Luo A, Qi M, Wang X, Wang X, Huang J, Zhang Y, Zhang H, Wang D, Shang H, Hong Q, Zhang R, Jia Q, Zheng Q, Qin G, Li Y, Zhang T, Jin W, Chen ZY, Wang H, Zhou Q, Meng A, Wei H, Yang S, Zhao J (2017) Creation of miniature pig model of human Waardenburg syndrome type 2A by ENU mutagenesis. Hum Genet 136(11-12):1463–1475. https://doi.org/10.1007/s00439-017-1851-2

Hammer RE, Pursel VG, Rexroad CE Jr, Wall RJ, Bolt DJ, Ebert KM, Palmiter RD, Brinster RL (1985) Production of transgenic rabbits, sheep and pigs by microinjection. Nature 315(6021):680–683. https://doi.org/10.1038/315680a0

Han X, Gao Y, Li G, Xiong Y, Zhao C, Ruan J, Ma Y, Li X, Li C, Zhao S, Xie S (2020) Enhancing the antibacterial activities of sow milk via site-specific knock-in of a lactoferrin gene in pigs using CRISPR/Cas9 technology. Cell Biosci 10(1):133. https://doi.org/10.1186/s13578-020-00496-y

Hatada L (2017) Genome editing in animals, vol 1630. Methods in molecular biology. Springer New York, New York. https://doi.org/10.1007/978-1-4939-7128-2

Hauschild J, Petersen B, Santiago Y, Queisser AL, Carnwath JW, Lucas-Hahn A, Zhang L, Meng X, Gregory PD, Schwinzer R, Cost GJ, Niemann H (2011) Efficient generation of a biallelic knockout in pigs using zinc-finger nucleases. Proc Natl Acad Sci U S A 108(29):12013–12017. https://doi.org/10.1073/pnas.1106422108

He Q, Yu D, Bao M, Korensky G, Chen J, Shin M, Kim J, Park M, Qin P, Du K (2020) High-throughput and all-solution phase African Swine Fever Virus (ASFV) detection using CRISPR-Cas12a and fluorescence based point-of-care system. Biosens Bioelectron 154:112068. https://doi.org/10.1016/j.bios.2020.112068

Hein R, Sake HJ, Pokoyski C, Hundrieser J, Brinkmann A, Baars W, Nowak-Imialek M, Lucas-Hahn A, Figueiredo C, Schuberth HJ, Niemann H, Petersen B, Schwinzer R (2020) Triple (GGTA1, CMAH, B2M) modified pigs expressing an SLA class I(low) phenotype- Effects on immune status and susceptibility to human immune responses. Am J Transplant 20(4):988–998. https://doi.org/10.1111/ajt.15710

Hinrichs A, Kessler B, Kurome M, Blutke A, Kemter E, Bernau M, Scholz AM, Rathkolb B, Renner S, Bultmann S, Leonhardt H, de Angelis MH, Nagashima H, Hoeflich A, Blum WF, Bidlingmaier M, Wanke R, Dahlhoff M, Wolf E (2018) Growth hormone receptor- deficient pigs resemble the pathophysiology of human Laron syndrome and reveal altered activation of signaling cascades in the liver. Molecular metabolism 11:113–128. https://doi.org/10.1016/j.molmet.2018.03.006

Hinrichs A, Riedel EO, Klymiuk N, Blutke A, Kemter E, Langin M, Dahlhoff M, Kessler B, Kurome M, Zakhartchenko V, Jemiller EM, Ayares D, Bidlingmaier M, Flenkenthaler F, Hrabe de Angelis M, Arnold GJ, Reichart B, Frohlich T, Wolf E (2020) Growth hormone receptor knockout to reduce the size of donor pigs for preclinical xenotransplantation studies. Xenotransplantation n/a (n/a):e12664. https://doi.org/10.1111/xen.12664

Hirata M, Tanihara F, Wittayarat M, Hirano T, Nguyen NT, Le QA, Namula Z, Nii M, Otoi T (2019) Genome mutation after introduction of the gene editing by electroporation of Cas9 protein (GEEP) system in matured oocytes and putative zygotes. Vitro Cell Dev Biol Anim 55(4):237–242. https://doi.org/10.1007/s11626-019-00338-3

Hirata M, Wittayarat M, Tanihara F, Sato Y, Namula Z, Le QA, Lin Q, Takebayashi K, Otoi T (2020) One-step genome editing of porcine zygotes through the electroporation of a CRISPR/Cas9 system with two guide RNAs. In Vitro Cell Dev Biol Anim 56 (8):614–621. https://doi.org/10.1007/s11626-020-00507-9

Hockemeyer D, Wang H, Kiani S, Lai CS, Gao Q, Cassady JP, Cost GJ, Zhang L, Santiago Y, Miller JC, Zeitler B, Cherone JM, Meng X, Hinkley SJ, Rebar EJ, Gregory PD, Urnov FD, Jaenisch R (2011) Genetic engineering of human pluripotent cells using TALE nucleases. Nat Biotechnol 29(8):731–734. https://doi.org/10.1038/nbt.1927

Hofmann A, Kessler B, Ewerling S, Weppert M, Vogg B, Ludwig H, Stojkovic M, Boelhauve M, Brem G, Wolf E, Pfeifer A (2003) Efficient transgenesis in farm animals by lentiviral vectors. EMBO Rep 4(11):1054–1060. https://doi.org/10.1038/sj.embor.embor7400007

Hooper M, Hardy K, Handyside A, Hunter S, Monk M (1987) HPRT-deficient (Lesch-Nyhan) mouse embryos derived from germline colonization by cultured cells. Nature 326(6110):292–295. https://doi.org/10.1038/326292a0

Hornak M, Hulinska P, Musilova P, Kubickova S, Rubes J (2009) Investigation of chromosome aneuploidies in early porcine embryos using comparative genomic hybridization. Cytogenet Genome Res 126(1-2):210–216. https://doi.org/10.1159/000245922

Hryhorowicz M, Lipinski D, Hryhorowicz S, Nowak-Terpilowska A, Ryczek N, Zeyland J (2020) Application of genetically engineered pigs in biomedical research. Genes (Basel) 11(6):670. https://doi.org/10.3390/genes11060670

Hu S, Qiao J, Fu Q, Chen C, Ni W, Wujiafu S, Ma S, Zhang H, Sheng J, Wang P, Wang D, Huang J, Cao L, Ouyang H (2015) Transgenic shRNA pigs reduce susceptibility to foot and mouth disease virus infection. eLife 4:e06951. https://doi.org/10.7554/eLife.06951

Huan YJ, Zhu J, Xie BT, Wang JY, Liu SC, Zhou Y, Kong QR, He HB, Liu ZH (2013) Treating cloned embryos, but not donor cells, with 5-aza-2′-deoxycytidine enhances the developmental competence of porcine cloned embryos. J Reprod Dev 59(5):442–449. https://doi.org/10.1262/jrd.2013-026

Huang L, Hua Z, Xiao H, Cheng Y, Xu K, Gao Q, Xia Y, Liu Y, Zhang X, Zheng X, Mu Y, Li K (2017) CRISPR/Cas9-mediated ApoE -/- and LDLR -/- double gene knockout in pigs elevates serum LDL-C and TC levels. Oncotarget 8(23)

Huang J, Wang A, Huang C, Sun Y, Song B, Zhou R, Li L (2020a) Generation of marker-free pbd-2 knock-in pigs using the CRISPR/Cas9 and Cre/loxP systems. Genes (Basel) 11(8):1–14. https://doi.org/10.3390/genes11080951

Huang X, Zou X, Xu Z, Tang F, Shi J, Zheng E, Liu D, Moisyadi S, Urschitz J, Wu Z, Li Z (2020b) Efficient deletion of LoxP-flanked selectable marker genes from the genome of transgenic pigs by an engineered Cre recombinase. Transgenic Res 29(3):307–319. https://doi.org/10.1007/s11248-020-00200-3

Hubner A, Petersen B, Keil GM, Niemann H, Mettenleiter TC, Fuchs W (2018) Efficient inhibition of African swine fever virus replication by CRISPR/Cas9 targeting of the viral p30 gene (CP204L). Scientific Rep 8(1):1449. https://doi.org/10.1038/s41598-018-19626-1

Hunter RHF, Polge C, Rowson LE (1967) The recovery, transfer and survival of blastocysts in pigs. J Reprod Fertil 14(3):501–502

Hwang SU, Eun K, Yoon JD, Kim H, Hyun SH (2018) Production of transgenic pigs using a pGFAP-CreER(T2)/EGFP (LoxP) inducible system for central nervous system disease models. J Vet Sci 19(3):434–445. https://doi.org/10.4142/jvs.2018.19.3.434

Im GS, Lai L, Liu Z, Hao Y, Wax D, Bonk A, Prather RS (2004) In vitro development of preimplantation porcine nuclear transfer embryos cultured in different media and gas atmospheres. Theriogenology 61(6):1125–1135. https://doi.org/10.1016/j.theriogenology.2003.06.006

Jeong YS, Yeo S, Park JS, Lee KK, Kang YK (2007) Gradual development of a genome-wide H3-K9 trimethylation pattern in paternally derived pig pronucleus. Dev Dyn 236(6):1509–1516. https://doi.org/10.1002/dvdy.21150

Jinek M, Chylinski K, Fonfara I, Hauer M, Doudna JA, Charpentier E (2012) A programmable dual-RNA-guided DNA endonuclease in adaptive bacterial immunity. Science 337(6096):816–821. https://doi.org/10.1126/science.1225829

Kang JT, Ryu J, Cho B, Lee EJ, Yun YJ, Ahn S, Lee J, Ji DY, Lee K, Park KW (2016) Generation of RUNX3 knockout pigs using CRISPR/Cas9-mediated gene targeting. Reprod Domes Anim 51(6):970–978. https://doi.org/10.1111/rda.12775

Kang J-D, Kim S, Zhu H-Y, Jin L, Guo Q, Li X-C, Zhang Y-C, Xing X-X, Xuan M-F, Zhang G-L, Luo Q-R, Kim YS, Cui C-D, Li W-X, Cui Z-Y, Kim J-S, Yin X-J (2017) Generation of cloned adult muscular pigs with myostatin gene mutation by genetic engineering. RSC Adv 7(21):12541–12549. https://doi.org/10.1039/c6ra28579a

Kikuchi K, Onishi A, Kashiwazaki N, Iwamoto M, Noguchi J, Kaneko H, Akita T, Nagai T (2002) Successful piglet production after transfer of blastocysts produced by a modified in vitro system. Biol Reprod 66(4):1033–1041

Kim YB, Komor AC, Levy JM, Packer MS, Zhao KT, Liu DR (2017) Increasing the genome- targeting scope and precision of base editing with engineered Cas9-cytidine deaminase fusions. Nat Biotechnol 35(4):371–376. https://doi.org/10.1038/nbt.3803

Koblan LW, Doman JL, Wilson C, Levy JM, Tay T, Newby GA, Maianti JP, Raguram A, Liu DR (2018) Improving cytidine and adenine base editors by expression optimization and ancestral reconstruction. Nat Biotechnol 36(9):843–846. https://doi.org/10.1038/nbt.4172

Koketsu Y, Tani S, Iida R (2017) Factors for improving reproductive performance of sows and herd productivity in commercial breeding herds. Porcine Health Manag 3:1. https://doi.org/10.1186/s40813-016-0049-7

Komor AC, Kim YB, Packer MS, Zuris JA, Liu DR (2016) Programmable editing of a target base in genomic DNA without double-stranded DNA cleavage. Nature 533(7603):420–424. https://doi.org/10.1038/nature17946

Komor AC, Zhao KT, Packer MS, Gaudelli NM, Waterbury AL, Koblan LW, Kim YB, Badran AH, Liu DR (2017) Improved base excision repair inhibition and bacteriophage Mu Gam protein yields C:G-to-T:A base editors with higher efficiency and product purity. Sci Adv 3 (8):eaao4774. https://doi.org/10.1126/sciadv.aao4774

Koppes EA, Redel BK, Johnson MA, Skvorak KJ, Ghaloul-Gonzalez L, Yates ME, Lewis DW, Gollin SM, Wu YL, Christ SE, Yerle M, Leshinski A, Spate LD, Benne JA, Murphy SL, Samuel MS, Walters EM, Hansen SA, Wells KD, Lichter-Konecki U, Wagner RA, Newsome JT, Dobrowolski SF, Vockley J, Prather RS, Nicholls RD (2020) A porcine model of phenylketonuria generated by CRISPR/Cas9 genome editing. JCI Insight 5(20). https://doi.org/10.1172/jci.insight.141523

Kumar BM, Maeng GH, Lee YM, Lee JH, Jeon BG, Ock SA, Kang T, Rho GJ (2013) Epigenetic modification of fetal fibroblasts improves developmental competency and gene expression in porcine cloned embryos. Vet Res Commun 37(1):19–28. https://doi.org/10.1007/s11259-012-9542-x
Kurome M, Kessler B, Wuensch A, Nagashima H, Wolf E (2015) Nuclear transfer and transgenesis in the pig. In: N., H., A. (eds) Nuclear reprogramming. Methods in molecular biology, vol 1222. Methods and protocols. Humana Press, New York, pp 37–59. https://doi.org/10.1007/978-1-4939-1594-1_4
Kurtz S, Lucas-Hahn A, Schlegelberger B, Gohring G, Niemann H, Mettenleiter TC, Petersen B (2021) Knockout of the HMG domain of the porcine SRY gene causes sex reversal in gene-edited pigs. Proc Natl Acad Sci U S A 118(2). https://doi.org/10.1073/pnas.2008743118
Kwon J, Namgoong S, Kim NH (2015) CRISPR/Cas9 as tool for functional study of genes involved in preimplantation embryo development. PLoS One 10(3):e0120501. https://doi.org/10.1371/journal.pone.0120501
Lai L, Kolber-Simonds D, Park KW, Cheong HT, Greenstein JL, Im GS, Samuel M, Bonk A, Rieke A, Day BN, Murphy CN, Carter DB, Hawley RJ, Prather RS (2002) Production of alpha-1,3-galactosyltransferase knockout pigs by nuclear transfer cloning. Science 295(5557):1089–1092. https://doi.org/10.1126/science.1068228
Lai S, Wei S, Zhao B, Ouyang Z, Zhang Q, Fan N, Liu Z, Zhao Y, Yan Q, Zhou X, Li L, Xin J, Zeng Y, Lai L, Zou Q (2016) Generation of knock-in pigs carrying Oct4-tdTomato reporter through CRISPR/Cas9-mediated genome engineering. PLoS One 11 (1):e0146562. https://doi.org/10.1371/journal.pone.0146562
Lamas-Toranzo I, Guerrero-Sanchez J, Miralles-Bover H, Alegre-Cid G, Pericuesta E, Bermejo-Alvarez P (2017) CRISPR is knocking on barn door. Reprod Domes Anim 52 Suppl 4:39–47. https://doi.org/10.1111/rda.13047
Lamas-Toranzo I, Fonseca Balvis N, Querejeta-Fernandez A, Izquierdo-Rico MJ, Gonzalez-Brusi L, Lorenzo PL, Garcia-Rebollar P, Aviles M, Bermejo-Alvarez P (2019a) ZP4 confers structural properties to the zona pellucida essential for embryo development. eLife 8:e48904. doi:https://doi.org/10.7554/eLife.48904
Lamas-Toranzo I, Galiano-Cogolludo B, Cornudella-Ardiaca F, Cobos-Figueroa J, Ousinde O, Bermejo-Alvarez P (2019b) Strategies to reduce genetic mosaicism following CRISPR- mediated genome edition in bovine embryos. Scientific reports 9(1):14900. https://doi.org/10.1038/s41598-019-51366-8
Lassnig C, Müller M (2015) Disease-resistant transgenic animals. In: Meyers (ed) Encyclopedia of sustainability science and technology. Springer New York, New York, pp 1–17. https://doi.org/10.1007/978-1-4939-2493-6_10-3
Lavitrano M, Forni M, Varzi V, Pucci L, Bacci ML, Di Stefano C, Fioretti D, Zoraqi G, Moioli B, Rossi M, Lazzereschi D, Stoppacciaro A, Seren E, Alfani D, Cortesini R, Frati L (1997) Sperm-mediated gene transfer: production of pigs transgenic for a human regulator of complement activation. Transplant Proc 29(8):3508–3509. https://doi.org/10.1016/s0041-1345(97)00998-6
Le QA, Hirata M, Nguyen NT, Takebayashi K, Wittayarat M, Sato Y, Namula Z, Nii M, Tanihara F, Otoi T (2020) Effects of electroporation treatment using different concentrations of Cas9 protein with gRNA targeting Myostatin (MSTN) genes on the development and gene editing of porcine zygotes. Anim Sci J Nihon chikusan Gakkaiho 91 (1):e13386. https://doi.org/10.1111/asj.13386
Lee K, Prather RS (2014) Cloning pigs by somatic cell nuclear transfer. In: Cibelli, Gurdon, Wilmut et al. (eds) Principles of cloning, 2nd edn. Academic, San Diego, pp 245–254. https://doi.org/10.1016/B978-0-12-386541-0.00019-9
Legge M (1995) Oocyte and zygote zona pellucida permeability to macromolecules. J Exp Zool 271(2):145–150. https://doi.org/10.1002/jez.1402710210
Li J, Rieke A, Day BN, Prather RS (1996) Technical note: porcine non-surgical embryo transfer. J Anim Sci 74 (9):2263–2268. https://doi.org/10.2527/1996.7492263x

Li L, Pang D, Wang T, Li Z, Chen L, Zhang M, Song N, Nie D, Chen Z, Lai L, Ouyang H (2009) Production of a reporter transgenic pig for monitoring Cre recombinase activity. Biochem Biophys Res Commun 382(2):232–235. https://doi.org/10.1016/j.bbrc.2009.02.146

Li S, Flisikowska T, Kurome M, Zakhartchenko V, Kessler B, Saur D, Kind A, Wolf E, Flisikowski K, Schnieke A (2014) Dual fluorescent reporter pig for Cre recombination: transgene placement at the ROSA26 locus. PLoS One 9(7):e102455. https://doi.org/10.1371/journal.pone.0102455

Li P, Estrada JL, Burlak C, Montgomery J, Butler JR, Santos RM, Wang ZY, Paris LL, Blankenship RL, Downey SM, Tector M, Tector AJ (2015) Efficient generation of genetically distinct pigs in a single pregnancy using multiplexed single-guide RNA and carbohydrate selection. Xenotransplantation 22(1):20–31. https://doi.org/10.1111/xen.12131

Li G, Zhang X, Zhong C, Mo J, Quan R, Yang J, Liu D, Li Z, Yang H, Wu Z (2017a) Small molecules enhance CRISPR/Cas9-mediated homology-directed genome editing in primary cells. Scientific Rep 7(1):8943. https://doi.org/10.1038/s41598-017-09306-x

Li Z, Yang HY, Wang Y, Zhang ML, Liu XR, Xiong Q, Zhang LN, Jin Y, Mou LS, Liu Y, Li RF, Rao Y, Dai YF (2017b) Generation of tryptophan hydroxylase 2 gene knockout pigs by CRISPR/Cas9-mediated gene targeting. Journal of biomedical research 31 (5):445- 452. https://doi.org/10.7555/JBR.31.20170026

Li M, Ouyang H, Yuan H, Li J, Xie Z, Wang K, Yu T, Liu M, Chen X, Tang X, Jiao H, Pang D (2018) Site-Specific Fat-1 Knock-In Enables Significant Decrease of n-6PUFAs/n- 3PUFAs Ratio in Pigs. G3 (Bethesda, Md) 8(5):1747–1754. https://doi.org/10.1534/g3.118.200114

Li L, Meng H, Zou Q, Zhang J, Cai L, Yang B, Weng J, Lai L, Yang H, Gao Y (2019a) Establishment of gene-edited pigs expressing human blood-coagulation factor VII and albumin for bioartificial liver use. J Gastroenterol Hepatol 34(10):1851–1859. https://doi.org/10.1111/jgh.14666

Li R, Miao J, Wang Z (2019b) Production of genetically engineered porcine embryos by handmade cloning. Methods Mol Biol 1874:347–360. https://doi.org/10.1007/978-1-4939-8831-0_20

Li G, Zhang X, Wang H, Mo J, Zhong C, Shi J, Zhou R, Li Z, Yang H, Wu Z, Liu D (2020a) CRISPR/Cas9-Mediated Integration of Large Transgene into Pig CEP112 Locus. G3 (Bethesda, Md) 10(2):467–473. https://doi.org/10.1534/g3.119.400810

Li R, Zeng W, Ma M, Wei Z, Liu H, Liu X, Wang M, Shi X, Zeng J, Yang L, Mo D, Liu X, Chen Y, He Z (2020b) Precise editing of myostatin signal peptide by CRISPR/Cas9 increases the muscle mass of Liang Guang Small Spotted pigs. Transgenic Res 29 (1):149–163. https://doi.org/10.1007/s11248-020-00188-w

Li W, Shi L, Zhuang Z, Wu H, Lian M, Chen Y, Li L, Ge W, Jin Q, Zhang Q, Zhao Y, Liu Z, Ouyang Z, Ye Y, Li Y, Wang H, Liao Y, Quan L, Xiao L, Lai L, Meng G, Wang K (2020c) Engineered Pigs Carrying a Gain-of-Function NLRP3 Homozygous Mutation Can Survive to Adulthood and Accurately Recapitulate Human Systemic Spontaneous Inflammatory Responses. J Immunol 205(9):2532–2544. https://doi.org/10.4049/jimmunol.1901468

Liu L, Lee C, Moor RM (1996) DNA synthesis, microtubule and nuclear dynamics in porcine parthenotes. Zygote 4(2):139–144. https://doi.org/10.1017/s0967199400003014

Liu L, Liu Y, Gao F, Song G, Wen J, Guan J, Yin Y, Ma X, Tang B, Li Z (2012) Embryonic development and gene expression of porcine SCNT embryos treated with sodium butyrate. J Exp Zool B Mol Dev Evol 318(3):224–234. https://doi.org/10.1002/jez.b.22440

Liu T, Dou H, Xiang X, Li L, Li Y, Lin L, Pang X, Zhang Y, Chen Y, Luan J, Xu Y, Yang Z, Yang W, Liu H, Li F, Wang H, Yang H, Bolund L, Vajta G, Du Y (2015) Factors determining the efficiency of porcine somatic cell nuclear transfer: data analysis with over 200,000 reconstructed embryos. Cell Reprogram 17(6):463–471. https://doi.org/10.1089/cell.2015.0037

Liu Y, Lucas-Hahn A, Petersen B, Li R, Hermann D, Hassel P, Ziegler M, Larsen K, Niemann H, Callesen H (2017) Developmental competence and epigenetic profile of porcine embryos produced by two different cloning methods. Cell Reprogram 19(3):171–179. https://doi.org/10.1089/cell.2016.0055

Liu X, Liu H, Wang M, Li R, Zeng J, Mo D, Cong P, Liu X, Chen Y, He Z (2019) Disruption of the ZBED6 binding site in intron 3 of IGF2 by CRISPR/Cas9 leads to enhanced muscle development in Liang Guang Small Spotted pigs. Transgenic Res 28(1):141–150. https://doi.org/10.1007/s11248-018-0107-9

Luo W, Li Z, Huang Y, Han Y, Yao C, Duan X, Ouyang H, Li L (2014a) Generation of AQP2- Cre transgenic mini-pigs specifically expressing Cre recombinase in kidney collecting duct cells. Transgenic Res 23(2):365–375. https://doi.org/10.1007/s11248-013-9774-8

Luo W, Li Z, Li P, Huang Y, Han Y, Yao C, Zhang Z, Yan H, Pang D, Ouyang H, Li L (2014b) Expression of Cre recombinase in alveolar epithelial cells of the AQP2-Cre transgenic mini-pigs. Cell Physiol Biochem 34(5):1597–1613. https://doi.org/10.1159/000366363

Luo L, Wang S, Zhu L, Fan B, Liu T, Wang L, Zhao P, Dang Y, Sun P, Chen J, Zhang Y, Chang X, Yu Z, Wang H, Guo R, Li B, Zhang K (2019) Aminopeptidase N-null neonatal piglets are protected from transmissible gastroenteritis virus but not porcine epidemic diarrhea virus. Scientific Rep 9(1):13186. https://doi.org/10.1038/s41598-019-49838-y

Lv L, Lu X, Feng T, Rehman S, Sun J, Wu Z, Shi D, Liu Q, Cui K (2020) Valproic acid enhances in vitro developmental competence of porcine handmade cloned embryos. Livestock Science 233:103957. https://doi.org/10.1016/j.livsci.2020.103957

Mali P, Yang L, Esvelt KM, Aach J, Guell M, DiCarlo JE, Norville JE, Church GM (2013) RNA- guided human genome engineering via Cas9. Science 339(6121):823–826. https://doi.org/10.1126/science.1232033

Mao J, Ajakaiye A, Lan Y, Olk DC, Ceballos M, Zhang T, Fan MZ, Forsberg CW (2008) Chemical structures of manure from conventional and phytase transgenic pigs investigated by advanced solid-state NMR spectroscopy. J Agric Food Chem 56(6):2131–2138. https://doi.org/10.1021/jf071588x

Marchal R, Feugang JM, Perreau C, Venturi E, Terqui M, Mermillod P (2001) Meiotic and developmental competence of prepubertal and adult swine oocytes. Theriogenology 56(1):17–29. https://doi.org/10.1016/s0093-691x(01)00539-8

Martinez EA, Martinez CA, Cambra JM, Maside C, Lucas X, Vazquez JL, Vazquez JM, Roca J, Rodriguez-Martinez H, Gil MA, Parrilla I, Cuello C (2019) Achievements and future perspectives of embryo transfer technology in pigs. Reprod Domes Anim 54 Suppl 4(S4):4–13. https://doi.org/10.1111/rda.13465

Martinez CA, Cambra JM, Gil MA, Parrilla I, Alvarez-Rodriguez M, Rodriguez-Martinez H, Cuello C, Martinez EA (2020) Seminal Plasma Induces Overexpression of Genes Associated with Embryo Development and Implantation in Day-6 Porcine Blastocysts. Int J Mol Sci 21 (10):3662. https://doi.org/10.3390/ijms21103662

McCreath KJ, Howcroft J, Campbell KH, Colman A, Schnieke AE, Kind AJ (2000) Production of gene-targeted sheep by nuclear transfer from cultured somatic cells. Nature 405(6790):1066–1069. https://doi.org/10.1038/35016604

Mehravar M, Shirazi A, Nazari M, Banan M (2019) Mosaicism in CRISPR/Cas9-mediated genome editing. Dev Biol 445(2):156–162. https://doi.org/10.1016/j.ydbio.2018.10.008

Miller JC, Holmes MC, Wang J, Guschin DY, Lee YL, Rupniewski I, Beausejour CM, Waite AJ, Wang NS, Kim KA, Gregory PD, Pabo CO, Rebar EJ (2007) An improved zinc-finger nuclease architecture for highly specific genome editing. Nat Biotechnol 25(7):778–785. https://doi.org/10.1038/nbt1319

Miyoshi K, Kawaguchi H, Maeda K, Sato M, Akioka K, Noguchi M, Horiuchi M, Tanimoto A (2016) Birth of cloned microminipigs derived from somatic cell nuclear transfer embryos that have been transiently treated with valproic acid. Cell Reprogram 18(6):390–400. https://doi.org/10.1089/cell.2016.0025

Mojica FJM, Díez-Villaseñor C, Soria E, Juez G (2000) Biological significance of a family of regularly spaced repeats in the genomes of Archaea, Bacteria and mitochondria. vol 36. https://doi.org/10.1046/j.1365-2958.2000.01838.x

Mojica FJ, Diez-Villasenor C, Garcia-Martinez J, Soria E (2005) Intervening sequences of regularly spaced prokaryotic repeats derive from foreign genetic elements. J Mol Evol 60(2):174–182. https://doi.org/10.1007/s00239-004-0046-3

Moore FE, Reyon D, Sander JD, Martinez SA, Blackburn JS, Khayter C, Ramirez CL, Joung JK, Langenau DM (2012) Improved somatic mutagenesis in zebrafish using transcription activator-like effector nucleases (TALENs). PLoS One 7(5):e37877. https://doi.org/10.1371/journal.pone.0037877

Murray JD, Maga EA (2016) Genetically engineered livestock for agriculture: a generation after the first transgenic animal research conference. Transgenic Res 25(3):321–327. https://doi.org/10.1007/s11248-016-9927-7

Murray D, Meidinger RG, Golovan SP, Phillips JP, O'Halloran IP, Fan MZ, Hacker RR, Forsberg CW (2007) Transgene and mitochondrial DNA are indicators of efficient composting of transgenic pig carcasses. Bioresource technology 98(9):1795–1804. https://doi.org/10.1016/j.biortech.2006.06.029

Naeimi Kararoudi M, Hejazi SS, Elmas E, Hellstrom M, Naeimi Kararoudi M, Padma AM, Lee D, Dolatshad H (2018) Clustered regularly interspaced short palindromic repeats/Cas9 gene editing technique in xenotransplantation. Front Immunol 9:1711. https://doi.org/10.3389/fimmu.2018.01711

Navarro-Serna S, Romar R, Dehesa-Etxebeste M, Lopes JS, Lopez de Munain A, Gadea J (2019) First production of Calpain3 KO pig embryo by CRISPR/Cas9 technology for human disease modelling: efficiency comparison between electroporation and intracytoplasmic microinjection. In: 35rd Scientific meeting Association of Embryo Technology in Europe (AETE). Murcia (Spain). 13-14th September 2019

Navarro-Serna S, Vilarino M, Park I, Gadea J, Ross PJ (2020) Livestock gene editing by one- step embryo manipulation. J Equine Vet Sci 89:103025. https://doi.org/10.1016/j.jevs.2020.103025

Navarro-Serna S, Hachem A, Canha-Gouveia A, Hanbashi A, Garrappa G, Lopes JS, Paris-Oller E, Sarrias-Gil L, Flores-Flores C, Bassett A, Sanchez R, Bermejo-Alvarez P, Matas C, Romar R, Parrington J, Gadea J (2021) Generation of nonmosaic, two-pore channel 2 Biallelic knockout pigs in one generation by CRISPR-Cas9 microinjection before oocyte insemination. CRISPR J 4(1):132–146. https://doi.org/10.1089/crispr.2020.0078

Niemann H, Petersen B (2016) The production of multi-transgenic pigs: update and perspectives for xenotransplantation. Transgenic Res 25(3):361–374. https://doi.org/10.1007/s11248-016-9934-8

Nishio K, Tanihara F, Nguyen TV, Kunihara T, Nii M, Hirata M, Takemoto T, Otoi T (2018) Effects of voltage strength during electroporation on the development and quality of in vitro-produced porcine embryos. Reprod Domes Anim 53 (2):313–318. https://doi.org/10.1111/rda.13106

Niu D, Wei HJ, Lin L, George H, Wang T, Lee IH, Zhao HY, Wang Y, Kan Y, Shrock E, Lesha E, Wang G, Luo Y, Qing Y, Jiao D, Zhao H, Zhou X, Wang S, Wei H, Güell M, Church GM, Yang L (2017) Inactivation of porcine endogenous retrovirus in pigs using CRISPR- Cas9. Science 357(6357):1303–1307. https://doi.org/10.1126/science.aan4187

Niu D, Ma X, Yuan T, Niu Y, Xu Y, Sun Z, Ping Y, Li W, Zhang J, Wang T, Church GM (2021) Porcine genome engineering for xenotransplantation. Adv Drug Delivery Rev 168:229–245. https://doi.org/10.1016/j.addr.2020.04.001

Novoselova TA, Meuwissen MPM, Jongbloed AW, Huirne RBM (2013) Expected economic performance of genetic modification in pork production. NJAS Wageningen J Life Sci 64-65:9–15. https://doi.org/10.1016/j.njas.2012.03.002

Onishi A, Iwamoto M, Akita T, Mikawa S, Takeda K, Awata T, Hanada H, Perry AC (2000) Pig cloning by microinjection of fetal fibroblast nuclei. Science 289(5482):1188–1190. https://doi.org/10.1126/science.289.5482.1188

Ostedgaard LS, Price MP, Whitworth KM, Abou Alaiwa MH, Fischer AJ, Warrier A, Samuel M, Spate LD, Allen PD, Hilkin BM, Romano Ibarra GS, Ortiz Bezara ME, Goodell BJ, Mather SE, Powers LS, Stroik MR, Gansemer ND, Hippee CE, Zarei K, Goeken JA, Businga TR, Hoffman EA, Meyerholz DK, Prather RS, Stoltz DA, Welsh MJ (2020) Lack of airway submucosal glands impairs respiratory host defenses. eLife 9:e59653. doi:https://doi.org/10.7554/eLife.59653

Palmiter RD, Brinster RL, Hammer RE, Trumbauer ME, Rosenfeld MG, Birnberg NC, Evans RM (1982) Dramatic growth of mice that develop from eggs microinjected with metallothionein-growth hormone fusion genes. Nature 300(5893):611–615. https://doi.org/10.1038/300611a0

Paris-Oller E, Navarro-Serna S, Soriano-Ubeda C, Lopes JS, Matas C, Ruiz S, Latorre R, Lopez-Albors O, Romar R, Canovas S, Coy P (2021) Reproductive fluids, used for the in vitro production of pig embryos, result in healthy offspring and avoid aberrant placental expression of PEG3 and LUM. J Anim Sci Biotechnol 12(1):32. https://doi.org/10.1186/s40104-020-00544-0

Park KW, Cheong HT, Lai LX, Im GS, Kuhholzer B, Bonk A, Samuel M, Rieke A, Day BN, Murphy CN, Carter DB, Prather RS (2001) Production of nuclear transfer-derived swine that express the enhanced green fluorescent protein. Animal Biotechnology 12(2):173–181

Park SJ, Park HJ, Koo OJ, Choi WJ, Moon JH, Kwon DK, Kang JT, Kim S, Choi JY, Jang G, Lee BC (2012) Oxamflatin improves developmental competence of porcine somatic cell nuclear transfer embryos. Cell Reprogram 14(5):398–406. https://doi.org/10.1089/cell.2012.0007

Park KE, Park CH, Powell A, Martin J, Donovan DM, Telugu BP (2016) Targeted gene knockin in porcine somatic cells using CRISPR/Cas ribonucleoproteins. Int J Mol Sci 17(6):810–810. https://doi.org/10.3390/ijms17060810

Park KE, Powell A, Sandmaier SE, Kim CM, Mileham A, Donovan DM, Telugu BP (2017) Targeted gene knock-in by CRISPR/Cas ribonucleoproteins in porcine zygotes. Scientific Rep 7:42458. https://doi.org/10.1038/srep42458

Peng J, Wang Y, Jiang J, Zhou X, Song L, Wang L, Ding C, Qin J, Liu L, Wang W, Liu J, Huang X, Wei H, Zhang P (2015) Production of human albumin in pigs through CRISPR/Cas9-mediated knockin of human cDNA into swine albumin locus in the zygotes. Scientific Rep 5:16705. https://doi.org/10.1038/srep16705

Perleberg C, Kind A, Schnieke A (2018) Genetically engineered pigs as models for human disease. Dis Models Mechan 11 (1):dmm030783. https://doi.org/10.1242/dmm.030783

Petersen B (2017) Basics of genome editing technology and its application in livestock species. Reprod Domes Anim 52 Suppl 3:4-13. doi:10.1111/rda.13012

Petersen B, Frenzel A, Lucas-Hahn A, Herrmann D, Hassel P, Klein S, Ziegler M, Hadeler KG, Niemann H (2016) Efficient production of biallelic GGTA1 knockout pigs by cytoplasmic microinjection of CRISPR/Cas9 into zygotes. Xenotransplantation 23(5):338–346. https://doi.org/10.1111/xen.12258

Petters RM, Shuman RM, Johnson BH, Mettus RV (1987) Gene transfer in swine embryos by injection of cells infected with retrovirus vectors. J Exp Zool 242(1):85–88. https://doi.org/10.1002/jez.1402420111

Pfeiffer CA, Meyer AE, Brooks KE, Chen PR, Milano-Foster J, Spate LD, Benne JA, Cecil RF, Samuel MS, Ciernia LA, Spinka CM, Smith MF, Wells KD, Spencer TE, Prather RS, Geisert RD (2020) Ablation of conceptus PTGS2 expression does not alter early conceptus development and establishment of pregnancy in the pigdagger. Biol Reprod 102(2):475–488. https://doi.org/10.1093/biolre/ioz192

Polejaeva IA, Chen SH, Vaught TD, Page RL, Mullins J, Ball S, Dai Y, Boone J, Walker S, Ayares DL, Colman A, Campbell KH (2000) Cloned pigs produced by nuclear transfer from adult somatic cells. Nature 407(6800):86–90. https://doi.org/10.1038/35024082

Prather RS, Sims MM, First NL (1989) Nuclear transplantation in early pig embryos. Biol Reprod 41(3):414–418. https://doi.org/10.1095/biolreprod41.3.414

Prather RS, Rowland RR, Ewen C, Trible B, Kerrigan M, Bawa B, Teson JM, Mao J, Lee K, Samuel MS, Whitworth KM, Murphy CN, Egen T, Green JA (2013) An intact sialoadhesin (Sn/SIGLEC1/CD169) is not required for attachment/internalization of the porcine reproductive and respiratory syndrome virus. J Virol 87(17):9538–9546. https://doi.org/10.1128/JVI.00177-13

Ramlee MK, Yan T, Cheung AMS, Chuah CTH, Li S (2015) High-throughput genotyping of CRISPR/Cas9-mediated mutants using fluorescent PCR-capillary gel electrophoresis. Scientific Rep 5:15587. doi:https://doi.org/10.1038/srep15587., https://www.nature.com/articles/srep15587#supplementary-information

Ramlee MK, Wang J, Cheung AMS, Li S (2017) Using a Fluorescent PCR-capillary Gel Electrophoresis Technique to Genotype CRISPR/Cas9-mediated Knockout Mutants in a High-throughput Format. J Vis Exp 122. https://doi.org/10.3791/55586

Redel BK, Spate LD, Lee K, Mao J, Whitworth KM, Prather RS (2016) Glycine supplementation in vitro enhances porcine preimplantation embryo cell number and decreases apoptosis but does not lead to live births. Mol Reprod Dev 83(3):246–258. https://doi.org/10.1002/mrd.22618

Robertson E, Bradley A, Kuehn M, Evans M (1986) Germ-line transmission of genes introduced into cultured pluripotential cells by retroviral vector. Nature 323(6087):445–448. https://doi.org/10.1038/323445a0

Rocadembosch J, Amador J, Bernaus J, Font J, Fraile LJ (2016) Production parameters and pig production cost: temporal evolution 2010-2014. Porcine Health Manag 2:11. https://doi.org/10.1186/s40813-016-0027-0

Romar R, Canovas S, Matas C, Gadea J, Coy P (2019) Pig in vitro fertilization: Where are we and where do we go? Theriogenology 137:113–121. https://doi.org/10.1016/j.theriogenology.2019.05.045

Sake HJ, Frenzel A, Lucas-Hahn A, Nowak-Imialek M, Hassel P, Hadeler KG, Hermann D, Becker R, Eylers H, Hein R, Baars W, Brinkmann A, Schwinzer R, Niemann H, Petersen B (2019) Possible detrimental effects of beta-2-microglobulin knockout in pigs. Xenotransplantation 26(6):e12525. https://doi.org/10.1111/xen.12525

Sakurai T, Watanabe S, Kamiyoshi A, Sato M, Shindo T (2014) A single blastocyst assay optimized for detecting CRISPR/Cas9 system-induced indel mutations in mice. BMC Biotechnol 14(1):69. https://doi.org/10.1186/1472-6750-14-69

Sato M, Miyoshi K, Nagao Y, Nishi Y, Ohtsuka M, Nakamura S, Sakurai T, Watanabe S (2014) The combinational use of CRISPR/Cas9-based gene editing and targeted toxin technology enables efficient biallelic knockout of the alpha-1,3-galactosyltransferase gene in porcine embryonic fibroblasts. Xenotransplantation 21(3):291–300. https://doi.org/10.1111/xen.12089

Sato M, Kosuke M, Koriyama M, Inada E, Saitoh I, Ohtsuka M, Nakamura S, Sakurai T, Watanabe S, Miyoshi K (2018) Timing of CRISPR/Cas9-related mRNA microinjection after activation as an important factor affecting genome editing efficiency in porcine oocytes. Theriogenology 108:29–38. https://doi.org/10.1016/j.theriogenology.2017.11.030

Schachtschneider KM, Schwind RM, Darfour-Oduro KA, De AK, Rund LA, Singh K, Principe DR, Guzman G, Ray CE, Jr., Ozer H, Gaba RC, Schook LB (2017) A validated, transitional and translational porcine model of hepatocellular carcinoma. Oncotarget 8 (38):63620-63634. https://doi.org/10.18632/oncotarget.18872

Schook LB, Collares TV, Hu W, Liang Y, Rodrigues FM, Rund LA, Schachtschneider KM, Seixas FK, Singh K, Wells KD, Walters EM, Prather RS, Counter CM (2015) A genetic porcine model of cancer. PLoS One 10(7):e0128864. https://doi.org/10.1371/journal.pone.0128864

Schook LB, Rund L, Begnini KR, Remiao MH, Seixas FK, Collares T (2016) Emerging technologies to create inducible and genetically defined porcine cancer models. Front Genet 7:28. https://doi.org/10.3389/fgene.2016.00028

Sheets TP, Park CH, Park KE, Powell A, Donovan DM, Telugu BP (2016) Somatic cell nuclear transfer followed by CRIPSR/Cas9 microinjection results in highly efficient genome editing in cloned pigs. Int J Mol Sci 17(12):2031. https://doi.org/10.3390/ijms17122031

Sheets TP, Park KE, Park CH, Swift SM, Powell A, Donovan DM, Telugu BP (2018) Targeted mutation of NGN3 gene disrupts pancreatic endocrine cell development in pigs. Scientific Rep 8(1):3582. https://doi.org/10.1038/s41598-018-22050-0

Shen B, Zhang W, Zhang J, Zhou J, Wang J, Chen L, Wang L, Hodgkins A, Iyer V, Huang X, Skarnes WC (2014) Efficient genome modification by CRISPR-Cas9 nickase with minimal off-target effects. Nat Methods 11(4):399–402. https://doi.org/10.1038/nmeth.2857

Shi J, Zhou R, Luo L, Mai R, Zeng H, He X, Liu D, Zeng F, Cai G, Ji H, Tang F, Wang Q, Wu Z, Li Z (2015) Influence of embryo handling and transfer method on pig cloning efficiency. Anim Reprod Sci 154:121–127. https://doi.org/10.1016/j.anireprosci.2015.01.006

Shi X, Tang T, Lin Q, Liu H, Qin Y, Liang X, Cong P, Mo D, Liu X, Chen Y, He Z (2020) Efficient generation of bone morphogenetic protein 15-edited Yorkshire pigs using CRISPR/Cas9dagger. Biol Reprod 103(5):1054–1068. https://doi.org/10.1093/biolre/ioaa138

Smedley D, Salimova E, Rosenthal N (2011) Cre recombinase resources for conditional mouse mutagenesis. Methods 53(4):411–416. https://doi.org/10.1016/j.ymeth.2010.12.027

Somfai T, Yoshioka K, Tanihara F, Kaneko H, Noguchi J, Kashiwazaki N, Nagai T, Kikuchi K (2014) Generation of live piglets from cryopreserved oocytes for the first time using a defined system for in vitro embryo production. PLoS One 9(5):e97731. https://doi.org/10.1371/journal.pone.0097731

Song Y, Lai L, Li L, Huang Y, Wang A, Tang X, Pang D, Li Z, Ouyang H (2016) Germ cell- specific expression of Cre recombinase using the VASA promoter in the pig. FEBS open bio 6(1):50–55. https://doi.org/10.1002/2211-5463.12005

Soppe JA, Lebbink RJ (2017) Antiviral goes viral: harnessing CRISPR/Cas9 to combat viruses in humans. Trends Microbiol 25(10):833–850. https://doi.org/10.1016/j.tim.2017.04.005

Sperandio S, Lulli V, Bacci ML, Forni M, Maione B, Spadafora C, Lavitrano M (1996) Sperm mediated gene transfer in bovine and swine species. Anim Biotech 7:59–77

Su X, Chen W, Cai Q, Liang P, Chen Y, Cong P, Huang J (2020) Effective generation of maternal genome point mutated porcine embryos by injection of cytosine base editor into germinal vesicle oocytes. Sci China Life Sci 63(7):996–1005. https://doi.org/10.1007/s11427-019-1611-1

Sun JM, Cui KQ, Li ZP, Lu XR, Xu ZF, Liu QY, Huang B, Shi DS (2017) Suberoylanilide hydroxamic acid, a novel histone deacetylase inhibitor, improves the development and acetylation level of miniature porcine handmade cloning embryos. Reprod Domes Anim 52(5):763–774. https://doi.org/10.1111/rda.12977

Tajima S, Uchikura K, Kurita T, Kikuchi K (2020) Insemination of recipient sows improves the survival to term of vitrified and warmed porcine expanded blastocysts transferred non- surgically. Anim Sci J Nihon chikusan Gakkaiho 91 (1):e13453. https://doi.org/10.1111/asj.13453

Tanihara F, Takemoto T, Kitagawa E, Rao S, Do LT, Onishi A, Yamashita Y, Kosugi C, Suzuki H, Sembon S, Suzuki S, Nakai M, Hashimoto M, Yasue A, Matsuhisa M, Noji S, Fujimura T, Fuchimoto D, Otoi T (2016) Somatic cell reprogramming-free generation of genetically modified pigs. Sci Adv 2(9):e1600803. https://doi.org/10.1126/sciadv.1600803

Tanihara F, Hirata M, Nguyen NT, Le QA, Hirano T, Takemoto T, Nakai M, Fuchimoto DI, Otoi T (2018) Generation of a TP53-modified porcine cancer model by CRISPR/Cas9- mediated gene modification in porcine zygotes via electroporation. PLoS One 13(10):e0206360. https://doi.org/10.1371/journal.pone.0206360

Tanihara F, Hirata M, Nguyen NT, Le QA, Hirano T, Takemoto T, Nakai M, Fuchimoto DI, Otoi T (2019a) Generation of PDX-1 mutant porcine blastocysts by introducing CRISPR/Cas9-system into porcine zygotes via electroporation. Anim Sci J Nihon chikusan Gakkaiho 90(1):55–61. https://doi.org/10.1111/asj.13129

Tanihara F, Hirata M, Nguyen NT, Le QA, Wittayarat M, Fahrudin M, Hirano T, Otoi T (2019b) Generation of CD163-edited pig via electroporation of the CRISPR/Cas9 system into porcine in vitro-fertilized zygotes. Anim Biotechnol:1–8. doi:https://doi.org/10.1080/10495398.2019.1668801

Tanihara F, Hirata M, Nguyen NT, Sawamoto O, Kikuchi T, Doi M, Otoi T (2020a) Efficient generation of GGTA1-deficient pigs by electroporation of the CRISPR/Cas9 system into in vitro-fertilized zygotes. BMC Biotechnol 20(1):40. https://doi.org/10.1186/s12896-020-00638-7

Tanihara F, Hirata M, Thi Nguyen N, Anh Le Q, Hirano T, Otoi T (2020b) Generation of viable PDX1 gene-edited founder pigs as providers of nonmosaics. Mol Reprod Dev 87 (4):471–481. doi:https://doi.org/10.1002/mrd.23335

Tanihara F, Hirata M, Nguyen NT, Sawamoto O, Kikuchi T, Otoi T (2021) One-step generation of multiple gene-edited pigs by electroporation of the CRISPR/Cas9 system into zygotes to reduce xenoantigen biosynthesis. Intl J Mol Sci 22(5):2249

Tao L, Yang M, Wang X, Zhang Z, Wu Z, Tian J, An L, Wang S (2016) Efficient biallelic mutation in porcine parthenotes using a CRISPR-Cas9 system. Biochem Biophys Res Commun 476(4):225–229. https://doi.org/10.1016/j.bbrc.2016.05.100

Taweechaipaisankul A, Kim GA, Jin JX, Lee S, Qasim M, Kim EH, Lee BC (2019) Enhancement of epigenetic reprogramming status of porcine cloned embryos with zebularine, a DNA methyltransferase inhibitor. Mol Reprod Dev 86(8):1013–1022. https://doi.org/10.1002/mrd.23178

Tu Z, Yang W, Yan S, Guo X, Li XJ (2015) CRISPR/Cas9: a powerful genetic engineering tool for establishing large animal models of neurodegenerative diseases. Mol Neurodegener 10(1):35. https://doi.org/10.1186/s13024-015-0031-x

Tu Z, Yang W, Yan S, Yin A, Gao J, Liu X, Zheng Y, Zheng J, Li Z, Yang S, Li S, Guo X, Li XJ (2017) Promoting Cas9 degradation reduces mosaic mutations in non-human primate embryos. Scientific reports 7(1):42081. https://doi.org/10.1038/srep42081

Ulloa Ullo CM, Yoshizawa M, Komoriya E, Mitsui A, Nagai T, Kikuchi K (2008) The blastocyst production rate and incidence of chromosomal abnormalities by developmental stage in in vitro produced porcine embryos. J Reprod Dev 54(1):22–29. https://doi.org/10.1262/jrd.19102

Vajta G, Zhang Y, Machaty Z (2007) Somatic cell nuclear transfer in pigs: recent achievements and future possibilities. Reprod Fertil Dev 19(2):403–423. https://doi.org/10.1071/rd06089

Van Eenennaam AL, Silva FF, Trott JF, Zilberman D (2020) Genetic engineering of livestock: the opportunity cost of regulatory delay. Ann Rev Anim Biosci 9 (1):null. https://doi.org/10.1146/annurev-animal-061220-023052

Vanamee ES, Santagata S, Aggarwal AK (2001) FokI requires two specific DNA sites for cleavage. J Mol Biol 309(1):69–78. https://doi.org/10.1006/jmbi.2001.4635

Wall RJ, Pursel VG, Hammer RE, Brinster RL (1985) Development of porcine ova that were centrifuged to permit visualization of pronuclei and nuclei. Biol Reprod 32(3):645–651. https://doi.org/10.1095/biolreprod32.3.645

Wang K, Ouyang H, Xie Z, Yao C, Guo N, Li M, Jiao H, Pang D (2015a) Efficient generation of myostatin mutations in pigs using the CRISPR/Cas9 system. Scientific Rep 5(1):16623. https://doi.org/10.1038/srep16623

Wang X, Zhou J, Cao C, Huang J, Hai T, Wang Y, Zheng Q, Zhang H, Qin G, Miao X, Wang H, Cao S, Zhou Q, Zhao J (2015b) Efficient CRISPR/Cas9-mediated biallelic gene disruption and site-specific knockin after rapid selection of highly active sgRNAs in pigs. Scientific reports 5:13348. https://doi.org/10.1038/srep13348

Wang Y, Du Y, Shen B, Zhou X, Li J, Liu Y, Wang J, Zhou J, Hu B, Kang N, Gao J, Yu L, Huang X, Wei H (2015c) Efficient generation of gene-modified pigs via injection of zygote with Cas9/sgRNA. Scientific reports 5(1):8256. https://doi.org/10.1038/srep08256

Wang X, Cao C, Huang J, Yao J, Hai T, Zheng Q, Wang X, Zhang H, Qin G, Cheng J, Wang Y, Yuan Z, Zhou Q, Wang H, Zhao J (2016) One-step generation of triple gene-targeted pigs using CRISPR/Cas9 system. Scientific reports 6(1):20620. https://doi.org/10.1038/srep20620

Wang HX, Li M, Lee CM, Chakraborty S, Kim HW, Bao G, Leong KW (2017a) CRISPR/Cas9-Based Genome Editing for Disease Modeling and Therapy: Challenges and Opportunities for Nonviral Delivery. Chem Rev 117(15):9874–9906. https://doi.org/10.1021/acs.chemrev.6b00799

Wang K, Tang X, Xie Z, Zou X, Li M, Yuan H, Guo N, Ouyang H, Jiao H, Pang D (2017b) CRISPR/Cas9-mediated knockout of myostatin in Chinese indigenous Erhualian pigs. Transgenic Res. https://doi.org/10.1007/s11248-017-0044-z

Wang H, Shen L, Chen J, Liu X, Tan T, Hu Y, Bai X, Li Y, Tian K, Li N, Hu X (2019a) Deletion of CD163 Exon 7 Confers Resistance to Highly Pathogenic Porcine Reproductive and Respiratory Viruses on Pigs. Int J Biol Sci 15(9):1993–2005. https://doi.org/10.7150/ijbs.34269

Wang J, Liu M, Zhao L, Li Y, Zhang M, Jin Y, Xiong Q, Liu X, Zhang L, Jiang H, Chen Q, Wang C, You Z, Yang H, Cao C, Dai Y, Li R (2019b) Disabling of nephrogenesis in porcine embryos via CRISPR/Cas9-mediated SIX1 and SIX4 gene targeting. Xenotransplantation 26(3):e12484. https://doi.org/10.1111/xen.12484

Wang H, Li G, Zhong C, Mo J, Sun Y, Shi J, Zhou R, Li Z, Wu Z, Liu D, Zhang X (2020) Generation of Multi-Transgenic Pigs Using PiggyBac Transposons Co-expressing Pectinase, Xylanase, Cellulase, beta-1.3-1.4-Glucanase and Phytase. Front Genet 11 (1521):597841. https://doi.org/10.3389/fgene.2020.597841

Wasmer M (2019) Roads forward for European GMO policy-uncertainties in wake of ECJ judgment have to be mitigated by regulatory reform. Front Bioeng Biotechnol 7 (JUN):132. doi:https://doi.org/10.3389/fbioe.2019.00132

Wells KD, Prather RS (2017) Genome-editing technologies to improve research, reproduction, and production in pigs. Mol Reprod Dev 84(9):1012–1017. https://doi.org/10.1002/mrd.22812

Whitelaw CB, Sang HM (2005) Disease-resistant genetically modified animals. Rev Sci Tech 24(1):275–283

Whitelaw CB, Sheets TP, Lillico SG, Telugu BP (2016) Engineering large animal models of human disease. J Pathol 238(2):247–256. https://doi.org/10.1002/path.4648

Whitworth KM, Prather RS (2017) Gene editing as applied to prevention of reproductive porcine reproductive and respiratory syndrome. Mol Reprod Dev 84(9):926–933. https://doi.org/10.1002/mrd.22811

Whitworth KM, Li R, Spate LD, Wax DM, Rieke A, Whyte JJ, Manandhar G, Sutovsky M, Green JA, Sutovsky P, Prather RS (2009) Method of oocyte activation affects cloning efficiency in pigs. Mol Reprod Dev 76(5):490–500. https://doi.org/10.1002/mrd.20987

Whitworth KM, Lee K, Benne JA, Beaton BP, Spate LD, Murphy SL, Samuel MS, Mao J, O'Gorman C, Walters EM, Murphy CN, Driver J, Mileham A, McLaren D, Wells KD, Prather RS (2014) Use of the CRISPR/Cas9 system to produce genetically engineered pigs from in vitro-derived oocytes and embryos. Biol Reprod 91(3):78. https://doi.org/10.1095/biolreprod.114.121723

Whitworth KM, Rowland RR, Ewen CL, Trible BR, Kerrigan MA, Cino-Ozuna AG, Samuel MS, Lightner JE, McLaren DG, Mileham AJ, Wells KD, Prather RS (2016) Gene-edited pigs are protected from porcine reproductive and respiratory syndrome virus. Nat Biotechnol 34(1):20–22. https://doi.org/10.1038/nbt.3434

Whitworth KM, Benne JA, Spate LD, Murphy SL, Samuel MS, Murphy CN, Richt JA, Walters E, Prather RS, Wells KD (2017) Zygote injection of CRISPR/Cas9 RNA successfully modifies the target gene without delaying blastocyst development or altering the sex ratio in pigs. Transgenic Res 26(1):97–107. https://doi.org/10.1007/s11248-016-9989-6

Whitworth KM, Cecil R, Benne JA, Redel BK, Spate LD, Samuel MS, Prather RS, Wells KD (2018) Zygote injection of RNA encoding Cre recombinase results in efficient removal of LoxP flanked neomycin cassettes in pigs. Transgenic Res 27(2):167–178. https://doi.org/10.1007/s11248-018-0064-3

Whitworth KM, Rowland RRR, Petrovan V, Sheahan M, Cino-Ozuna AG, Fang Y, Hesse R, Mileham A, Samuel MS, Wells KD, Prather RS (2019) Resistance to coronavirus infection in amino peptidase N-deficient pigs. Transgenic Res 28(1):21–32. https://doi.org/10.1007/s11248-018-0100-3

Whyte JJ, Meyer AE, Spate LD, Benne JA, Cecil R, Samuel MS, Murphy CN, Prather RS, Geisert RD (2018) Inactivation of porcine interleukin-1beta results in failure of rapid conceptus elongation. Proc Natl Acad Sci U S A 115(2):307–312. https://doi.org/10.1073/pnas.1718004115

Wolfe SA, Nekludova L, Pabo CO (2000) DNA recognition by Cys2His2 zinc finger proteins. Annu Rev Biophys Biomol Struct. 29. https://doi.org/10.1146/annurev.biophys.29.1.183

Wu J, Vilarino M, Suzuki K, Okamura D, Bogliotti YS, Park I, Rowe J, McNabb B, Ross PJ, Belmonte JCI (2017) CRISPR-Cas9 mediated one-step disabling of pancreatogenesis in pigs. Scientific reports 7(1):10487. https://doi.org/10.1038/s41598-017-08596-5

Wu H, Liu Q, Shi H, Xie J, Zhang Q, Ouyang Z, Li N, Yang Y, Liu Z, Zhao Y, Lai C, Ruan D, Peng J, Ge W, Chen F, Fan N, Jin Q, Liang Y, Lan T, Yang X, Wang X, Lei Z, Doevendans PA, Sluijter JPG, Wang K, Li X, Lai L (2018) Engineering CRISPR/Cpf1 with tRNA promotes genome editing capability in mammalian systems. Cell Mol Life Sci 75(19):3593–3607. https://doi.org/10.1007/s00018-018-2810-3

Xie Z, Pang D, Wang K, Li M, Guo N, Yuan H, Li J, Zou X, Jiao H, Ouyang H, Li Z, Tang X (2017) Optimization of a CRISPR/Cas9-mediated Knock-in strategy at the Porcine Rosa26 locus in porcine foetal fibroblasts. Scientific reports 7(1):3036. https://doi.org/10.1038/s41598-017-02785-y

Xie Z, Pang D, Yuan H, Jiao H, Lu C, Wang K, Yang Q, Li M, Chen X, Yu T, Chen X, Dai Z, Peng Y, Tang X, Li Z, Wang T, Guo H, Li L, Tu C, Lai L, Ouyang H (2018) Genetically modified pigs are protected from classical swine fever virus. PLoS pathogens 14(12):e1007193. https://doi.org/10.1371/journal.ppat.1007193

Xie J, Ge W, Li N, Liu Q, Chen F, Yang X, Huang X, Ouyang Z, Zhang Q, Zhao Y, Liu Z, Gou S, Wu H, Lai C, Fan N, Jin Q, Shi H, Liang Y, Lan T, Quan L, Li X, Wang K, Lai L (2019) Efficient base editing for multiple genes and loci in pigs using base editors. Nat Commun 10(1):2852. https://doi.org/10.1038/s41467-019-10421-8

Xie F, Zhou X, Lin T, Wang L, Liu C, Luo X, Luo L, Chen H, Guo K, Wei H, Wang Y (2020a) Production of gene-edited pigs harboring orthologous human mutations via double cutting by CRISPR/Cas9 with long single-stranded DNAs as homology-directed repair templates by zygote injection. Transgenic Res 29(5-6):587–598. https://doi.org/10.1007/s11248-020-00218-7

Xie Z, Jiao H, Xiao H, Jiang Y, Liu Z, Qi C, Zhao D, Jiao S, Yu T, Tang X, Pang D, Ouyang H (2020b) Generation of pRSAD2 gene knock-in pig via CRISPR/Cas9 technology. Antiviral Res 174:104696. https://doi.org/10.1016/j.antiviral.2019.104696

Xu K, Zhou Y, Mu Y, Liu Z, Hou S, Xiong Y, Fang L, Ge C, Wei Y, Zhang X, Xu C, Che J, Fan Z, Xiang G, Guo J, Shang H, Li H, Xiao S, Li J, Li K (2020) CD163 and pAPN double- knockout pigs are resistant to PRRSV and TGEV and exhibit decreased susceptibility to PDCoV while maintaining normal production performance. eLife 9:e57132. doi:https://doi.org/10.7554/eLife.57132

Yamashita S, Kogasaka Y, Hiradate Y, Tanemura K, Sendai Y (2020) Suppression of mosaic mutation by co-delivery of CRISPR associated protein 9 and three-prime repair exonuclease 2 into porcine zygotes via electroporation. J Reprod Dev 66(1):41–48. https://doi.org/10.1262/jrd.2019-088

Yang H, Wu Z (2018) Genome Editing of Pigs for Agriculture and Biomedicine. Front Genet 9:360. https://doi.org/10.3389/fgene.2018.00360

Yang L, Guell M, Niu D, George H, Lesha E, Grishin D, Aach J, Shrock E, Xu W, Poci J, Cortazio R, Wilkinson RA, Fishman JA, Church G (2015) Genome-wide inactivation of porcine endogenous retroviruses (PERVs). Science 350(6264):1101–1104. https://doi.org/10.1126/science.aad1191

Yang H, Zhang J, Zhang X, Shi J, Pan Y, Zhou R, Li G, Li Z, Cai G, Wu Z (2018) CD163 knockout pigs are fully resistant to highly pathogenic porcine reproductive and respiratory syndrome virus. Antiviral Res 151:63–70. https://doi.org/10.1016/j.antiviral.2018.01.004

Yao J, Huang J, Zhao J (2016) Genome editing revolutionize the creation of genetically modified pigs for modeling human diseases. Hum Genet 135(9):1093–1105. https://doi.org/10.1007/s00439-016-1710-6

Yao J, Zeng H, Zhang M, Wei Q, Wang Y, Yang H, Lu Y, Li R, Xiong Q, Zhang L, Chen Z, Xing G, Cao X, Dai Y (2019) OSBPL2-disrupted pigs recapitulate dual features of human hearing loss and hypercholesterolaemia. J Genetics Genom Yi chuan xue bao 46(8):379–387. https://doi.org/10.1016/j.jgg.2019.06.006

Yin Y, Hao H, Xu X, Shen L, Wu W, Zhang J, Li Q (2019) Generation of an MC3R knock-out pig by CRSPR/Cas9 combined with somatic cell nuclear transfer (SCNT) technology. Lipids in health and disease 18(1):122. https://doi.org/10.1186/s12944-019-1073-9

Yoshioka K, Suzuki C, Tanaka A, Anas IM, Iwamura S (2002) Birth of piglets derived from porcine zygotes cultured in a chemically defined medium. Biol Reprod 66(1):112–119. https://doi.org/10.1095/biolreprod66.1.112

Yoshioka K, Suzuki C, Itoh S, Kikuchi K, Iwamura S, Rodriguez-Martinez H (2003) Production of piglets derived from in vitro-produced blastocysts fertilized and cultured in chemically defined media: effects of theophylline, adenosine, and cysteine during in vitro fertilization. Biol Reprod 69(6):2092–2099. https://doi.org/10.1095/biolreprod.103.020081

Yoshioka K, Noguchi M, Suzuki C (2012) Production of piglets from in vitro-produced embryos following non-surgical transfer. Anim Reprod Sci 131(1-2):23–29. https://doi.org/10.1016/j.anireprosci.2012.01.018

Youngs CR (2001) Factors influencing the success of embryo transfer in the pig. Theriogenology 56(8):1311–1320. https://doi.org/10.1016/s0093-691x(01)00632-x

Yu HH, Zhao H, Qing YB, Pan WR, Jia BY, Zhao HY, Huang XX, Wei HJ (2016) Porcine Zygote Injection with Cas9/sgRNA Results in DMD-Modified Pig with Muscle Dystrophy. Int J Mol Sci 17(10). https://doi.org/10.3390/ijms17101668

Yuan Y, Spate LD, Redel BK, Tian Y, Zhou J, Prather RS, Roberts RM (2017) Quadrupling efficiency in production of genetically modified pigs through improved oocyte maturation. Proc Natl Acad Sci U S A 114(29):E5796–e5804. https://doi.org/10.1073/pnas.1703998114

Yuan H, Yu T, Wang L, Yang L, Zhang Y, Liu H, Li M, Tang X, Liu Z, Li Z, Lu C, Chen X, Pang D, Ouyang H (2020) Efficient base editing by RNA-guided cytidine base editors (CBEs) in pigs. Cell Mol Life Sci 77(4):719–733. https://doi.org/10.1007/s00018-019-03205-2

Yuk IH, Zhang JD, Ebeling M, Berrera M, Gomez N, Werz S, Meiringer C, Shao Z, Swanberg JC, Lee KH, Luo J, Szperalski B (2014) Effects of copper on CHO cells: insights from gene expression analyses. Biotechnol Prog 30(2):429–442. https://doi.org/10.1002/btpr.1868

Zettler S, Renner S, Kemter E, Hinrichs A, Klymiuk N, Backman M, Riedel EO, Mueller C, Streckel E, Braun-Reichhart C, Martins AS, Kurome M, Kessler B, Zakhartchenko V, Flenkenthaler F, Arnold GJ, Frohlich T, Blum H, Blutke A, Wanke R, Wolf E (2020) A decade of experience with genetically tailored pig models for diabetes and metabolic research. Anim Reprod 17(3):e20200064. https://doi.org/10.1590/1984-3143-AR2020-0064

Zhai Y, Li W, Zhang Z, Cao Y, Wang Z, Zhang S, Li Z (2018) Epigenetic states of donor cells significantly affect the development of somatic cell nuclear transfer (SCNT) embryos in pigs. Mol Reprod Dev 85(1):26–37. https://doi.org/10.1002/mrd.22935

Zhang R, Wang Y, Chen L, Wang R, Li C, Li X, Fang B, Ren X, Ruan M, Liu J, Xiong Q, Zhang L, Jin Y, Zhang M, Liu X, Li L, Chen Q, Pan D, Li R, Cooper DKC, Yang H, Dai Y (2018a) Reducing immunoreactivity of porcine bioprosthetic heart valves by genetically- deleting three major glycan antigens, GGTA1/beta4GalNT2/CMAH. Acta Biomater 72:196–205. https://doi.org/10.1016/j.actbio.2018.03.055

Zhang X, Li Z, Yang H, Liu D, Cai G, Li G, Mo J, Wang D, Zhong C, Wang H, Sun Y, Shi J, Zheng E, Meng F, Zhang M, He X, Zhou R, Zhang J, Huang M, Zhang R, Li N, Fan M, Yang J, Wu Z (2018b) Novel transgenic pigs with enhanced growth and reduced environmental impact. eLife:7. https://doi.org/10.7554/eLife.34286

Zhang B, Wang C, Zhang Y, Jiang Y, Qin Y, Pang D, Zhang G, Liu H, Xie Z, Yuan H, Ouyang H, Wang J, Tang X (2020) A CRISPR-engineered swine model of COL2A1 deficiency recapitulates altered early skeletal developmental defects in humans. Bone 137:115450. https://doi.org/10.1016/j.bone.2020.115450

Zhao J, Ross JW, Hao Y, Spate LD, Walters EM, Samuel MS, Rieke A, Murphy CN, Prather RS (2009) Significant improvement in cloning efficiency of an inbred miniature pig by histone deacetylase inhibitor treatment after somatic cell nuclear transfer. Biol Reprod 81(3):525–530. https://doi.org/10.1095/biolreprod.109.077016

Zheng Q, Lin J, Huang J, Zhang H, Zhang R, Zhang X, Cao C, Hambly C, Qin G, Yao J, Song R, Jia Q, Wang X, Li Y, Zhang N, Piao Z, Ye R, Speakman JR, Wang H, Zhou Q, Wang Y, Jin W, Zhao J (2017) Reconstitution of UCP1 using CRISPR/Cas9 in the white adipose tissue of pigs decreases fat deposition and improves thermogenic capacity. Proc Natl Acad Sci U S A 114(45):E9474–E9482. https://doi.org/10.1073/pnas.1707853114

Zhou X, Xin J, Fan N, Zou Q, Huang J, Ouyang Z, Zhao Y, Zhao B, Liu Z, Lai S, Yi X, Guo L, Esteban MA, Zeng Y, Yang H, Lai L (2015) Generation of CRISPR/Cas9-mediated gene- targeted pigs via somatic cell nuclear transfer. Cell Mol Life Sci 72(6):1175–1184. https://doi.org/10.1007/s00018-014-1744-7

Zhou X, Wang L, Du Y, Xie F, Li L, Liu Y, Liu C, Wang S, Zhang S, Huang X, Wang Y, Wei H (2016) Efficient Generation of Gene-Modified Pigs Harboring Precise Orthologous Human Mutation via CRISPR/Cas9-Induced Homology-Directed Repair in Zygotes. Hum Mutat 37(1):110–118. https://doi.org/10.1002/humu.22913

Zhu XX, Zhong YZ, Ge YW, Lu KH, Lu SS (2018) CRISPR/Cas9-mediated generation of Guangxi Bama minipigs harboring three mutations in alpha-synuclein causing Parkinson's disease. Scientific Rep 8(1):12420. https://doi.org/10.1038/s41598-018-30436-3

Zhu X, Wei Y, Zhan Q, Yan A, Feng J, Liu L, Tang D (2020a) CRISPR/Cas9-Mediated Biallelic Knockout of IRX3 Reduces the Production and Survival of Somatic Cell-Cloned Bama Minipigs. Animals (Basel) 10(3):501–501. https://doi.org/10.3390/ani10030501

Zhu XX, Zhan QM, Wei YY, Yan AF, Feng J, Liu L, Lu SS, Tang DS (2020b) CRISPR/Cas9- mediated MSTN disruption accelerates the growth of Chinese Bama pigs. Reprod Domes Anim 55(10):1314–1327. https://doi.org/10.1111/rda.13775

Zou Y, Li Z, Zou Y, Hao H, Li N, Li Q (2018) An FBXO40 knockout generated by CRISPR/Cas9 causes muscle hypertrophy in pigs without detectable pathological effects. Biochem Biophys Res Commun 498(4):940–945. https://doi.org/10.1016/j.bbrc.2018.03.085

Zou X, Ouyang H, Yu T, Chen X, Pang D, Tang X, Chen C (2019) Preparation of a new type 2 diabetic miniature pig model via the CRISPR/Cas9 system. Cell Death Dis 10(11):823. https://doi.org/10.1038/s41419-019-2056-5

Zuo E, Cai YJ, Li K, Wei Y, Wang BA, Sun Y, Liu Z, Liu J, Hu X, Wei W, Huo X, Shi L, Tang C, Liang D, Wang Y, Nie YH, Zhang CC, Yao X, Wang X, Zhou C, Ying W, Wang Q, Chen RC, Shen Q, Xu GL, Li J, Sun Q, Xiong ZQ, Yang H (2017) One-step generation of complete gene knockout mice and monkeys by CRISPR/Cas9-mediated gene editing with multiple sgRNAs. Cell Res 27(7):933–945. https://doi.org/10.1038/cr.2017.81

Chapter 4
Dietary Anti-nutritional Factors and Their Roles in Livestock Nutrition

Salma H. Abu Hafsa, Ayman A. Hassan, Mona M. M. Y. Elghandour, Alberto Barbabosa-Pliego, Miguel Mellado, and Abdelfattah Z. M. Salem

Abstract Nutrition is widely recognized as one of the chief factors driving profitability, efficiency, and development of livestock production. Plant-derived feedstuffs are high in macronutrients and micronutrients, but they also possess anti-nutritional factors (ANFs). Anti-nutritional factors are secondary compounds that lower the nutrient content of forages and reduce forage feed intake by livestock. Protease inhibitors, amylase inhibitors, lectins, tannins, mimosine, phytic acid, gossypol, oxalates, cyanogens, saponins, nitrates, alkaloids, and anti-vitamins are some of the most common ANFs found in livestock feed. The ANFs block or interfere with how the animal's body absorbs other nutrients, resulting in reduced bioavailability of various legumes and cereal components. Thus, ANFs may cause micronutrient malnutrition and mineral deficiencies. Different traditional techniques and methods are used alone or in combination to reduce the ANFs content in livestock feed, such as fermentation, germination, debarking, sterilization, steam sterilization, and soaking. The majority of ANFs found in livestock feeds offer potential health advantages or risks for livestock.

Keywords Plant-based diet · Anti-nutrients · Potential health benefits · Adverse health effects · Livestock

S. H. Abu Hafsa
Livestock Research Department, Arid Lands Cultivation Research Institute, City of Scientific Research and Technological Applications, Alexandria, Egypt
e-mail: hashim_salma@yahoo.com

A. A. Hassan
Animal Production Research Institute, Agricultural Research Centre, Dokki, Gizza, Egypt
e-mail: aymanan19@hotmail.com

Mona M. M. Y. Elghandour · A. Barbabosa-Pliego · A. Z. M. Salem (✉)
Facultad de Medicina Veterinaria y Zootecnia, Universidad Autónoma del Estado de México, Toluca, Mexico
e-mail: mmohamede@uaemex.mx; salem@uaemex.mx

M. Mellado
Department of Animal Nutrition, Autonomous Agrarian University Antonio Narro, Saltillo, Mexico

V. K. Yata et al. (eds.), *Sustainable Agriculture Reviews 57*, Sustainable Agriculture Reviews 57, https://doi.org/10.1007/978-3-031-07496-7_4

4.1 Introduction

Plants grown as animal feed are the basic feedstuff that provides nutrients to animals. Various shrubs, cereals, legumes, roughs, trees, herbs, and other non-conventional feeds for animals contain anti-nutritional factors (ANFs). These ANFs limit the usefulness of edible leaves, twigs, and pods of shrubs and trees as livestock feed. The ANFs are substances that, either by themselves or through their metabolic products, interfere with feed absorption and utilization, reduce nutrient intake and digestion, affect animal health and reproduction, or may produce other adverse effects (Akande et al. 2010). When consumed by animals in quantities above a critical threshold, even at a minimum level, ANFs reduce animal productivity, reproduction efficiency, and the quality of their products (milk, meat, and eggs) and cause toxicity. There is a wide distribution of non-lethal toxic factors throughout the plant kingdom, especially in plants used as animal feed (Igile 1996; D'Mello 2000). Several ANFs with potential toxicity for farm animals have been identified and are either heat-labile or heat-stable. These factors include protease inhibitors, amylase inhibitors, lectins, tannins, mimosine, phytic acid, gossypol, oxalates, cyanogens, saponins, nitrates, alkaloids, and anti-vitamin agents. These ANFs, found in plant-derived feeds, cause nutritional and animal health problems. Recently, the knowledge that these factors produce toxins and elicit beneficial biological responses has led to numerous investigations regarding their possible physiological implications in different biological systems (Igile 1996). Some of these factors are known as 'secondary metabolites,' which are widely applied in nutrition and as pharmacologically active agents as antioxidants and reduce inflammation (Soetan 2008; Petroski and Minich 2020). Proper precautions, including physical, chemical, and biotechnical treatments, and the quantities and methods of use can aid in destroying or reducing the ANF content in unconventional feeds before feeding to livestock and may help to overcome the deleterious actions of ANFs and to make them useful for livestock (Amaefule and Onwudike 2000; Balogun 2013).

4.2 Anti-nutritional Factors

Anti-nutritional factors (ANFs) are chemicals that interfere with the absorption and utilisation of feed and affect animal productivity and health by themselves or their metabolic products. ANFs are also referred to as anti-nutrients, secondary substances, or plant secondary metabolites. Many ANFs with potential toxicity for livestock have been identified and can be either heat-labile or heat-stable (Table 4.1).

Table 4.1 Heat-labile and/or heat-stable types of anti-nutritional factors in livestock feed

Heat-stable anti-nutritional factors	Heat-labile anti-nutritional factors
Maintained at high temperature Phytic acid, polyphenolic compounds (such as condensed tannins), Alkaloids, Saponins, and non-protein amino acids (Mimosine), etc.	Sensitive to standard temperature and lost at high temperature Lectins, Cyanogenic Glycosides, and Protease inhibitors, etc.

Adapted from Felix and Mello (2000)

4.3 Classification of Anti-nutritional Factors

Anti-nutritional factors (ANFs) in plants can be classified based on their chemical composition, properties, mechanisms of action (Aletor and Adeogun 1995), effects on the nutritional value of feedstuffs, and biological effects on the overall animal health (Huisman and Tolman 2001). ANFs which are frequently found in animal feed can be grouped as follows:

The major ANFs commonly found in plant-derived feedstuffs used in animal feed are summarised in Tables 4.2 and 4.3.

4.3.1 Direct and Indirect Factors Affect on Protein Digestion and Metabolism

4.3.1.1 Enzyme Inhibitors

Protease Inhibitors

Proteinases are enzymes that have diverse effects in improving the functional and nutritional properties of different protein molecules (Salas et al. 2018; Samtiya et al. 2020). Protease inhibitors are natural plant inhibitors. They have been amply studied due to their proteolytic action (reduces enzyme activity by protein–protein interactions), inflammatory response, ability to coagulate blood, and role in numerous hormone processing pathways (Gomes et al. 2011). They are widely distributed within the plant kingdom. For instance, protease inhibitors are present in seeds of most leguminous crops, and their presence prevents the utilization of the seeds as livestock feed, which may lead to reduced mineral bioavailability as well as reduced digestion and nutrient absorption (Bajpai et al. 2005; Yasmin et al. 2008) (Table 4.6). Compared with legumes, cereals contain much less of these digestive inhibitors, particularly those that act against proteases and amylases (Nikmaram et al. 2017). Protease inhibitors are concentrated in the outer portion of cereal cotyledons, which are the most common areas containing anti-nutritional factors in plants, and they can inhibit the activity of proteolytic enzymes secreted in the digestive system of animals (Nørgaard et al. 2019) by blocking the active site of the enzymes through a catalytic means. The N- and C-terminal and the exposed protease inhibitors are

Table 4.2 Classification of major anti-nutritional factors present in the plant-derived feedstuffs used in livestock feed

Anti-nutritional factors	Plant-derived nutrient source	Means of alleviation
Interaction with protein nutrition		
Enzyme inhibitors	Soybean, sunflower oil cake, rapeseed meal, lupin seed meal, sesame meal, pea seed meal, Jatropha kernel meal, Rapeseed, mustard oil cake	Heat, autoclaving, boiling, soaking
Lectins (Heamagglutinins)	Soybean, pea seed meal, Jatropha kernel meal	Heat, autoclaving
Saponins	Peas, Jatropha kernel meal, sunflower oil cake, lupin seed meal, pea seed meal	Soaking,
Tannins	Sorghum, mustard oil Cake, Jatropha kernel meal, pea seed meal, Rapeseed, mustard oil cake	Soaking, germination followed by dehulling, genetic modification
Mimosine	*Leucaena leucocephala*	Heat and chemical treatments, supplementation with amino acids or with metal ions
Interaction with mineralavailability		
Phytic acid	Soybean, pea seed meal, cottonseed meal, Jatropha kernel meal, sesame meal, Rapeseed, mustard oil cake	Supplementation, use of phytase, roasting, soaking, autoclaving, fermentation, germination, genetic modification
Oxalic acid	Leaf proteins	Heat treatment, Boiling
Gossypol	Cottonseed meal	Genetic modification, fermentation, use of iron salts
Interaction with vitamin availability		
Cyanogens,	Cassava, sorghum, pea seed meal	Heat treatment, boiling, simmering, blanching
Alkaloids	Lupin seed meal	Heat treatment, soaking
Antivitamins	Cottonseed meal, soybean meal, pea seed meal	Heat treatment

Francis et al. (2001)

often considered structural features necessary for inhibiting enzyme activity (Otlewski et al. 2005). They bind proteases, which resist digestion in the small intestine; thus, ensuring their removal through excretion (Fig. 4.1). Because of their protein-particular nature, protease inhibitors can be easily denatured by heat treatment, although some residual activity may remain in commercially produced products. The anti-nutrient activity of protease inhibitors is related to growth suppression and pancreatic hypertrophy (Chunmei et al. 2010). There are two types of protease inhibitors, the Kunitz inhibitor (inhibits trypsin only) and the Bowman-Burk inhibitor (inhibits trypsin and chymotrypsin) (Ramteke et al. 2019), commonly found in soybeans and cannot be quickly inactivated by heat treatment due to the presence of disulfide bridges (Liu 1997; Van Der Ven et al. 2005). The trypsin inhibitor in

Table 4.3 The types of anti-nutritional factors in forage crops

Anti-nutritional substances	Crops/species
Non-protein amino acids	
Mimosine	*Leucaena leucocephala*
Indospecine	*Indigofera spicta*
Glycosides	
Cyanogens	*Acacia giraffae* *Acacia sieberiana* *Acacia Cunninghamii* *Barteria fistulosa* *Bambusa bambos* *Manihot esculenta*
Saponins	*Albizia stipulate* *Sesbania sesban* *Bassia latifolia*
Phytohemagglutinins	
Ricin	*Bauhinia purpurea* *Robinia pseudoacacia*
Robin	*Ricinus communis*
Polyphenolic compounds	
Tannins	All vascular plants
Lignins	All vascular plants
Alkaloids	
N-methyl-B-phen	*Acacia berlandieri*
Ethylamine	*Sesbania vesiceria*
Sesbanine	*Sesbania punicea* *Sesbania drummondii*
Triterpenes	
Azadirachtin	*Azadirachta indica*
Limonin	*Azadirachta indica*
Oxalate	*Acacia aneura*

soybean interferes with methionine availability from raw soybean and forms non-digestible complexes with dietary protein in the gastrointestinal tract (Ramteke et al. 2019). These complexes are not digestible even in large amounts of digestive enzymes (Thakur et al. 2019). Chicks fed raw soybeans develop pancreatic hypertrophy, but this is not observed in pigs and calves (Ramteke et al. 2019). The presence of trypsin inhibitors in the diet creates an irreversible condition known as the enzyme-trypsin inhibitor complex, which leads to a reduction of trypsin in the intestine and a decrease in protein digestion, slowing down the animal growth. Several enzyme inhibitors are found in plant-derived feeds, but those that affect trypsin and α-amylase activity are the two main types found in all cereals and legume-based feeds. The factors controlling the destruction of protease inhibitors are heat treatment, duration of heating, particle size, and moisture level (Vaz Patto et al. 2015).

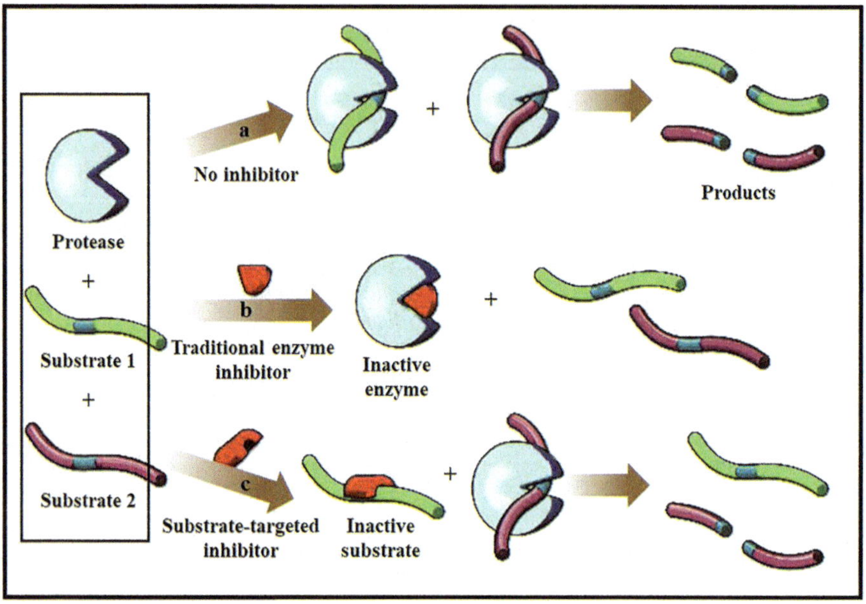

Fig. 4.1 Potential Target molecules of protease inhibitors

Amylase Inhibitors

α-amylase regulates the breakdown of carbohydrates, such as the breakdown of polysaccharides into oligosaccharides. Amylase inhibitors are known as starch blockers because they contain substances that prevent the absorption of dietary starches. Therefore, α-amylase inhibitors increase the time for carbohydrate absorption by delaying carbohydrate digestion, thus decreasing the rate of glucose absorption and affecting the average postprandial plasma glucose concentration (Bhutkar and Bhise 2012). These inhibitors are heat-labile and are active in the pH range of 4.5–9.5 (Marshall and Lauda 2007). Amylase inhibitors do not inhibit bacterial, fungal, or endogenous amylase but can inhibit bovine pancreatic amylase. This inhibitor's instability in the gastrointestinal tract leads to reduced insulin responses and increased caloric production from food when the inhibitor is used in starch blocking tablets (Giri and Kachole 1998).

4.3.1.2 Lectins (Haemagglutinins)

Lectins are sugar-binding proteins that readily bind to red blood cells to cause agglutination and are found in most plants, especially seeds such as grains and beans, tubers like potatoes, and raw meat (Hamid et al. 2013). Grains and legumes generally contain lectins, which are glycoproteins. Lectin activity has been determined in more than 800 legumes; 2–10% of the total legume seed proteins are lectins in soybean and ricin (castor bean), the latter is toxic and causes severe inflammation in the intestine, kidney, thyroid gland, etc. (Ramteke et al. 2019). In addition, the transport and hydrolytic functions of intestinal cells can be impaired by the consumption of foods containing lectins (Krupa 2008). Lectins impair the absorption of nutrients by binding directly to the intestinal mucosa, interacting with enterocytes, and resulting in severe intestinal damage, which disrupts digestion, causes nutrient deficiencies and epithelial lesions within the intestine, and allows bacterial populations to come in contact with the bloodstream (Muramoto 2017) (Fig. 4.2). In a study by Bardocz et al. (1995), the epithelium had an increased density of goblet cells and a marked decrease or absence of absorptive vacuoles; the microvilli of the intestinal cells were shortened with an increase in microvillar

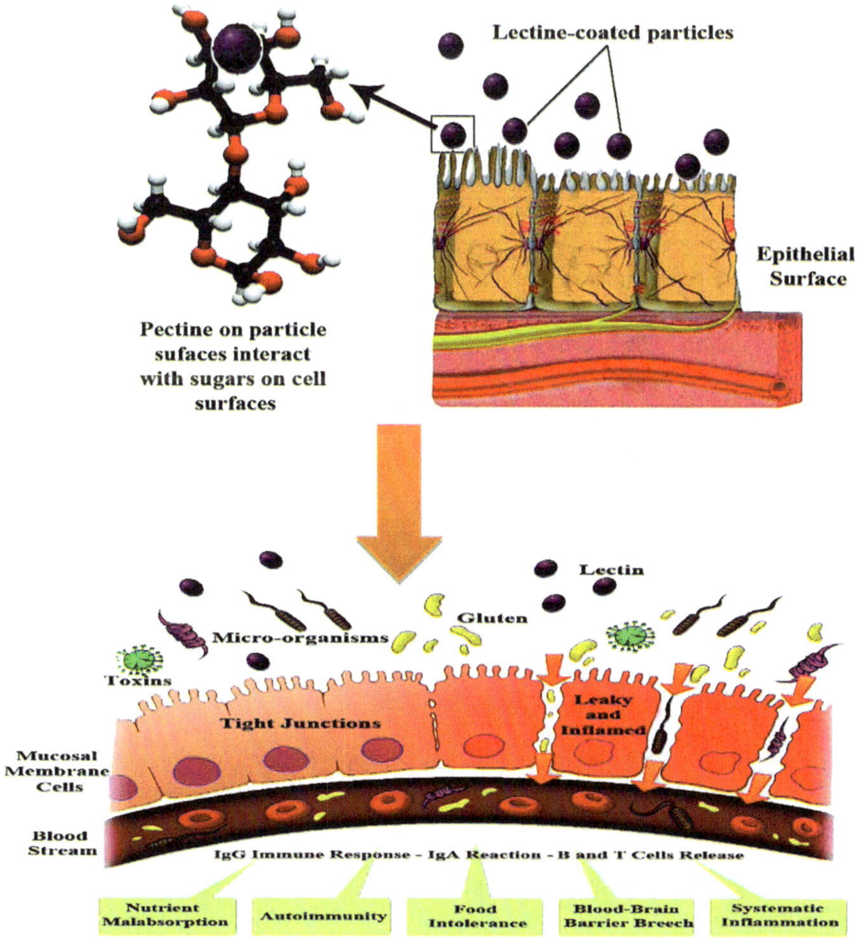

Fig. 4.2 Binding of particle-lectin conjugates with sugar residues of the cell membrane

vesicle formation and the number of intestinal crypt cells. The increase in the density of the goblet cells may have been due to the production of hypertrophic mucus in the intestine as a result of the irritation by lectins; thus, preventing the absorption of digestion end products in the small intestine. Lectins may also have resisted digestion through pancreatic juice (Ramteke et al. 2019). Lectins from soybeans, known as soybean agglutinins, impair animal growth, cause enlargement of the small intestine, damage the small intestine's epithelium, and stimulate hypertrophy and pancreas hyperplasia (Grant 1991) (Table 4.6). Lectins also bind with the glycoprotein components of erythrocytes, causing cell agglutination. Lectins have some interesting chemical and biological properties, such as interacting with specific blood groups, performing various mitotic division functions, destroying cancer cells, and having toxic effects in some animals. Since they bind with different sugar groups, lectins that attach to the intestinal wall may vary depending on sugar type. Dietary lectins are important because they are resistant to digestion and are not hydrolyzed in the intestine (Fig. 4.2). Although lectins are proteins, they are partially resistant to proteolytic degradation in the intestine. Soybean lectins can bind to brush border surfaces, particularly in the small intestine's distal part (Grant 1991; Dublecz 2011). Lectins selectively bind carbohydrates and, most importantly, the carbohydrate moieties of glycoproteins present on most animal cell surfaces. Lectins act as protein antigens that simultaneously bind to surface glycoproteins or glycolipids in red blood cells and immune factors, causing haemagglutination and anemia (Sauvion et al. 2004). They are present in small amounts in 30% of foods and in higher quantities in whole-grain diets. Haemagglutination of red blood cells is commonly used to measure lectin activity (Dublecz 2011; Fereidoon 2014). Consumption of feed-containing lectins may result in endogenous loss of nitrogen and reduced protein utilization. Undigested and unabsorbed proteins and carbohydrates in the small intestine reach the colon, where the bacterial flora ferments them into short-chain fatty acids and gases. These may, in turn, contribute to some digestive symptoms related to the intake of raw beans or purified lectins. The gastrointestinal mucosal disruption caused by lectins may allow bacteria and their endotoxins to enter the bloodstream and cause toxicity. Lectins can also be absorbed directly and cause systemic effects such as increased protein catabolism, breakdown of stored fats and glycogen, and mineral metabolism disturbances (Fereidoon 2014).

4.3.1.3 Tannins

Tannins are astringent and bitter plant polyphenols with molecular weights higher than 500 Da. One of the properties of these compounds is their ability to precipitate proteins and various other organic compounds, including amino acids and alkaloids. Tannins are secondary compounds found in plants' leaves, fruits, and bark (Timotheo and Lauer 2018). They are also found in cereals such as sorghum (containing up to 5% condensed tannin) and barley (Serrano et al. 2009; Morzelle et al. 2019), food crops and legumes such as lima beans, fava beans, sunflower seed meal, and rapeseed, in the foliage of many trees and shrubs, and many seeds and agro-industrial by-products (Dube et al. 2001) (Table 4.4). Tannins usually affect protein digestion by forming reversible and non-reversible tannin-protein complexes between the hydroxyl group of tannins and the carbonyl group of proteins, reducing essential amino acids (Lampart-Szczapa et al. 2003; Patra and Saxena 2010; Raes et al. 2014). In nature, there are two types of tannins: hydrolyzable (e.g., gallotannins and ellagitannins) and condensed (e.g., proanthocyanidins) (Patel et al. 2013) (Fig. 4.3). The two types differ in their molecular weight, structure, and nutritional and toxic effects on herbivorous animals, especially in ruminants that ingest tannin-rich forages (Fig. 4.3).

Condensed tannins (CT) are the most common type of tannins present in legumes, seeds, trees, and stems (Barry and McNabb 1999). They are extensively distributed in legume pasture species, several Acacia species, seeds, and other plant species (Degan et al. 1995). The CT consists of flavonoid units (flavan-3-ol) linked by carbon–carbon bonds, which influence its physical and biological properties (Hassanpour et al. 2011). The complexity of CT relies on flavonoid units that vary

Table 4.4 Distribution of tannins in selected feedstuffs

Feed ingredients	Tannin concentration (%)[a]	References
Sorghum grain (white)	0.55	Gowda et al. (1994)
Sorghum grain (yellow)	0.2–2.0	Fuller et al. (1996)
Sorghum grain (red)	1.54–7.44	Medugu et al. (2010)
Sunflower cake	2.36	Jacob et al. (1996)
Sesame seed cake	2.15	Jacob et al. (1996)
Mango seed kernel	5.47	Diarra et al. (2008)
Mango seed kernel	0.08–0.10	Bala et al. (2013)
Soybean meal	2.47	Jacob et al. (1996)
Pigeon pea	4.3–11.4	Jambunathan et al. (1988)
Chick pea	1.9–6.1	Jambunathan et al. (1988)
Mucuna beans	0.80	Akinmutimi (2007)

[a]Dry matter basis

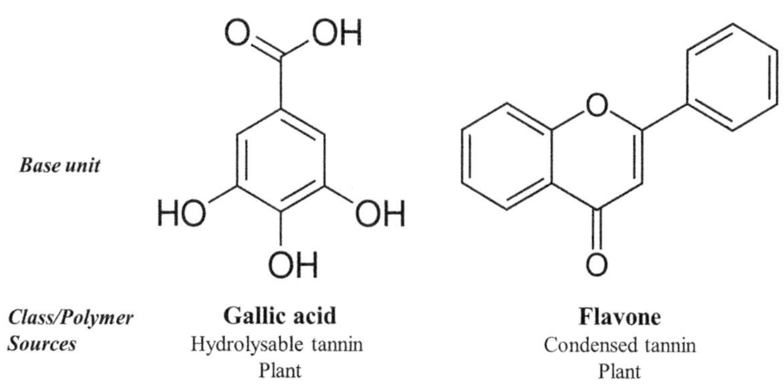

Fig. 4.3 Types of tannins and their primary structures

between components and within sites to form interflavan bonds. Hydrolyzable tannins (HT) are usually found at lower concentrations in plants than CT, and they are further divided into taragallotannins (gallic and quinic acid) and caffetannins (caffeic and quinic acid) (Mangan 1988). The HT are easily hydrolyzed during the digestion process by tanninase enzymes that engage in ester-bond hydrolysis. They can form compounds such as pyrogallol, which is toxic to ruminants. Poisonous compounds from more than 20% HT in the diet can cause kidney damage, proximal tubular necrosis, liver necrosis, lesions related to haemorrhagic gastroenteritis, and high mortality in sheep and cattle (Reed 1995). Previous studies have shown that cattle and sheep are sensitive to these tannins, while goats are resistant (D'Mello 2000; Bhattarai et al. 2016; Smeriglio et al. 2017). Tannins mainly accumulate on the seed coat of legumes; when ingested, they form protein-containing complexes that disrupt various digestive enzymes and reduce protein digestion (Joye 2019). In non-ruminants, HT can reduce growth rates, protein utilization, cause damage to the mucosa of the digestive tract and increase the excretion of protein and amino acids (Hassanpour et al. 2011). CT strongly reduces hydrolyzable tannin's digestibility, while HT causes varied toxic manifestations due to hydrolysis in the rumen (Akande et al. 2010). Tannins are the most common anti-nutritional factors found in plants. Their anti-nutritional effects depend on their chemical structure and concentration. They can inhibit trypsin, chymotrypsin, amylase, and lipase activities, reduce dietary protein quality, and interfere with dietary iron absorption (Lumen and Salamat 1980; Rao and Desothale 1998). Tannins also form insoluble complexes with proteins, which may explain the anti-nutritional effects of feeds containing tannins (Gemede and Ratta 2014) (Table 4.6). Tannins interfere with digestion by displaying anti-amylase activity and forming a complex with vitamin B (Liener 1980). Other adverse nutritional effects of tannins include intestinal damage and a possible carcinogenic effect, depression of feed intake, growth rate, feed efficiency, and microbial enzyme activities, including cellulose and intestinal digestion, as well as increased endogenous protein excretion, digestive tract malfunctioning, and toxicity of absorbed tannins or their metabolites. Tannins may form small digestive complexes with the feed antagonistic to arginine, interfere with RNA proteins, bind and inhibit endogenous proteins such as digestive enzymes, make proteins partially unavailable, and increase faecal nitrogen (Kumar and Singh 1984) (Fig. 4.4). Tannin-protein complexes include both hydrogen bonding and hydrophobic interactions. The protein–tannin complex's precipitation depends on the pH, molecular size, and ionic strength of tannins (Fig. 4.5). Both protein precipitation and incorporation of tannins in the precipitate increase as the tannins' molecular weight exceeds 5000 Da, and the tannins become insoluble and lose their ability to precipitate protein. The degree of polymerization then becomes necessary to assess the role of tannins in ruminant nutrition. CTs are responsible for the test-linked trypsin inhibitor activity of fava beans (Helsper et al. 1993).

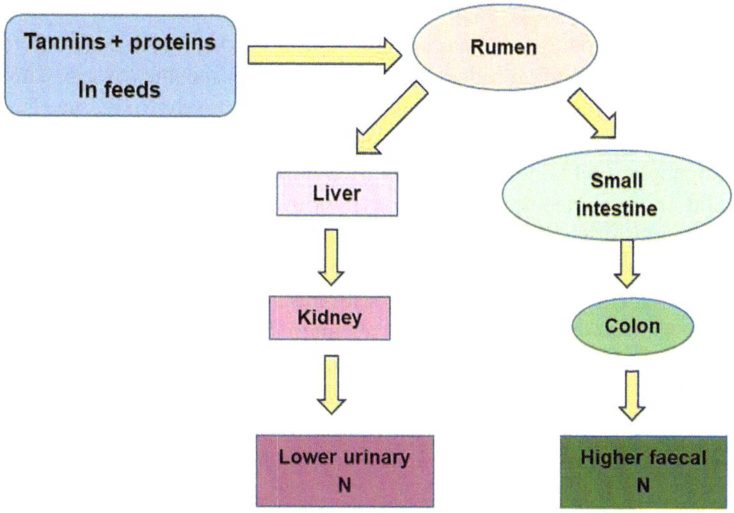

Fig. 4.4 Tannins that bind to dietary protein increase the nitrogen flux from the rumen to the small intestine

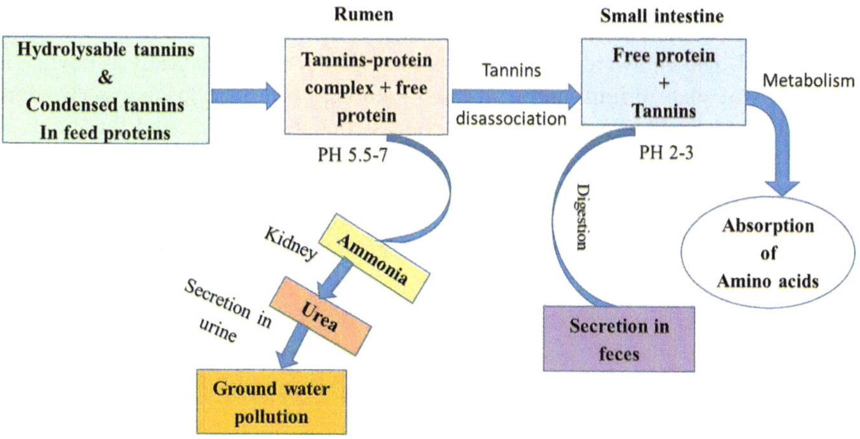

Fig. 4.5 Protection of feed proteins through tannin-protein complexes

4.3.1.4 Non-protein Amino Acids

Mimosine

Mimosine is a non-protein amino acid that is structurally similar to tyrosine. It is present in *Leucaena leucocephala*, in which the leaf mimosine level is approximately 2–6% and varies depending on season and maturity of leaves and stems. The main clinical symptoms of toxicity in non-ruminants include poor growth, reproductive problems, eye cataracts, and alopecia. When *L. leucocephala* is used as a feed meal for poultry, rabbit, or pigs, more than 5–10% of the meal generally causes poor growth and reproduction. In ruminants, mimosine toxicity causes poor body growth, poor wool development, depressed serum thyroxine levels, goiters, alopecia, dullness, swollen and raw coronets above the hooves, lesions in the mouth and oesophagus, and lameness (Table 4.6). Symptoms may be due to the mimosine metabolite in the rumen or 3,4-dihydroxypyridine.

Additionally, Jones et al. (1989) observed a diminished calving percentage in cows fed *L. leucocephala*. The problems caused by mimosine can be solved by genetically selecting *Leucaena* species with low mimosine content, but it has been noted that the low-mimosine species are unproductive and have low vigor; this problem can be solved by producing feed containing *Leucaena* mixed with other forages and concentrates. Hiremat (1981) suggested that the use of *Leucaena* as fodder can be limited to 50% of green forage for goats and 30% for cattle and buffalo. This strategy results in better livestock growth and production.

4.3.2 Factors Interfering with Minerals Utilisation

4.3.2.1 Phytic Acid

Phytic acid, also known as inositol hexakisphosphate, occurs naturally as phytate in feedstuffs of plant origin, and it acts as a storage form of phosphorus (Bedford 2000) (Fig. 4.6). Phytic acid is a phosphorus-containing compound that binds to

Fig. 4.6 Phytic acid and its basic structure

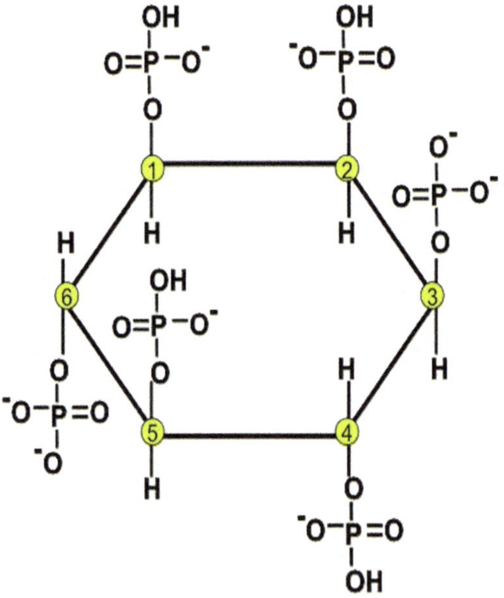

minerals and inhibits mineral absorption, resulting in decreased bioavailability of essential minerals, eventually turning them into insoluble compounds that are less readily absorbed and digested in the small intestine (Desphande and Cheryan 1984; Lott et al. 2000; Raboy 2000). Phytic acid is a ubiquitous secondary compound ranging from 0.1% to 6.0% among plant species, especially seeds, legumes, and cereals (Lolas 1976; García-Estepa et al. 1999; Lori et al. 2001; Loewus 2002; Margier et al. 2018). Phytic acid is primarily present as a salt of the mono- and divalent cations K^+, Mg^{2+}, and Ca^{2+}, and it accumulates in seeds during ripening (Maenz 2001).

Phytic acid is generally a negatively charged structure that binds to positively charged metal ions such as zinc, calcium, magnesium, and iron to form complexes and reduce these ions' bioavailability through lowered absorption rates. Phytic acid is one of the most effective anti-nutrients in animal feeds due to its chelating property. Its presence causes mineral ion deficiency in animal and human nutrition (Walter et al. 2002; Bora 2014; Grace et al. 2017) (Table 4.6). Since phytic acid, one of the strongest ANFs in plant feedstuffs, accumulates in seed storage sites, behaves as a chelating ligand with minerals and forms complex salt phytates; and can act as potent chelators that form protein and mineral-phytic acid complexes in a reduced bioavailability of protein and minerals (Erdman 1979). Most of the phosphorus contained within phytic acid is unavailable mainly to non-ruminants due to the absence of phytase in these animals' gastrointestinal tract (GIT). In chickens, there is a significant inverse relationship between phytic acid availability and the availability of phosphorus, magnesium, zinc, and calcium in feedstuffs, such as rapeseed, palm kernel seed, soybean meal, and cottonseed meal. Phytic acid, a highly negatively charged ion, works in a broad pH range and binds nutritionally important divalent cations in the diet such as iron, zinc, copper, magnesium, calcium, and molybdenum and endogenous GIT secretions such as digestive enzymes and mucins. This binding leads to the formation of insoluble complexes that are not readily absorbed by the GIT and increase the endogenous secretion of nutrients (Frontela et al. 2008; Woyengo and Nyachoti 2013) (Fig. 4.7). It also inhibits the action of GI tyrosinase, trypsin, pepsin, lipase, and amylase. Phytic acid is poorly hydrolyzed by non-ruminants (Woyengo and Nyachoti 2011, 2013). Most poultry does not have endogenous enzymes to break down phytate and release nutrients; thus, phytate transits undigested through the GIT (Fig. 4.8). This is also why high proportions of valuable nutrients from plant sources are not utilized by non-ruminants and are wasted in the excreta (Mueller 2001). Phosphorus bound to phytate is not bioavailable to non-ruminants. Ruminants, such as sheep and cows, chew, swallow, and then regurgitate their food; this regurgitated food is known as cud and is chewed a second time. Due to the phytase activity of rumen microorganisms, these animals can separate and process phosphorus into phytates (Haese 2017).

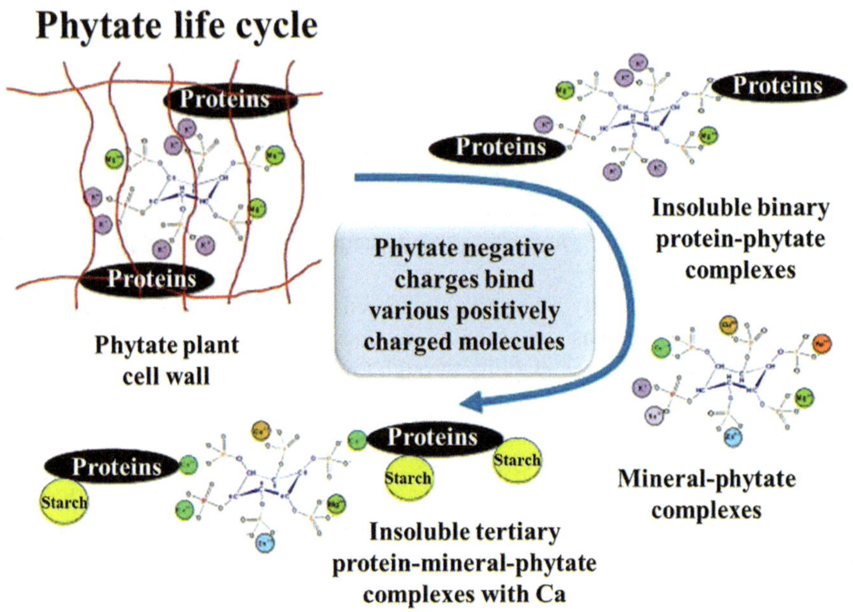

Fig. 4.7 Phytate life cycle

4.3.2.2 Gossypol

Gossypol is a phenolic compound with a yellow pigment, which is a complex of esters and ethers of various carbohydrates found in the pigment glands of plants of the genus Gossypium, family Malvaceae. It is present in two forms: free gossypol or bound form (Abbas 2020). Free gossypol contains aldehyde and phenolic groups, making it more reactive and toxic (Leeson and Summer 2001) (Fig. 4.9). Bound gossypol (BG) is not absorbed and is non-toxic. Gossypol is found in higher concentrations in cotton seeds (0.4–2.4%), and the average content of free gossypol in cottonseed meal is 0.01% (Liener 1980). Whole cotton seeds contain the highest amount of free gossypol. Cottonseed meal is a by-product of extracting cottonseed oil from whole seeds. Different extraction techniques significantly impact on the amount of free gossypol contained in cottonseed meal. The screw-press method uses heat that increases protein binding, thus converting more free gossypol (toxic form) into BG (non-toxic). Solvent extraction is widely used because more oil can be extracted. However, because heat is not used in solvent extraction, the amount of free gossypol content in cottonseed meal is approximately ten times higher than that in cottonseed meal processed by the screw-press method. This can be a considerable difference if there is much gossypol in the seed. This switch to solvent extraction explains the increase in gossypol toxicity in the past decades (Morgan 1989). Free gossypol is the most common anti-nutritional factor in cottonseed meal, primarily affecting the heart, liver, reproductive tract, and kidneys (Nagalakshmi et al. 2007). During cotton seed oil extraction, free gossypol binds to the epsilon amino group of

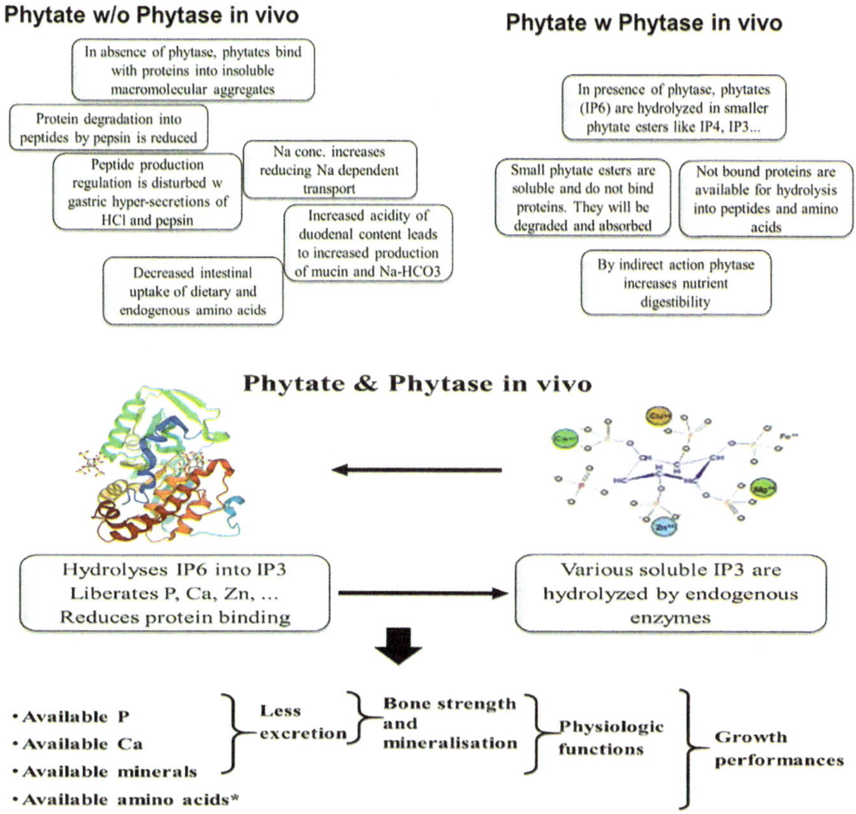

Fig. 4.8 Phytate with/without phytase in vivo

Fig. 4.9 Gossypol and its basic structure

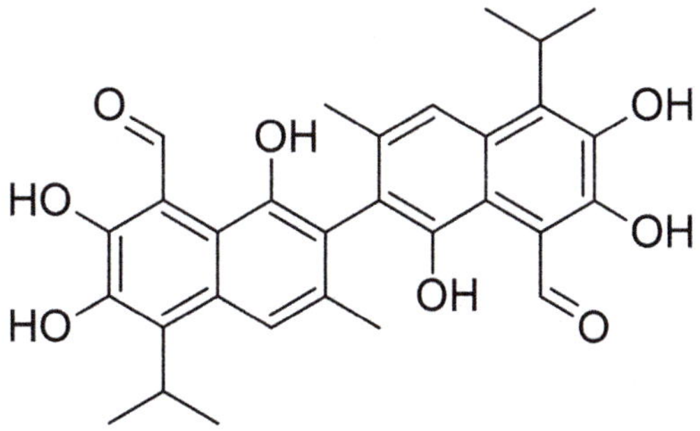

lysine, resulting in BG, which reduces lysine availability to animals. The amount of free gossypol in cottonseed meal can be variable. Many factors influence its content, such as cotton plant species, climatic and soil conditions, oil extraction methods, kernel to husk ratio, and seed coat (Nagalakshmi et al. 2007). This makes it impossible to know how much free gossypol the cottonseed meal contains without testing it. Hence, the quality of cottonseed meal is restricted by the free gossypol, its unbalanced digestible amino acid, and its high fluctuation in free gossypol concentration. Moreover, methionine, lysine, threonine, and valine deficiency in cottonseed meal protein causes a decrease in its digestibility, probably due to high cell wall components, which provokes a faster digestion passage rate due to gossypol binding to the soluble protein (Nagalakshmi et al. 2007). Gossypol makes an insoluble chelate with many essential elements such as iron and amino acids, hence reducing these nutrients' availability (Church 1991; Robinson 1991). Gossypol may reduce protein digestibility by binding to the free epsilon amino group of lysine during heat treatment, and the gossypol protein complex formed in cottonseed meal may render the adjacent peptides unavailable for proteolytic action. Gossypol inhibits the activity of important enzymes by binding to their free epsilon amino groups (Sharma et al. 1978). Non-ruminants have long been known to be susceptible to gossypol toxicity. Ruminants such as cattle and sheep can tolerate higher free gossypol levels because gossypol binds to proteins in the rumen. However, young calves and lambs are pretty susceptible to gossypol toxicosis. Although they are ruminants, their rumen is not fully functional and cannot bind as much free gossypol as the rumen of adult animals. General signs of gossypol toxicity are reduced appetite and the productive performance of animals, and causes contraception and infertility in animals (Leeson and Summer 2001), inhibition of haemoglobin synthesis by iron-binding, inhibition of respiratory enzymes resulting in difficulty breathing and cardiac arrhythmias (Ferguson et al. 1959; Skutches et al. 1973), reduction in the oxygen-carrying capacity of hemoglobin, and a decrease in the ratio of hemoglobin to red blood cells and decreased serum protein concentration. Dietary gossypol may also cause diarrhea, oedema of the body cavities, liver discoloration, and degeneration of myocardium, liver, and spleen (Church 1991; Olomu 1995). In poultry, free gossypol reduces production performance and causes leg weakness (Lordelo et al. 2007) and egg yolk mottling (i.e., olive green discoloration of yolk) (Davis et al. 2002) due to the interaction between gossypol and yolk iron, and may also harm blood biochemistry variables (Adeyemo 2008) (Table 4.6). Cotton seeds are rich in gossypol and can thus produce severe toxicity to farm animals; however, the cumulative effects of dietary gossypol and toxicity can occur after an ingestion period of 1–3 months (Patton et al. 1985; Kerr 1989; Soto-Blanco 2008; Gadelha et al. 2011). Gossypol toxicity has been reported in many species, including broiler chicks (Henry et al. 2001), pigs (Haschek et al. 1989), goats (East et al. 1994), and sheep (Morgan et al. 1988). Non-ruminants are more susceptible to gossypol toxicity than ruminants (Alexander et al. 2008; Kenar 2006; Randel et al. 1992; Zhang et al. 2007). Moreover, young ruminants are more sensitive to gossypol than adult ruminants (Soto-Blanco 2008) because gossypol is not bound during rumen fermentation, as it is in animals with fully functional rumen. However, if gossypol intake overwhelms the rumen's detoxification capacity, free gossypol may be absorbed in hazardous concentrations even in adult ruminants (Willard et al. 1995). The rate of gossypol absorption is inversely proportional to the amount of iron in the diet (Haschek et al.

1989); thus, dietary supplementation with ferrous sulfate inhibits free gossypol (Barraza et al. 1991). In ruminants, microbial fermentation in the rumen binds dietary free gossypol with proteins (Schneider et al. 2002). However, it is unknown whether the intestine can absorb the BG form or if microorganisms can release free gossypol from the bound form, as absorbed gossypol accumulates in the liver (Lindsey et al. 1980) and kidneys (Kim et al. 1996). The primary route of gossypol excretion is through the bile; it is then eliminated through faeces after conjugation with glucuronides and sulfates (Abou-Donia et al. 1989). Small amounts of gossypol are also excreted in expired air (Soto-Blanco 2008), and some gossypol is excreted in the milk (Lindsey et al. 1980).

4.3.2.3 Oxalates

Oxalate (oxalic acid) is a substance that can form insoluble salts with minerals such as Ca, K, Na, Mg, and Fe. These compounds are found in small amounts in both plants and mammals (Petroski and Minich 2020). Under normal conditions, oxalate is confined to separate compartments, but when it is processed and or digested, it comes into contact with the nutrients in the digestive system (Noonan and Savage 1999). When released, it binds with nutrients, rendering them unavailable to the body. If feed with excessive amounts of oxalic acid is consumed regularly, a nutritional deficiency is likely and severe irritation of the gut lining (Liebman and Al-Wahsh 2011). Strong bonds are formed between oxalic acid and many other minerals, such as Ca, K, Na, and Mg (Fig. 4.10). These chemical combinations lead to the formation of oxalate salts found in plants' soluble and insoluble forms. Soluble salts are formed when oxalate binds to Mg, Na, and K, while insoluble salts are produced when oxalate binds to Fe and Ca. Oxalate affects Ca and Mg metabolism and interacts with proteins to form complexes that inhibit digestion. The high content of soluble oxalate content prevents the absorption of soluble Ca ions, as oxalate binds to Ca ions to form insoluble Ca-oxalate complexes (Hamid et al. 2017). This renders Ca unavailable for maintaining healthy bones, as a cofactor in enzymatic reactions, the transmission of nerve impulses, and as a clotting factor in the blood (Table 4.6). Ca loss leads to bone deterioration, impaired blood clotting, and a disturbance in the absorbed Ca:P ratio, which leads to bone mineral mobilization to alleviate hypocalcemia; therefore, prolonged mobilization of bone minerals

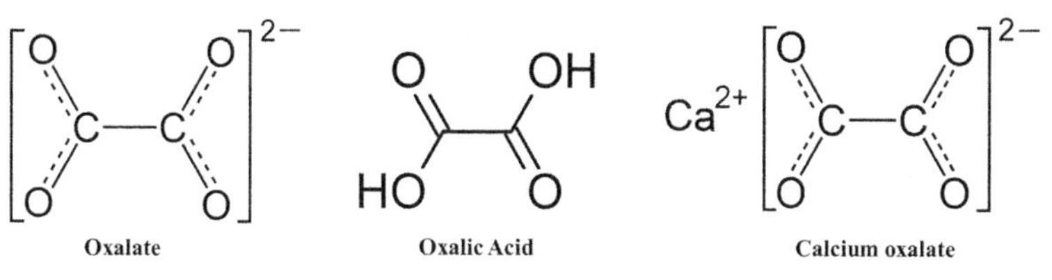

Fig. 4.10 Show the different chemical structure between oxalate, oxalic acid and calcium oxalate

leads to osteodystrophy fibrosa or hyperparathyroidism (Rahman and Kawamura 2011). In ruminants, oxalate is marginally significant as ANF because rumen microflora efficiently metabolize soluble oxalates (Gemede and Ratta 2014) and, to a lesser extent, even insoluble Ca-oxalate. If large quantities of oxalate-rich plants are ingested, the rumen is overwhelmed and cannot metabolize oxalates, poisoning the animal. A soluble oxalate level of 2% or more in forage grasses may cause severe poisoning in ruminants, but in non-ruminants, a level $<0.5\%$ is safe. However, these proposed safe levels of soluble oxalate should be considered preliminary (Rahman et al. 2013). Various tropical grasses, including pangola and buffel grasses, kikuyugrass, and *Setaria* grasses, contain soluble oxalates in sufficient concentrations to induce Ca deficiency in grazing animals. Young plants contain more oxalate than older ones (Jones and Ford 1972). There is a rapid rise in oxalate content during the early stages of growth, followed by a decrease as the plant matures (Davis 1981). The highest oxalate content in grasses occurs during rapid growth, reaching concentrations up to 6% of the dry weight (Cheeke 1996). Additionally, oxalate content can be manipulated by varying the harvesting interval, decreasing with an increased harvest interval (Rahman et al. 2009; Patel et al. 2013).

4.3.3 Anti-vitamins

Some anti-vitamin factors are found in plants, especially leguminous plants. Antivitamins are organic compounds that destroy specific vitamins, combine and form non-absorbable complexes, or interfere with digestive and/or metabolic functions (Ramteke et al. 2019) (Table 4.6). Antivitamin A in raw soybeans contains lipoxygenase enzymes that oxidize carotene, a precursor of vitamin A. Heating soybeans can destroy it for 5 min at atmospheric pressure. Antivitamin D is a rachitogenic factor in isolated soy protein (unheated). It interferes with the absorption of Ca and P in pigs and chicks, and it is destroyed by autoclaving. Antivitamin E is present in soybeans and alfalfa, and it causes muscle dystrophy and liver necrosis in lambs and chicks by reducing plasma vitamin E. It is similarly destroyed by autoclaving. Antivitamin K in sweet clover causes a fatal haemorrhagic condition in cattle known as sweet clover disease. Dicoumarol reduces the levels of prothrombin in the blood and affects blood clotting. Other anti-vitamins include anti-thiamine, also called thiaminase, which is found in cotton seeds, linseed, mustard seed, and mung bean, and anti-niacin, which is found in sorghum, maize, and wheat bran and causes perosis (chondrodystrophy) and growth depression. Additionally, anti-pyridoxine, also called linatine, has been identified as 1-amino-D-proline, and is naturally occurring with glutamic acid as a peptide, and can be destroyed after water treatment and autoclaving. Finally, anti-vitamin B12 is found in raw soybeans (Ramteke et al. 2019).

4.3.4 *Miscellaneous*

4.3.4.1 Cyanogens

When consumed, several plant species produce hydrogen cyanide from cyanogenic glycosides, which are sugar glycosides or polysaccharides that combine with cyanide and contain aglycone (Fig. 4.11). More than 2500 plant species have been reported to contain cyanogenic glucosides, including important staple foods such as sorghum, cassava, white clover, and linseed (Rosling 1987; Vennesland et al. 1982). Cyanogenic glucosides, or cyanoglycosides, represent approximately 90% of the broadest group of plant toxins known as cyanogens. The main feature of these toxins is cyanogenes, the formation of free hydrogen cyanide, which binds to cyanohydrins that have been stabilized by glycosylation (binding of polysaccharides) to form cyanogenic glycosides. In addition, cyanogenic glucosides are classified as phytoanticipins. Their function in plants depends on activation by glucosidases to release toxic volatile hydrogen cyanide and aldehydes or ketones to repel herbivory and pathogens (Zagrobelny et al. 2004). Hydrogen cyanide inhibits the cytochrome oxidase enzyme in the mitochondria of cells by binding to the Fe^{3+}/Fe^{2+} present in the enzyme, resulting in decreased tissue O_2 utilization, causing increased levels of blood glucose and lactic acid, and reducing the ATP/ADP ratio, indicating a shift from aerobic to anaerobic metabolism.

Moreover, many enzyme systems inhibit growth by interfering with certain essential amino acids and utilizing associated nutrients (Table 4.6) and causing severe poisoning, neuropathy, and death (Osuntokun 1972). Cyanide activates the glycogenolysis process and converts glucose in the pentose phosphate pathway, which reduces the rate of glycolysis and inhibits the tricarboxylic acid cycle. Cyanide then reduces the energy availability in all cells, but its effect is immediate on the respiratory system and heart. Since cyanide is highly toxic, it inhibits cytochrome oxidase, which is the last step in electron transport, thus inhibiting ATP synthesis. The most obvious symptom is death, but before dying, symptoms include faster and deeper respiration, a faster irregular and weaker pulse, salivation and frothing mouth, muscular spasms, dilation of the pupils, and bright red mucous membranes. The potential toxicity of cyanogen depends mainly on the potential concentration of hydrogen cyanide that may be released upon consumption. When

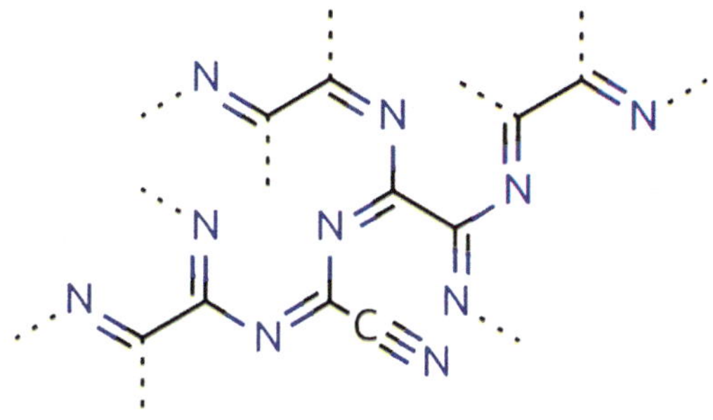

Fig. 4.11 Cyanogens and their basic structure

a cyanogenic plant is consumed, β-glycosidase is released during digestion and follows the known cyanide metabolic pathways and toxicokinetic processes in animals and humans (Oke 1979, 1980); it remains active until inactivated by a low gastric pH. After the catabolic intracellular enzyme β-glucosidase is released upon contact with glycosides, this enzyme breaks down cyanogenic glycosides releasing hydrogen cyanide, glucose, ketones, or benzaldehydes (Gonzales and Sabatini 1989; Rosling 1987; WHO 2010). The hydrolytic reaction can occur in the rumen by microbial activity. Hence, ruminants are more susceptible to cyanide toxicity than non-ruminants (Patel et al. 2013). After its absorption, cyanide is rapidly distributed in the animal body through the blood. It is known to combine with Fe in both methemoglobin and haemoglobin found in red blood cells, leading to an increased cyanide concentration in red blood cells compared with that of plasma. Cyanide is detoxified in the liver by the enzyme rhodanese, forming thiocyanate, which is excreted in the urine (Oke 1979, 1980). There is a range for the lethal dose of hydrogen cyanide in animals for various species (0.66–15 mg/kg body weight). For cattle and sheep, the range is 2.0–4.0 mg/kg body weight (Robson 2007).

Furthermore, a level greater than 50 mg/kg is harmful to poultry (Udedibie et al. 2004). Decreased growth and egg production rates have been observed in hens offered fed containing cyanide (Okafor and Okorie 2006), and the acute toxicity of hydrogen cyanide in rabbits occurs at 0.66 mg/kg body weight (EPA 1990). The presence of cyanogens in feed can also deplete sulfur-containing amino acids in birds, resulting in reduced protein synthesis and growth since an adequate amount of amino acids is one of the basic requirements for protein synthesis. The need to supplement cassava feed with methionine and cysteine (sulfur-containing amino acids) has been demonstrated for non-ruminant species (Maner and Gomez 1973).

4.3.4.2 Saponins

Saponins are a heterogeneous group of foam-producing triterpenes or steroidal glycosides that are widely distributed in nature, occurring primarily in the plant kingdom, including in pulses (chickpeas, beans, lentils, among others), oilseeds, groundnuts, lupin beans, sunflower, and alfalfa, with minor levels in other legumes such as soybeans, rapeseed, and various varieties of peas. The name 'saponin' is derived from the Latin word *sapo,* meaning 'soap' because saponin molecules form soap-like foams when shaken in water (Fig. 4.12). Saponins consist of non-polar aglycones coupled with one or more monosaccharide moieties (Oleszek 2002). This combination of polar and non-polar structural elements in their molecules explains their soap-like behavior in aqueous solutions. The structural complexity of saponins results in physical, chemical, and biological properties such as sweetness and bitterness, foaming and emulsifying, pharmacological and medicinal, hemolytic, antimicrobial, insecticidal and molluscicidal activities (Sparg et al. 2004; Gemede and Ratta 2014). Saponins have been recognized as ANF constituents because of their

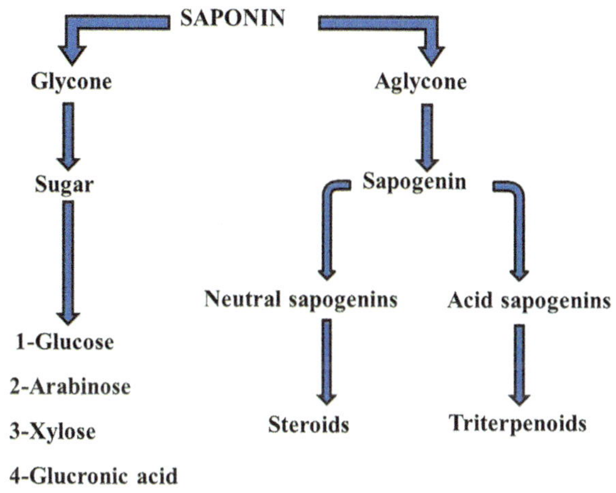

Fig. 4.12 Pathways of saponins synthesis

adverse health effects, including impaired growth and reduced feed intake owing to bitterness and throat irritation that they cause (Shi et al. 2004). Dietary saponins are poorly absorbed because they can form complexes with sterols, leading to harmful biological effects in the digestive system (Cheeke 1996). Saponins increase the permeability of intestinal mucosal cells, prevent active mucosal transport, and facilitate the uptake of substances that are not customarily absorbed (Johnson et al. 1986). In addition, they reduce the bioavailability of nutrients, decrease enzyme activity, and affect protein digestibility by inhibiting various digestive enzymes such as trypsin and chymotrypsin (Simee 2011). Saponins reduce the absorption of certain nutrients, including glucose and cholesterol, in the gut through an intra-lumenal physicochemical interaction (Table 4.6); hence, the effects of hypocholesterolaemia have been reported (Umaru et al. 2007). Additionally, saponins can form complexes with dietary Fe, rendering it unavailable for absorption (Southon et al. 1988). In fact, saponins have lytic action (the specific ability to form pores in membranes) on erythrocyte membranes, causing hemolysis in red blood cells (Seeman 1974). Since they reduce the surface tension of blood in cold-blooded animals, saponins have a highly toxic effect. They reduce growth performance in both poultry and swine. In chickens, saponins reduce growth and feed efficiency and interfere with the absorption of dietary lipids, vitamin A, and vitamin E (Jenkins and Atwal 1994). Compared with non-ruminants, poultry is more sensitive to saponins. Saponins increase the digestibility of carbohydrate-rich foods through a detergent-like activity that reduces viscosity, preventing the regular blocking action of such foods in the intestine. In general, saponins are of minor concern in non-ruminants because they are only present at low levels in common feedstuffs (Dublecz 2011).

4.3.4.3 Nitrates

Nitrates are ANF whose toxicity is associated with consumption of plants with high levels of nitrates, which causes health problems similar to bloating, sweet clover poisoning, or grass/winter tetany. Nitrate accumulates in plants damaged by frost, hail, drought, or even sudden cold and cloudy weather conditions. Also, high nitrates occur in forages subjected to excessive fertilization (Basso and Ritchie 2005). Affected plants must be grazed or harvested to avoid an adverse effect on livestock, especially ruminants. Nitrate toxicity is caused by excess nitrates in feeds, leading to a dangerous condition in ruminants due to a lack of O_2 in the bloodstream; death may result if not treated immediately. It is possible to treat nitrate toxicity, but it is challenging to identify animals with symptoms of this condition. Some forages such as Sudan grass, pearl millet (Andrews and Kumar 1992), and oats can accumulate nitrates at potentially toxic levels. Most nitrate is accumulated in the stem, followed by leaves, and very little in grains (Singh et al. 2000). Nitrate toxicity occurs mainly in ruminants when animals consume feed containing excess nitrate (NO_3^-), which is converted into nitrite (NO_2^-) by rumen bacteria (Fig. 4.13). Then, NO_2^- crosses the rumen wall and enters the bloodstream, where it combines with hemoglobin to form methemoglobin, which hampers the ability of red blood cells to carry O_2 into body tissues. NO_3^- at low levels in forages is converted into ammonia by bacteria in the rumen (Lee and Beauchemin 2014). This process is one of detoxification because NO_2^- is ten times more toxic than NO_3^-. This detoxification process occurs more slowly than the conversion of NO_3^- into NO_2^-. When microbes that convert NO_2^- to ammonia are overwhelmed by high NO_2^- levels in the rumen, toxicity will occur. NO_3^- and NO_2^- pass through the rumen wall and interfere with Fe ions in the red blood cells, and the ferrous Fe of hemoglobin turns into the ferric form; thus, forming methemoglobin. This metalloprotein, in which the iron in the heme group is in the Fe^{3+} (ferric) state, not the Fe^{2+} (ferrous), does not have the same O_2 carrying capacity as hemoglobin. So the tissues do not get enough O_2 and, thus, suffer from O_2 deprivation (Fig. 4.14). The blood turns to bluish chocolate-brown color rather than the usual bright red. An animal that dies from NO_3^- (NO_2^-) poisoning actually dies from a lack of O_2 (asphyxiation) (Benjamin 2006) (Table 4.6). When forages contain an unusually high level of nitrate, animals cannot complete the conversion process, consequently accumulating the nitrite (Table 4.5). A positive feedback loop occurs if the animals consistently have access to a high-nitrate

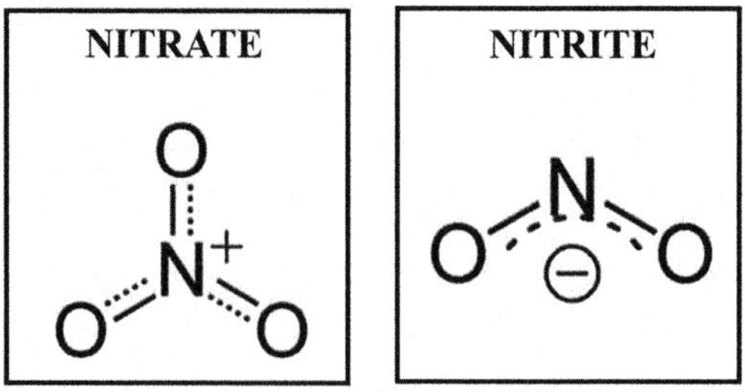

Fig. 4.13 Show the different chemical structures between nitrate and nitrite

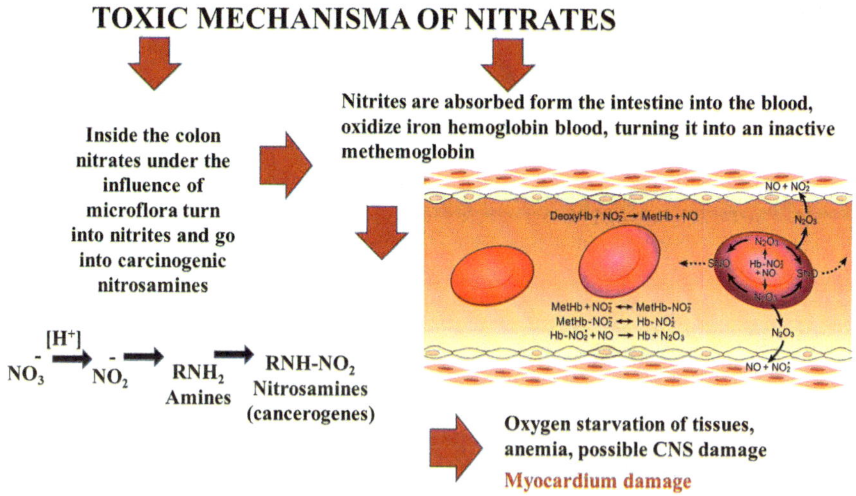

Fig. 4.14 Mechanism of nitrate toxicity

Table 4.5 Level of nitrate in forage (as DM basis) and potential effects on animals

Nitrate content (ppm)	Effect on animals
0–1000	This level is considered safe to feed under all conditions
1000–1500	This level should be safe to feed to non-pregnant animals under all conditions. It may be best to limit its use to pregnant animals to 50% of the total ration
1500–2000	Feeds are fed safely if limited to 50% of the ration
2000–3500	Feeds should be limited to 35–40% of the ration. Feeds containing over 2000 ppm nitrate- nitrogen should not be used for pregnant animals
3500–4000	Feeds should be limited to 25% of the ration. Do not use it for pregnant animals
> 4000	Feeds containing over 4000 ppm are potentially toxic. Do not feed

Adapted from Andrae (2008)

feed. While NO_3^- in the bloodstream, which does not initially cause toxicity problems, can be recycled back into the rumen via saliva or intestinal secretions and converted into NO_2^-. High-nitrate feed exacerbates the problems because NO_3^- is constantly being flooded into the system and either rapidly turns into NO_2^- in the rumen or enters the bloodstream to be recycled back into the rumen and reabsorbed into the blood as NO_2^-. The rate and quantity of forage consumption, type of forage, energy level, or diet adequacy are the factors that affect the severity of nitrate poisoning. Sheep, goats, and other ruminants are likely to suffer from NO_3^- toxicity as cattle. Sheep are the least sensitive to NO_3^- toxicity compared with cattle, which are the most sensitive. Sheep convert methemoglobin to hemoglobin and NO_2^- to ammonia more efficiently than cattle; therefore they can be safely fed feeds with a higher NO_3^- content (Benjamin 2006). Non-ruminants are unlikely to be affected by nitrate poisoning because NO_3^- is primarily converted into NO_2^- in the intestine,

Table 4.6 Deleterious and beneficial effects of some anti-nutrients on livestock

Anti-nutritional factors	Deleterious effects	Beneficial effects
Enzyme's inhibitors	Reduce protein digestion and absorption Disturb digestive functions Reduce bioavailability of minerals Decrease the growth Cause diarrhea or excessive gas Pancreatic hypertrophy Animal growth depression	Decrease the incidences of pancreatic cancer Act as anti-carcinogenic agents
Lectins (Haemagglutinins)	Impair animal growth Caused damage to the epithelium of the small intestine Reduced nutrient absorption Cause damage to the gastrointestinal tract Cause endogenous loss of nitrogen and protein utilization Increased protein catabolism, Breakdown of stored fats and glycogen Disturbance in mineral metabolism Stimulate hypertrophy and hyperplasia of the pancreas Cause hemagglutination and anemia Allow bacteria and their endotoxins to enter the bloodstream and cause a toxic response	
Tannins	Decreased growth rate Decrease bioavailability of amino acids Reduce protein digestibility Decreased palatability and feed intake A less digestible complex with dietary proteins Inhibit the endogenous protein such as digestive enzymes Interfere with dietary iron absorption Have the ability to complex with vitamin B	Show antioxidant, antibacterial, anti-diarrhea, free-radical scavenging, anti-proliferative activity Reduce protein degradation during ensilage Increase protein utilization efficiency Reduce parasite burden on gastrointestinal tract Prevent bloating Reduce N emissions into the environment Reduce methane emissions Increase animal product quality Improve live weight gain, reproductive efficiency, and wool production in sheep Increase amino acids absorption in the small intestine

(continued)

Table 4.6 (continued)

Anti-nutritional factors	Deleterious effects	Beneficial effects
Mimosine	Act as an amino acid Lead to inactivation of the catalytic, transaminases, Complicated with metal such as zinc Cause poor body growth, poor wool development Depress serum thyroxine level and goiter alopecia, dullness, swollen and raw coronets above the hooves, lesions of the mouth and esophagus, and lameness	
Phytic acid	Chelating of mineral cofactors or interacting with a protein Reduce mineral bioavailability and inhibit their absorption Inhibit the action of gastrointestinal tyrosinase, pepsin, trypsin, amylase, and lipase	A powerful natural antioxidant Reducing lipid peroxidation Reduce cholesterol Protect against cancer Against foodborne pathogen Coronary heart disease
Gossypol	Reduce protein digestion Reduce the availability of lysine to the non-ruminants Decreased appetite, weight gain, leg weakness, egg production, and egg size, decreased egg hatchability, caused egg yolk mottling, olive-green discoloration of egg yolk in poultry Decrease hemoglobin, total red blood cell count, protein and albumin to globulin ratio in blood serum	
Oxalates	Form insoluble calcium oxalate Negatively affect the absorption and utilization of calcium in the animal body Cause severe irritation to the lining of the gut	
Cyanogens	Depress growth Interfering with essential amino acids and utilization of nutrients Inhibit many enzyme systems Cause severe toxicity, neuropathy, and death Reduce energy availability in all cells Inhibit cytochrome oxidase Depression in birds growth and small eggs production Deplete the sulfur-containing amino acids for birds resulting in reduced protein synthesis Respiratory inhibitors	

(continued)

Table 4.6 (continued)

Anti-nutritional factors	Deleterious effects	Beneficial effects
Saponins	Reduce growth rate, bloat (ruminants) Reduce the absorption of nutrients (monosaccharide, glucose, and lipids) Hemolysis in erythrocytes Low blood and liver cholesterol Inhibition of smooth muscle activity Alter the integrity of intestinal epithelial cells Inhibit microbial fermentation and synthesis in the rumen	Antibacterial and antiprotozoal, antioxidants, antitumor property Lowering cholesterol Immune potentiating Bind ammonia and hydrogen sulfide, thus improving air quality in poultry houses Reduced risk of coronary heart diseases Form of first collagen that has a role in the wound-healing process (hydrocarbon ointment)
Nitrate	Nitrite poisoning Convert haemoglobin in the blood to methaemoglobin (blood turns to a chocolate brown color) Animal death from asphyxiation	
Alkaloids	Cause gastrointestinal and neurological disorders Cause infertility	
Anti-vitamins	From non-absorbable complexes Interfere with digestive and/or metabolic functions Anti-vitamin A Anti-vitamin D interferes with the absorption of calcium and phosphorous in chicks and pigs, Anti-vitamin E causing liver necrosis and muscle dystrophy in chicks and lambs Anti-vitamin K causes a fatal haemorrhagic condition in cattle (known as sweet clover disease) Anti-niacin causes Perosis and growth depression	

closer to the end of the digestive system, thus reducing the chance for NO_2^- being absorbed into the bloodstream. However, NO_3^- poisoning in non-ruminants is still possible, but it is not as severe as in ruminants (Okafor and Okorie 2006). In ruminants, NO_3^- toxicity is most commonly reported in ruminants grazing fresh herbage. Due to insufficient data on nitrite levels in the most common livestock diet feeds, reliable exposure estimates can be calculated. The highest estimated dietary exposure of cattle to NO_3^- from feed was for beef cattle fed a grass silage-based diet (53 mg/kg body weight/day). For sheep and goats, the categories 'lactating sheep' and 'goats for fattening' had the highest estimates of exposure to NO_3^- from the silage-based diet, at 46 and 60 mg/kg body weight/day, respectively. In non-ruminants, exposure estimates are low (from the average upper limit 5.6 mg/kg body weight/day in laying hens). However, this may be underestimated due to the lack of data on key ingredients in their diets (EFSA 2020). Risks of nitrate poisoning exist, mainly when animals are not accustomed to consuming nitrate-containing feeds. When animals are introduced to these feeds slowly over time, they can slowly adapt to a feed intake with at least 1% nitrates; it is important to note that this introduction must be prolonged for feeds high in nitrates. Acute toxicity signs and symptoms include a rapid and weak heartbeat, an abnormal body temperature, muscle tremors, weakness, and ataxia. Additionally, brown/bluish-grey mucous membranes, excessive salivation, lacrimation, labored and rapid breathing, frequent urination, vomiting (more common in non-ruminants), diarrhea or scouring, and an inability to get back up from laying down are all common symptoms. Death follows within a few hours of feeding cattle with a nitrate-rich forage. Animals die from asphyxiation due to a lack of O_2 in the body tissues. Subacute nitrate poisoning often corresponds to decreased weight, decreased feed intake, decreased milk production, increased susceptibility to infections and diseases, and reproductive problems such as silent heats and reduced fertility (Lee and Beauchemin 2014).

4.3.4.4 Alkaloids

Alkaloids are common groups of chemical compounds synthesised by plants from amino acids. They are generally found as salts such as malic, oxalic, citric, or tartaric acids (Yilkal 2015), which are small organic molecules found in 15–20% of all plants. Usually, alkaloids consist of several carbon rings with side chains replaced by one or more carbon or nitrogen atoms. Decarboxylation of amino acids produces amines that react with amine oxides to form aldehydes. The distinctive heterocyclic ring in alkaloids undergoes Mannich-type condensation of aldehyde and amine groups (Felix and Mello 2000; Yilkal 2015). The chemical type of their nitrogen ring provides how the alkaloids are sub-classified; for example, glycoalkaloids (aglycone fraction) glycosylated with a carbohydrate moiety are formed as metabolic by-products. Plants repel insects and herbivores with alkaloids due to their potential toxicity and bitter taste (Fereidoon 2012, 2014; Yilkal 2015). Lupins contain high alkaloids, specifically quinolizidine alkaloids, while soybeans and linseed may be contaminated by *Datura stramonium*. Linseed also contains scopolamine and hyoscyamine alkaloids (Dublecz 2011).

Potato tubers naturally contain the two toxic and bitter glycoalkaloids, a-solanine, and a-chaconine. The levels are usually low and have no adverse effects on food safety or culinary quality. However, consumption of potato tubers that are unusually high in glycoalkaloids has sometimes been associated with severe toxicity, including gastrointestinal and neurological disturbances, and the disruption or inappropriate augmentation of electrochemical transmission (Fernando et al. 2012). In tubers, glycoalkaloid levels are inheritable and vary significantly between different species. Environmental factors to which tubers are exposed during germination, growth, harvest, and storage may further affect glycoalkaloid levels (Jadhav et al. 2009). Indeed, the physiological effects of alkaloids on humans and animals are quite evident (Gemede and Ratta 2014). Consuming a high dose of tropane alkaloids accelerates the heartbeat, causes paralysis and fatality. Ingesting a high dose of tryptamine alkaloids causes a staggering gait and death (Fernando et al. 2012; Gemede and Ratta 2014). Alkaloids are oxidized in the liver to produce metabolites, such as dehydrosparteine, responsible for the observed toxicity. The level of toxicity is affected by the alkaloid composition. There is a high degree of variation in the ability of different animal species to deal with these compounds. Toxicity by alkaloids and their metabolites is mainly mediated through the nervous system, although they also stimulate the liver cells to absorb copper, resulting in copper toxicity. Pigs appear to be more sensitive to alkaloids than poultry. Glycoalkaloids significantly inhibit cholinesterase, and this also causes symptoms of neurological disorders. Alkaloids cause gastro intestinal and neurological disorders (Aletor 1993; Stegelmeier et al. 2020). Coumarins, which are feed components, have been associated with hemorrhagic disease in cattle that consume spoiled sweet clover. Alkaline pH conditions generally improve glycoalkaloid absorption, as binding to sterols in cell membranes leads to extra disruption. Nicotine (tobacco), cocaine (leaves of coca plant), caffeine, quinine (cinchona bark), morphine (dried latex of opium poppy), and solanine (unripe potatoes and potato sprouts), and strychnine are well-known examples of alkaloids. Pyrrolizidine alkaloid toxicity causes liver damage in chickens that initially might not show any clinical signs. The symptoms may appear very vague and are often confused with other diseases. Toxicity occurs in chickens of all ages and breeds, but not all flock members show signs of liver damage. Clinical symptoms of pyrrolizidine alkaloid poisoning include loss of appetite, lethargy, neurobehavioral abnormalities, and excessive drinking.

4.4 Mechanism of ANFs Toxicity

Protease inhibitors reduce protein digestion, so when legumes are eaten either raw or without being properly cooked, they disturb digestive functions and cause diarrhea or excessive gas (Thakur et al. 2019). Feeding animals with raw soybean or isolated soybean inhibitors causes a pancreatic enlargement in susceptible animals, which can be characterized histologically as hypertrophy (i.e., an increase in the size of pancreatic acinar cells) (Friedman and Brandon 2001); this is accompanied by an increase in the secretion of digestive enzymes including trypsin, chymotrypsin, and elastase. This supports the hypothesis that the growth inhibition caused by trypsin inhibitors results from an endogenous loss of amino acids in the form of enzymes secreted by a hyperactive pancreas. Pancreatic enzymes, such as trypsin and chymotrypsin, are particularly rich in sulfur-containing amino acids. Hence, the outcome of pancreatic hyperactivity is converting these amino acids from body tissue protein synthesis to the synthesis of these enzymes, which are subsequently lost in feces (Friedman and Brandon 2001). Trypsin inhibitor-induced pancreatic hypertrophy/hyperplasia observed in susceptible animal species has been explained by an adverse reaction mechanism where enzyme secretion is inversely related to the level of trypsin present in the small intestine (Liener 1994). Therefore, when the level of active trypsin in the gut is reduced due to the inhibitor's presence, the pancreas compensates by producing more enzymes. Mimosine's function is unclear, but it may act as an amino acid, lead to inactivation of the catalyst (transaminase), or bind to a mineral such as zinc (Hiremat 1981). To overcome the mimosine problem, *Leucaena leucocephala* should be restricted to 50% green forage for goats and 30% for cattle and buffalo (Hegarty 1987). The effects of tannins come from their ability to form strong H-bonds with nutrients, resulting in the inhibition of digestive enzymes and microbial activity in the rumen (Kumar and Singh 1984). These effects can be significantly increased with an increase in tannin molecules. It is well known that tannins are potential protein precipitants (Ahmed et al. 1991) and reduce animal protein digestibility (Salunkhe et al. 1990; Jansman 1993). The increase in faecal nitrogen associated with the ingestion of tannin-containing feed is largely due to interactions between tannin and dietary proteins or tannin and digestive enzymes, or both (Jansman 1993; Kelln et al. 2021). In a study by Ahmed et al. (1991), diets containing tannins, mostly hydrolyzable gallotannins, were fed to broilers at 13.5, 25, and 50 g/kg to verify their effects on enzymes in the pancreas, intestinal lumen, and intestinal mucosa. Pancreatic weight showed a significant increase with an increased dietary tannin level. The activities of trypsin and α-amylase in the pancreas of birds fed the highest level of tannins were more than double those of birds fed a tannin-free diet. In the intestinal lumen, the inhibition of trypsin activity increases with an increase in dietary tannin level. Likewise, dipeptidase and sucrose α-glucosidase were inhibited by tannins in the intestinal mucosa. Protein digestion and bird growth were negatively affected by tannin-containing diets. Similarly, feeding pigs with fava bean hulls high in tannins significantly reduced aminopeptidase activity in the jejunal mucosal homogenates in young piglets (Van Leeuwen et al. 1995). Low aminopeptidase activity was associated with decreased protein digestibility. Pancreatic enlargement caused by diets containing tannins may be mediated by hormones transported in the blood (Ahmed et al. 1991). The pancreatic enlargement caused by dietary tannins has also been reported in response to dietary trypsin inhibitors (Liener 1994). This might indicate a common mode of action for these ANFs, at least on the cellular level. The consumption of tannin-rich sorghum, CT, which has been isolated and purified from sorghum, or tannic acid, increase the size of the parotid glands, synthesis and secretion of proline-rich proteins (Mehansho et al. 1992), and the synthesis of proline-rich proteins secreted with saliva and associated with dietary tannins in the oral cavity to protect dietary protein. The association of tannins with dietary and endogenous proteins, such as digestive enzymes and proteins located on the luminal side of the intestine, has been used to explain the reduced digestion of protein in diets containing tannins (Jansman et al. 1994). There is no clear evidence for systemic effects in animals after they have been intensively fed with CT. It is hypothesized that CTs are resistant to intestinal degradation and are poorly absorbed due to forming a less digestible compound with dietary proteins. They may bind and inhibit endogenous proteins, including digestive enzymes.

These compounds are astringent and negatively affect animal feed intake (Patel et al. 2013). CT concentration above 4% of diet is toxic to ruminants because they are resistant to microbial attack and are harmful to various microorganisms (Waghorn 2008), resulting in reduced palatability, feed intake, growth rate, utilization, and iron absorption (Roeder 1995). Phytate can adversely affect digestive enzyme activity by chelating mineral cofactors or interacting with proteins at an acidic or alkaline pH (Ryden and Selvendran 1993; Khare 2000). For full activity, some digestive enzymes require metal cofactors, such as calcium, zinc, copper, magnesium, iron, and molybdenum. For example, these enzymes include α-amylase, carboxypeptidases, aminopeptidases, and alkaline phosphatase. Phytate binding to proteins may directly form phytate-protein complexes or indirectly form phytate-cation-protein complexes. These processes may differ according to pH, time, and relative concentrations (Ryden and Selvendran 1993). At the low pH in the stomach, the positively charged side-groups of protein essential amino acids can bind to the negatively charged phytate due to strong electrostatic interactions (Cheryan 1980; Deshpande and Cheryan 1984). Above its isoelectric point, the protein carries a net negative charge. A multivalent cation bridge (which includes calcium) appears to be involved in the complex formation between phytate and proteins. Phytate–cation–protein interactions are expected to be predominant at high pH in the small intestine (Selle et al. 2000). Another indirect mechanism for phytate inhibition of digestive enzyme activity measured in vitro has been proposed to include complex interactions between phytate, digestive enzymes, and other proteins present in solution (Li et al. 1993). These interactions also inhibit the action of gastrointestinal tyrosinase, pepsin, trypsin, amylase, and lipase (Khare 2000). Oxalic acid binds with calcium to form insoluble calcium oxalate, which negatively affects the absorption and utilization of calcium in the bodies of animals (Akande et al. 2010). Gossypol binds to proteins and/or to a group of free essential amino acids. In particular, gossypol binds to lysine in cottonseed meal, resulting in BG, which is less toxic to non-ruminants than is free gossypol. The free and BG content in the meal varies with the cultivar and the type of treatment. Gossypol reduces protein digestion in two ways. First, by binding to free epsilon amino group of lysine during heat treatment and the gossypol-protein complex formed in the meal makes the adjacent peptides unavailable for proteolytic action. Second, gossypol may directly affect certain enzymes in the gastrointestinal tract, such as pepsinogen, pepsin, and trypsin, by binding to the free epsilon amino groups (Sharma et al. 1978). Gossypol toxicity in poultry results in decreased appetite, weight loss, leg weakness, reduced egg production, and egg size, egg yolk discoloration, and decreased egg hatchability, hemoglobin, total red blood cell count, protein, and albumin-to-globulin ratio in blood serum (Waldroup and Goodner 1973; Suryawanshi et al. 1993). Saponins decrease the absorption of certain nutrients, including monosaccharides, glucose, and cholesterol in the gut through intraluminal physical and chemical interactions; thus, they have been reported to have hypocholesterolemic effects (Umaru et al. 2007). Additionally, saponins have distinctive foaming properties, as observed in white clover and alfalfa. Saponins can negatively affect animal performance and metabolism in several ways: hemolysis in erythrocytes, low blood and liver cholesterol, reduced growth rate and bloat in ruminants, inhibition of enzyme and smooth muscle activity, reduced absorption of nutrients, and inhibition of microbial fermentation and synthesis in the rumen (Akande et al. 2010). However, saponins have diverse biological effects due to structural differences in sapogenin fractions. Finally, some plant alkaloids have been reported to cause digestive and nervous disorders and infertility (Aletor 1993; Olayemi 2010), while glycoalkaloids (solanine and chaconine), found in potatoes and *Solanum* spp. (Saito et al. 1990; Aletor and Adeogun 1995) are toxic to humans (Table 4.6).

4.5 Methods of Reducing the Deleterious Effects of ANFs

The abundance of ANFs and consequent toxic effects in the plant-based diets of animals is a cause for concern, and ways to reduce their levels should be explored (Soetan and Oyewole 2009). Removing undesirable components is essential for improving a plant-based diet's nutritional quality and effectively utilizing its full potential as a feed source for livestock. It is widely accepted that simple and inexpensive processing techniques effectively achieve desirable changes in feed ingredient composition (Akande and Fabiyi 2010). Uhegbu et al. (2009) reported the effects of several methods tested to overcome ANFs and their harmful effects in various browses, grains, seeds, and agro-industrial by-products. These methods include mechanical or physical techniques (e.g., processing, wilting, and ensiling) and biological or chemical techniques (e.g., treatment with alkalis, organic solvents, and precipitants). In general, of the different processing methods that are used to reduce levels of various ANFs (soaking, boiling, and toasting) (Balogun 2013), soaking is one of the most common methods to lower trypsin inhibitors (from 13.8 to 9.4 TIU/100 g), phytates (from 0.18% to 0.09%), tannins (from 0.23% to 0.09%), saponins (from 0.42% to 0.24%), hydrogen cyanide (from 8.6% to 5.7%), and alkaloids (from 0.34% to 0.28%) -(Nwosu 2010). Boiling, simmering, and blanching reduce the cyanide content in *Moringa oleifera* leaves by 88%, 81%, and 62%, respectively (Sallau et al. 2012). Owing to the water solubility of oxalates, wet treatment methods such as boiling and steaming produce the highest oxalate reduction (Mada et al. 2012; Petroski and Minich 2020). Autoclaving seeds for 20 min reduce phytic acid by 24.7%, while roasting can reduce phytic acid content by 23.1–28.6% (Embaby 2011). Amaefule and Onwudike (2000) and Kankuka et al. (2000) reported that cooking legumes improve their nutritional value by destroying most ANFs and improving protein and energy availability. Autoclave treatment or boiling also reduces the content of protease inhibitors. Ramteke et al. (2019) reported that the trypsin inhibitor activity of soybean meal was destroyed by autoclaving under specific conditions (i.e., 5, 10, and 15 psi for 45, 30, and 20 min, respectively) or by exposure to steam for 60 min. Likewise, a longer boiling time (40 min), autoclaving (20 min), and microwaving (12 min) cause complete disruption of trypsin inhibitor activities (Sallau et al. 2012). Although lectins are usually degradable, their stability varies among plant species, as many lectins are resistant to dry heat inhibition and require moisture for destruction (Boehm and Huck 2009; Ramteke et al. 2019) or can quickly disintegrate by hydrothermal treatment (100 °C for 10 min) or autoclaving (Grant 1991). Physical treatments such as heat and chemical treatments or supplementation with amino acids or mineral ions such as Zn, Fe, and Al reduce mimosine's toxicity (Hiremat 1981). Other processing methods, like germination followed by dehulling, reduce tannins by 43–52% and phytic acid by 47–52% (Ghavidel and Prakash 2006). Polyethylene glycol is the most frequently used reagent to neutralize the secondary compounds in tannin-rich diets for animals. It can be used in various ways, such as applied in concentrate or feed blocks, sprayed on feed, or dissolved in water. Additionally, polyethylene glycol is an effective supplement for increasing feed intake, digestion, daily weight gain, and the synchronized, fractionated, and balanced supply of essential nutrients for rumen microflora and host animals fed on diets rich in tannins (Mueller 2001; Patel et al. 2013). Fermentation can reduce some anti-nutrients in feed, such as phytic acid and tannins (Sarangthem and Singh 2013; Singh et al. 2017). This method could also improve the nutritional value of cottonseed meal and increase the lysine and methionine content of cottonseed meal when fermented with Aspergillus oryzae NRRL 506 *Aspergillus Janus* NRRL 1935 for 48 h (Nagalakshmi et al. 2007). One of the recent trends in reducing free gossypol content in cottonseed meal is to produce varieties of cotton plants through genetic modification; however, other cotton seed processing methods such as pelleting, extrusion, cooking, and $Ca(OH)_2$ can be used to improve the nutritional value of cottonseed meal for poultry (Nagalakshmi et al. 2007). Treatment with iron 1: 1 ratio can remove 80–99% of gossypol; moreover, high protein content in the meal is also helpful in reducing the effect of gossypol (Leeson and Summer 2001). Moreover, using exogenous phytases to enhance phosphorous digestibility is now common practice where animal agriculture's contribution to environmental pollution is of concern. The phytase enzyme releases phosphorus, bound minerals, and amino acids from phytate, increasing nutrient utilization. Evidence indicates that phytases enhance ileal amino acids, which increase the body's nitrogen, calcium, and phosphorus retention and increase fecal energy digestibility in poultry (Woyengo and Nyachoti 2011). The use of gamma irradiation to reduce anti-nutrients in canola meal causes a significant increase in its nutritional value for broilers (Gharaghani et al. 2008). Finally, the use of the electron beam radiation method reduces hydrocyanic, phytic, and tannic acids.

4.6 Beneficial Effects of Anti-nutritional Factors

The potential beneficial effects of protease inhibitors remain unclear. However, a decrease in the incidence of pancreatic cancer has been observed in a population where the consumption of soybean and its products was high (Giri and Kachole 1998) (Table 4.6). Additionally, protease inhibitors have been associated with pancreatic cancer in animal studies, and they may act as anti-carcinogenic agents (Chunmei et al. 2010). It is not necessary to completely eliminate anti-nutritional factors from plant-based diets such as that of *Moringa oleifera* leaves because low amounts of ANFs, such as tannins and hydrolyzable phenols, may act as antioxidants in animal feed and may not only help improve meat quality but also reduce methane emissions (Su and Chen 2020) and gastrointestinal nematodes (Naumann et al. 2017) from ruminants. The benefits of tannins in animal feed and health include increased protein utilization efficiency, amino acid absorption (Hervás et al. 2003), antioxidants, antibacterial and anti-diarrhoeal effects, free-radical scavenging, and anti-proliferative activity (Corder 2006). Also, tannins reduce the impact of gastrointestinal parasitism, protein degradation during ensilage (Reed et al. 1990), prevent bloating, increase animal product quality, reduce N emission to the environment and promote rumen defaunation as a CH_4 mitigation strategy (Animut et al. 2008; Boadi et al. 2004), and reduce methane emissions from rumen fermentation (Waghorn 2008) (Table 4.6). CTs help to improve weight gain, reproductive efficiency, and wool production in sheep, reduces the effect of gastrointestinal parasitism, and lessens methane emissions from rumen fermentation (Waghorn 2008), allows dietary protein to bypass the rumen for digestion in the lower gastrointestinal tract (Hassanpour et al. 2011), and shows a bacteriostatic effect on *Salmonella enteritidis* infection in broiler chickens and can reduce excretion of these bacteria (Redondo et al. 2013a). Both HT and CT have antimicrobial activity against *Campylobacter jejuni* (Anderson et al. 2012; Gutierrez-Banuelos et al. 2011). The *in vivo* effects of tannins in the necrotic enteritis model reduce the incidence and severity of gross lesions and improve broilers' productive performance (Redondo et al. 2013b). Tannic acid can treat diarrhea because it causes constipation without inflammation (Phytolab 2007).

Phytic acid has been suggested as a store of cations, high-energy phosphoryl groups, and free iron chelates, making phytic acid a potent natural antioxidant (Mueller 2001). In addition, phytic acid can reduce cholesterol and protect against iron-induced intestinal cancer. Furthermore, phytic acid exhibits natural antioxidant properties, providing other benefits such as reducing lipid peroxidation (Table 4.6). Of the many plant compounds, saponins have beneficial biological effects, including antibacterial, antiprotozoal (Avato et al. 2006), antioxidant, antitumor, cholesterol-lowering, immune-potentiating, and anticancer (Blumert and Liu 2003) effects. Furthermore, saponins reduce the risk of coronary heart diseases (Ferri 2009) and the probability of forming collagen, a protein with a role in wound healing that can also be used as a hydrocarbon ointment. Saponins attract considerable interest due to their beneficial effects in the poultry industry; they can bind ammonia and hydrogen sulfide, thus improving air quality in poultry houses (Tab*le* 4.6).

4.7 Conclusions

Anti-nutritional factors (ANFs) in plant-based protein diets for livestock may reduce their full utilization. Thus, to justify the nutritional value of any plant-based protein source, it is imperative to appropriately assess the nature, type, and concentration of the ANFs present in the diet and the bioavailability of the required nutrients. Basic information about the most common factors found in plants used for animal feed includes protease inhibitors, amylase inhibitors, lectins, tannins, mimosine, phytic acid, gossypol oxalates, cyanogens, saponins, nitrates, alkaloids, and anti-vitamin agents. These ANFs interfere with the nutritional value of feed by reducing digestion, absorption, and utilization of proteins and minerals that cause toxicity and lead to undesirable effects on animal health if their consumption is excessive. However, these ANFs may have beneficial health effects if they are present in low amounts. Risk factors associated with ANFs include a lack of knowledge on tolerance levels of these compounds in livestock, limited information on the degree of variability of individual risk, and lack of knowledge regarding the influence of environmental factors on the detoxification capacity of livestock feed. Thus, using appropriate and effective techniques and methods alone or in combinations can help eliminate or reduce the harmful effects of these ANFs in plant-based diets and, therefore, improve their nutritional value. Several strategies are used to counteract the effects of ANFs, including physical and chemical treatments, supplementation with enzymes, amino acids or mineral ions, germination, fermentation, and genetic modification.

References

Abbas G (2020) An augmented review about anti-nutrients and toxins of feedstuff and their control strategies, a step toward sustainable resource utilization. Int J Agric Innov Res 8:2319–1473

Abou-Donia MB, Othman MA, Obih P (1989) Interspecies comparison of pharmacokinetic profile and bioavailability of (±)-gossypol in male Fischer-344 rats and male B6C3F mice. Toxicology 55:37–51. https://doi.org/10.1016/0300-483x(89)90173-x

Adeyemo GO (2008) Effects of cottonseed cake-based diets on haematology and serum biochemistry of egg-type chickens. Int J Poult Sci 7:23–27

Ahmed AE, Smithard R, Ellis M (1991) Activities of enzymes of the pancreas, and the lumen and mucosa of the small intestine in growing broiler cockerels fed on tannin-containing diets. Br J Nutr 65:189–197. https://doi.org/10.1079/bjn19910080

Akande KE, Fabiyi EF (2010) Effect of processing methods on some anti-nutritional factors in legume seeds for poultry feeding. Int J Poultry Sci 9:996–1001. https://doi.org/10.3923/ijps.2010.996.1001

Akande KE, Doma UD, Agu HO, Adamu HM (2010) Major anti-nutrients found in plant protein sources: their effect on nutrition. Pak J Nutr 9:827–832. https://doi.org/10.3923/pjn.2010.827.832

Akinmutimi AH (2007) Effect of cooking periods on the nutrient composition of velvet beans (*Mucuna pruscens*). In: Proceeding of the 32nd annual conference of the Nigerian society for animal production march 18th–21st. University of Calabar, Calabar, pp 223–236

Aletor VA (1993) Allelochemicals in plant foods and feeding Stuffs. Part I. Nutritional, biochemical and physiopathological aspects in animal production. Vet Hum Toxicol 35:57–67

Aletor VA, Adeogun OA (1995) Nutrient and anti-nutrient components of some tropical leafy vegetables. Food Chem 53:375–379. https://doi.org/10.1016/0308-8146(95)99830-S

Alexander J, Benford D, Cockburn A et al (2008) Gossypol as undesirable substance in animal feed. EFSA J 908:1–55. https://doi.org/10.2903/j.efsa.2009.908

Amaefule KU, Onwudike OC (2000) Comparative evaluation of the processing methods of pigeon pea seeds (*Cajanus cajan*) as protein source for broilers. J Sustain Agric Environ 1:134–136
Anderson RC, Vodovnik M, Min BR, Pinchak WE, Krueger NA, Harvey RB et al (2012) Bactericidal effect of hydrolysable and condensed tannin extracts on Campylobacter jejuni *in vitro*. Folia Microbiol 57:253–258. https://doi.org/10.1007/s12223-012-0119-4
Andrae J (2008) Preventing prussic acid poisoning, Forage Leaflet, 13. Clemson University
Andrews DJ, Kumar KA (1992) Pearl millet for food, feed and forage. Adv Agron 48:89–139. https://doi.org/10.1016/S0065-2113(08)60936-0
Animut G, Puchala R, Goetsch AL, Patra AK, Sahlu T, Varel VH, Wells J (2008) Methane emission by goats consuming diets with different levels of condensed tannins from lespedeza. Anim Feed Sci Technol 144: 212–227
Avato P, Bucci R, Tava A, Vitali C, Rosato A, Bialy Z, Jurzysta M (2006) Antimicrobial activity of saponins from *Medicago* spp.: Structure-activity relationship. Phytother Res 20:454–457. https://doi.org/10.1002/ptr.1876
Bala SM, Ameh DA, Omage JJ, Bugaje SM, Makeri MU, Dikko H, Usman MS (2013) Influence of processing method on the nutrients and antinutrients content of kernels of three nigerian local mango (mangifera indica l.) varieties. Journal of Applied Agricultural Research 5:203–213
Bajpai S, Sharma A, Gupta MN (2005) Removal and recovery of antinutritional factors from soybean flour. Food Chem 89:497–501. https://doi.org/10.1016/j.foodchem.2004.02.055
Balogun BI (2013) Evaluation of the nutritional potentials of fermented oil beans seed Pentaclethra macrophyllah Benth. PAT Nsuka J 9:73–87
Bardocz S, Grant G, Ewen SW, Duguid TJ, Brown DS, Englyst K, Pusztai A (1995) Reversible effect of phytohaemagglutinin on the growth and metabolism of rat gastrointestinal tract. Gut 37:353–360. https://doi.org/10.1136/gut.37.3.353
Barry TN, McNabb WC (1999) The implications of condensed tannins on the nutritive value of temperate forages fed to ruminants. Br J Nutr 81:263–272
Barraza ML, Coppock CE, Brooks KN, Wilks DL, Saunders RG, Latimer GW Jr (1991) Iron sulfate and feed pelleting to detoxify free gossypol in cottonseed diets for dairy cattle. J Dairy Sci 74:3457–3467. https://doi.org/10.3168/jds.S0022-0302(91)78536-6
Basso B, Ritchie JT (2005) Impact of compost, manure and inorganic fertilizer on nitrate leaching and yield for a 6-year maize–alfalfa rotation in Michigan. Agric Ecosyst Environ 108:329–341. https://doi.org/10.1016/j.agee.2005.01.011
Bedford MR (2000) Exogenous enzymes in monogastric nutrition their current value and future benefits. Anim Feed Sci Technol 86:113
Benjamin DN (2006) Effect of fertilizer application and cutting interval on nitrate accumulation in bermudagrass. M.Sc. (Agri.), Thesis, University Tennesse, Martin, TN
Bhattarai S, Coulman B, Biligetu B (2016) Sainfoin (*Onobrychis viciifolia* Scop.): renewed interest as a forage legume for western Canada. Can J Plant Sci 96:748–756. https://doi.org/10.1139/cjps-2015-0378
Bhutkar MA, Bhise SB (2012) *In vitro* assay of alpha amylase inhibitory activity of some indigenous plants. Int J Chem Sci 10:457–462. https://doi.org/10.31031/MAPP.2018.01.000518
Blumert M, Liu J (2003) Jiaogulan (*Gynostemma pentaphyllum*), China's Immortality Herb, 3rd edn. Torchlight
Boadi D, Benchaar C, Chiquette J, Massé D (2004) Mitigation strategies to reduce enteric methane emissions from dairy cows: update review. Can J Anim Sci 84:319–335
Boehm S, Huck S (2009) Presynaptic inhibition by concanavalin: are alphalatrotoxin receptors involved in action potential-dependent transmitter release? J Neurochem 71:2421–2430
Bora P (2014) Anti-nutritional factors in foods and their effects. J Acad Ind Res 3:285–290
Cheeke PR (1996) Biological effects of feed commercial use and forage saponins and their impact on animal production. In: Saponins used in food and agriculture. Plenum Press, New York, pp 377–386. https://doi.org/10.1007/978-1-4613-0413-5_32
Cheryan M (1980) Phytic acid interactions in food systems. CRC Crit Rev Food Sci Nutr 13:297–335. https://doi.org/10.1080/10408398009527293

Chunmei G, Hongbin P, Zewei S, Guixin Q (2010) Effect of soybean variety on anti-nutritional factors content, and growth performance and nutrients metabolism in rat. Int J Mol Sci 11:1048–1056. https://doi.org/10.3390/ijms11031048

Church DC (1991) Livestock feeds and feeding, 3rd edn. Prentice Hall Incorporation, New Jersey, p 546

Corder R (2006) Tannins in some European Reds have Greater Protective Effects on Heart than Other Red Wines. Nature Online Advance Edition. News Release, Queen Mary University of London. http://www.cbsnews.com/stories/2006/11/29/health/webind/main2215429.html

D'Mello JPF (2000) Anti-nutritional factors and mycotoxins. In: Farm animal metabolism and Soybeans. CAB International, Wallingford, pp 383–403. https://doi.org/10.1079/9780851993782.0383

Davis AM (1981) The oxalate, tannin, c rude fiber, and crude protein composition of young plants of some Atriplex species. J Range Manag 34:329–331. https://doi.org/10.2307/3897862

Davis AJ, Lordelo MM, Dale N (2002) The use of cottonseed meal with or without added soapstock in laying hen diets. J Appl Poult Res 11:127–133. https://doi.org/10.1093/JAPR/11.2.127

Deshpande SS, Cheryan M (1984) Effects of Phytic acid, divalent cations, and their interactions on α-amylase activity. J Food Sci 49:516–524. https://doi.org/10.1111/j.1365-2621.1984.tb12456.x

Degan AA, Becker K, Makkar HPS, Borowy N (1995) Acacia saligna as fodder tree for desert livestock and the interaction of its tannins with fiber fractions. J Sci Food Agric 68:65–71

Diarra SS, Usman BA, Kwari ID, Yisa A (2008) Effect of processing methods on the antinutrional factor and the nutritional composition of sesame seed. Sahel J Vet Sci 7:16–20. https://www.ajol.info/index.php/sjvs/article/view/41483

Dube JS, Reed JD, Ndlovu LR (2001) Proanthocyanidins and other phenolics in Acacia leaves of Southern Africa. Anim Feed Sci Technol 91: 59–67. https://doi.org/https://doi.org/10.1016/S0377-8401(01)00229-2

Dublecz K (2011) Animal nutrition. http://www.tankonyvtar.hu/en/tartalom/tamop425/0010_1A_Book_angol_04_takarmanyozastan/index.html

East NE, Anderson M, Lowenstine LJ (1994) Apparent gossypol-induced toxicosis in adult dairy goats. J Am Vet Med Assoc 204:642–643

EFSA CONTAM Panel, Schrenk D, Bignami M, Bodin L, Chipman JK, del Mazo J, Grasl Kraupp B, Hoogenboom LR, Leblanc JC, Nebbia CS, Nielsen E, Ntzani E, Petersen A, Sand S, Schwerdtle T, Vleminckx C, Wallace H, Bampidis V, Cottrill B, Frutos MJ, Fürst P, Parker A, Christodoulidou A, Gergelova P, Hogstrand C (2020) Risk assessment of nitrate and nitrite in feed. EFSA J 18:6290, 110 pp. https://doi.org/10.2903/j.efsa.2020.6290

Embaby HE-S (2011) Effect of heat treatments on certain Antinutrients and in vitro protein digestibility of Peanut and sesame seeds. Food Sci Technol Res 17:31–38. https://doi.org/10.3136/fstr.17.31

EPA (1990) Summary review of health effects associated with hydrogen cyanide, Health Issue Assessment Environmental Criteria and Assessment Office, Office of Health and Environmental Assessment Office of Research and Development, US Environmental Protection Agency Research Triangle Park, North Carolina, USA

Erdman JW (1979) Oilseed phytates: Nutritional implications. J Am Oil Chem Soc 56: 736–741. https://doi.org/https://doi.org/10.1007/BF02663052

Felix JP, Mello D (2000) Farm animal metabolism and nutrition. CABI

Fereidoon S (2012) Antinutrients and Phytochemicals in Food. Developed from a symposium sponsored by the division of agricultural and food chemistry at the 210th National Meeting of the American Chemical Society, Chicago, Illinoi. ACS symposium series, ISSN 0097-6156; 2

Fereidoon S (2014) Beneficial health effects and drawbacks of Antinutrients and phytochemicals in foods. Appl Microbiol Biotechnol 97:45–55

Ferguson TM, Couch JR, Rigdon RH (1959) Histopathology of animal reactions to pigment compounds-chickens. In: Proceedings of the conference on chemistry structure and reactions of gossypol and non-gossypol pigments of cottonseed. National Cottonseed Products Association, Memphis, pp 131–141

Fernando R, Pinto MDP, Pathmeswaran A (2012) Goitrogenic food and prevalence of goitre in Sri Lanka. J Food Sci 41:1076–1081. https://doi.org/10.5923/j.ijim.20120102.02
Ferri M (2009) Binahong (*Anredera cordifolia*) Sebagai Obat. Bull Warta 15(1)., April 2009. Penelitian dan Pengembangan Tanaman Industri. Badan Penelitian dan Pengembangan Pertanian
Francis G, Makkar HPS, Becker K (2001) Antinutritional factors present in plant-derived alternate fish feed ingredients and their effects in fish. Aquaculture 199:197–227. https://doi.org/10.1016/S0044-8486(01)00526-9
Friedman M, Brandon DL (2001) Nutritional and health benefits of soy proteins. J Agric Food Chem 49:1069–1086. https://doi.org/10.1021/jf0009246
Frontela C, Garcıa-Alonso FJ, Ros G, Martınez C (2008) Phytic acid and inositol phosphates in raw flours and infant cereals: The effect of processing. J Food Compos Anal 21:343–350
Fuller JB, Patterson CEP, Hester K, Stringer DY (1996) A quantitative review of research on charismatic leadership. Psychol Rep 78:271–287. https://doi.org/10.2466/pr0.1996.78.1.271
Gadelha ICN, do Nascimento Rangel AH, Silva AR, Soto-Blanco B (2011) Efeitos do gossipol na reprodução animal. Acta Vet Bras 5:129–135. https://doi.org/10.21708/avb.2011.5.2.2343
García-Estepa RM, Guerra-Hernández E, García-Villanova B (1999) Phytic acid content in milled cereal products and breads. Food Res Int 32:217–221. https://doi.org/10.1016/S0963-9969(99)00092-7
Gemede HF, Ratta N (2014) Antinutritional factors in plant foods: potential health benefits and adverse effects. Int J Nutr Food Sci 3:284–289. https://doi.org/10.11648/j.ijnfs.20140304.18
Ghavidel RA, Prakash J (2006) Effect of germination and dehulling on functional properties of legume flours. J Sci Food Agric 86:1189–1195. https://doi.org/10.1002/jsfa.2460
Gharaghani H, Zaghari M, Shahhoseini G, Moravej H (2008) Effect of gamma irradiation on antinutritional factors and nutritional value of canola meal for broiler chickens. Asia Aust J Anim Sci 21:1479–1485
Giri AP, Kachole MS (1998) Amylase inhibitors of pigeon pea (*Cajanus cajan*) seeds. Phytochemistry 47:197–202. https://doi.org/10.1016/s0031-9422(97)00570-0
Gomes MTR, Oliva ML, Lopes MTP, Salas CE (2011) Plant proteinases and inhibitors: an overview of biological function and pharmacological activity. Curr Protein Pept Sci 12:417–436. https://doi.org/10.2174/138920311796391089
Gonzales I, Sabatini S (1989) Cyanide poisoning: pathophysiology and current approaches to therapy. Int J Artif Organs 12:347–355. PMID: 2674018
Gowda NKS, Katiyar RC, Sastry VRB (1994) Detoxified Mahua (*Bassia latifolia*) seed cake as protein source for high growth potential animals. Indian J Anim Nutr 11:1–6
Grace ZH, Sudipta S, Das B, Hoang ML, Karbaschi MR, Alexander AT, Williams B, Sagadevan G, Mundree SG (2017) Finger on pulse: iron pumping into chickpea. Front Plant Sci:1–17. https://doi.org/10.3389/fpls.2017.01755
Grant G (1991) Lectins. In: D'Mello JPF, Duffus CM (eds) Toxic substances in crop plants. The Royal Society of Chemistry, pp 49–67. https://www.worldcat.org/title/toxic-substances-in-crop-plants/oclc/28224827
Gutierrez-Banuelos H, Pinchak WE, Min BR, Carstens GE, Anderson RC, Tedeschi LO et al (2011) Effects of feed-supplementation and hide-spray application of two sources of tannins on enteric and hide bacteria of feed lot cattle. J Environ Sci Health B Pest Food Contamin Agric Wastes 46:360–365. https://doi.org/10.1080/03601234.2011.559419
Haese EB (2017) Studies on the extent of ruminal degradation of phytate from different feedstuffs. Thesis PhD in Agricultural Science, Faculty of Agricultural Science, Weingarten, Germany. http://opus.uni-hohenheim.de/volltexte/2017/1343/
Hassanpour S, Maheri-Sis N, Eshratkhah B, Mehmandar FB (2011) Plants and secondary metabolites (Tannins): A Review. International Journal of Forest, Soil and Erosion 1(1):47–53
Helsper JPFG, Kolodziej H, Hoogendijk JM, van Norel A (1993) Characterization and trypsin inhibitor activity of proanthocyanidins from Vicia faba. Phytochemistry 34:1255–1260

Hamid R, Masood A, Wani IH, Rafiq S (2013) Lectins: proteins with diverse applications. J Appl Pharm Sci 3:S93–S103. https://doi.org/10.7324/JAPS.2013.34.S18

Hamid, Thakur NS, Kumar P (2017) Anti-nutritional factors, their adverse effects and need for adequate processing to reduce them in food. Agric Int 4:56–60. https://doi.org/10.5958/2454-8634.2017.00013.4

Haschek WM, Beasley VR, Buck WB, Finnell JH (1989) Cottonseed meal (gossypol) toxicosis in a swine herd. J Am Vet Med Assoc 195:613–615

Hassanpour S, Maheri-Sis N, Eshratkhah B, Mehmandar FB (2011) Plants and secondary metabolites (Tannins): a review. Int J For Soil Eros 1:47–53

Hegarty (1987) Toxic amino acids of plant origin. In: Effect of poisonous plants on livestock. Academic

Henry MH, Pesti GM, Brown TP (2001) Pathology and histopathology of gossypol toxicity in broiler chicks. Avian Dis 45:598–604. PMID: 11569732

Hervás G, Frutos P, Giráldez FJ, Mantecón AR, Álvarez Del Pino MC (2003) Effect of different doses of quebracho tannins extract on rumen fermentation in ewes. Anim Feed Sci Technol 109:65–78. https://doi.org/10.1016/S0377-8401(03)00208-6

Hiremat NB (1981) Subabul (*Leucaena leucocephala*)- a wonder plant. Indian Dairy Man 33:351–356. https://agris.fao.org/agris-search/search.do?recordID=US201302634066

Huisman J, Tolman GH (2001) Antinutritional factors in the plant proteins of diets for non-ruminants. In: Recent developments in pig nutrition. Netherlands, 3. Elsevier, pp 261–291. https://www.cabdirect.org/cabdirect/abstract/20013120943

Igile GO (1996) Phytochemical and Biological studies on some constituents of *Vernonia amygdalina* (compositae) leaves. Ph.D thesis, Department of Biochemistry, University of Ibadan, Nigeria

Jacob JP, Mitaru BN, Mbugua PN, Blair R (1996) The effect of substituting Kenyan serena sorghum for maize in broiler starter diets with different dietary crude protein and methionine levels. Anim Feed Sci Technol 61:27–39. https://doi.org/10.1016/0377-8401(96)00955-8

Jadhav SJ, Sharma RP, Salunke DK (2009) Naturally occurring toxic alkaloids in foods. CRC Crit Rev Toxicol 9:21–104

Jambunathan R, Singh U, Subramanian V (1988) Grain quality of sorghum, pearl millet, pigeon pea and chick pea. In: Achaya KT (ed) Interfaces between Agriculture, Nutrition and Food Science. Proceedings, workshop on Hyberabad, India, November, 10–12th 1981, Food and Nutrition Bulletin supplement, 9, Tokyo, pp 47–50

Jansman AJM (1993) Tannins in feedstuffs for simple-stomached animals. Nutr Res Rev 6:209–236. https://doi.org/10.1079/NRR19930013

Jansman AJM, Frohlich AA, Marquardt RR (1994) Production of proline-rich proteins by the parotid glands of rats is enhanced by feeding diets containing tannins from faba beans (*Vicia faba L.*). J Nutr 124:249–258. https://doi.org/10.1093/jn/124.2.249

Jenkins KJ, Atwal AS (1994) Effects of dietary saponins on fecal bile acids and neutral sterols, and availability of vitamins A and E in the chick. J Nutr Biochem 5:134–138. https://doi.org/10.1016/0955-2863(94)90084-1

Johnson IT, Gee JM, Price K, Curl C, Fenwick GR (1986) Influence of saponins on gut permeability and active nutrient transport in vitro. J Nutr 116:2270–2277. https://doi.org/10.1093/jn/116.11.2270

Jones RJ, Ford CW (1972) Some factors affecting the oxalate content of the tropical grass (*Setaria sphacelata*). Aust J Exp Agric Anim Hus 12:400–406

Jones RM, McLennan MW, Dowsett KF (1989) The effect of *Leucaena leucocephla* on the beef cattle reproduction of beef cattle grazing leucaena-grass pastures (reduced calving rates) 23, 108–114. https://agris.fao.org/agris-search/search.do?recordID=AU9000106

Joye I (2019) Protein digestibility of cereal products. Foods 8:199. https://doi.org/10.3390/foods8060199

Kankuka FG, Balogun TF, Yaakugh IDI (2000) Effect of duration of cooking of full fat soybean on dry matter digestibility and energy utilization by young pigs. In: Ukachukwu SN,

Ibeawuchi JA, Ibe SN, Ezekwe AG, Abasiekong SF (eds). Proceedings of Annual Conference of the Nigerian Society for Animal Production. Umudike, Abia State, pp 191–193. 19 – 23rd March, 2000
Kelln BM, Penner GB, Acharya SN, McAllister TA, Lardner HA (2021) Impact of condensed tannin-containing legumes on ruminal fermentation, nutrition, and performance in ruminants: a review. Can J Anim Sci 101:210–223. https://doi.org/10.1139/cjas-2020-0096
Kenar JA (2006) Reaction chemistry of gossypol and its derivatives. J Am Oil Chem Soc 83:269–302. https://doi.org/10.1007/s11746-006-1203-1
Kerr LA (1989) Gossypol toxicosis in cattle. Compend Contin Educ Pract Vet 11:1139–1146
Khare SK (2000) Application of immobilized enzymes in soybean processing. In: The Third International Soybean Processing and Utilization Conference (ISPCRC III): 2000 of the Innovation Era for Soybeans. 15–20, October, 2000, Tsukuba, Ibaraka, Japan, pp 381–382
Kim HL, Calhoun MC, Stipanovic RD (1996) Accumulation of gossypol enantiomers in ovine tissues. Comp Biochem Physiol B: Biochem Mol Biol 113:417–420. https://doi.org/10.1016/0305-0491(95)02061-6
Krupa U (2008) Main nutritional and antinutritional compounds of bean seeds-a review. Polish J Food Nutr Sci 58:149–155. https://agris.fao.org/agrissearch/search.do?recordID=PL2008000713
Kumar R, Singh M (1984) Tannins their adverse role in ruminant nutrition. J Agric Food Chem 32:447–453. https://doi.org/10.1021/jf00123a006
Lampart-Szczapa E, Siger A, Trojanowska K, Nogala-Kalucka M, Malecka M, Pacholek B (2003) Chemical composition and antibacterial activities of lupin seeds extracts. Food Nahrung 47:286–290. https://doi.org/10.1002/food.200390068
Lee C, Beauchemin KA (2014) A review of feeding supplementary nitrate to ruminant animals: nitrate toxicity, methane emissions, and production performance. Can J Anim Sci 94:557–570
Leeson S, Summers JD (2001) Scott's nutrition of the chicken, 4th edn. University Books, Guelph
Li Z, Alli I, Kermasha S (1993) In-vitro α-amylase inhibitor activity-phytate relationships in proteins from Phaseolus beans. Food Res Int 26:195–201. https://doi.org/10.1016/0963-9969(93)90053-L
Liebman M, Al-Wahsh IA (2011) Probiotics and other key determinants of dietary oxalate absorption. Adv Nutr 2:254–260. https://doi.org/10.3945/an.111.000414
Liener IE (1980) In: Summerfield RJ, Bunting AH (eds) Advances in legume science. Academic, New York/London
Liener IE (1994) Implications of antinutritional components in soybean foods. Crit Rev Food Sci Nutr 34:31–67. https://doi.org/10.1080/10408399409527649
Lindsey TO, Hawkins GE, Guthrie LD (1980) Physiological responses of lactating cows to gossypol from cottonseed meal rations. J Dairy Sci 63:562–573. https://doi.org/10.3168/jds.S0022-0302(80)82972-9
Liu K (1997) Chemistry and nutritional value of soybean components. In: Soybeans. Springer, Boston, pp 25–113
Lolas GM (1976) The phytic acid-total phosphorus relationship in barley, oats, soybeans, and wheat. Cereal Chem 53:867–871. https://agris.fao.org/agrissearch/search.do?recordID=US19770146612
Lordelo MM, Calhoun MC, Dale NM, Dowd MK, Davis AJ (2007) Relative toxicity of gossypol enantiomers in laying and broiler breeder hens. Poult Sci 86:582–590. https://doi.org/10.1093/ps/86.3.582
Loewus FA (2002) Biosynthesis of phytate in food grains and seeds. In: Reddy NR, Sather SK (eds) Food Phytate. CRC Press, Boca Raton, Florida, pp 53–61
Lori O, Thava V, James HH (2001) Phytic acid. Food Rev Int 17: 419–431. https://doi.org/10.1081/FRI-100108531
Lott JN, Ockenden I, Raboy V, Batten GD (2000) Phytic acid and phosphorus in crop seeds and fruits: a global estimate. Seed Sci Res 10:11–33. https://doi.org/10.1017/S0960258500000039
Lumen BD, Salamat LA (1980) Trypsin inhibitor activity in winged beans (Psophocarpus tetragnolobus) and the posible role of tannins. J Agric Food Chem 28:533–536

Mada SB, Garba A, Mohammed A, Muhammad A, Olagunju A, Mohammed HA (2012) Effects of boiling and roasting on antinutrients and proximate composition of local and some selected improved varieties of *Arachis hypogaea* L (Groundnut). Int J Food Nutr Saf 1:45–53. Corpus ID: 100127073

Maenz DD (2001) Enzymatic characteristics of phytases as they relate to their use in animal feeds. In: Bedford MR, Partridge GG (eds) Enzymes in Farm Animal Nutrition. CABI, Wallingford, UK, pp 61–84

Maner HJ, Gomez G (1973) Chronic cassava toxicity. In: Proceedings of an interdisciplinary workshop, London January, 29–30

Mangan JL (1988) Nutritional effects of tannins in animal feeds. Nutr. Res. Reviews 1:209–231

Margier M, Georgé S, Hafnaoui N, Remond D, Nowicki M, Du Chaffaut L, Amiot MJ, Reboul E (2018) Nutritional composition and bioactive content of legumes: characterization of pulses frequently consumed in France and effect of the cooking method. Nutrients 10:1668. https://doi.org/10.3390/nu10111668

Marshall JJ, Lauda CM (2007) Purification and properties of Phaseolus vulgaris. J Biol Chem 250:8030–8037

Medugu CI, Kwari ID, Igwebuike J, Nkama I, Mohammed ID, Hamaker B (2010) Performance and economics of production of broiler chickens fed sorghum or millet as replacement for maize in the semi-arid zone of Nigeria. Agric Biol J N Am 1:321–325

Mehansho H, Asquith TN, Butler LG, Rogler JC, Carlson DM (1992) Tannin-mediated induction of proline-rich protein-synthesis. J Agric Food Chem 40:93–97

Min BR, Barry TN, Attwood GT, McNabb WC (2003) The effect of condensed tannins on the nutrition and health of ruminants fed fresh temperate forages: a review. Anim Feed Sci Technol 106:1–4

Morgan SE (1989) Gossypol as a toxicant in livestock. In: Burrows GE (ed) Vet Clin North Am Food Anim Pract 5: 251–263

Morgan S, Stair EL, Martin T, Edwards WC, Morgan GL (1988) Clinical, clinicopathologic, pathologic, and toxicologic alterations associated with gossypol toxicosis in feeder lambs. Am J Vet Res 49:493–499. PMID: 3377308

Morzelle MC, Salgado JM, Massarioli AP, Bachiega P, de Oliveira Rios A, Alencar SM et al (2019) Potential benefits of phenolics from pomegranate pulp and peel in Alzheimer's disease: antioxidant activity and inhibition of acetylcholinesterase. J Food Bioact 5:136–141. https://doi.org/10.31665/JFB.2019.5181

Mueller I (2001) Analysis of hydrolysable tannins. Anim Feed Sci Technol 91:3–20. https://doi.org/10.1016/S0377-8401(01)00227-9

Muramoto K (2017) Lectins as bioactive proteins in foods and feeds. Food Sci Technol Res 23:487–494. https://doi.org/10.3136/fstr.23.487

Nagalakshmi D, Rama RSV, Panda AK, Sastry VRB (2007) Cottonseed meal in poultry diets: a review. J Poult Sci 44:119–134. https://doi.org/10.2141/jpsa.44.119

Naumann HD, Tedeschi LO, Zeller WE, Huntley NF (2017) The role of condensed tannins in ruminant animal production: advances, limitations and future directions. Rev Bras Zoot 46:929–949

Nikmaram N, Leong SY, Koubaa M, Zhu Z, Barba FJ, Greiner R et al (2017) Effect of extrusion on the anti-nutritional factors of food products: an overview. Food Control 79:62–73. https://doi.org/10.1016/j.foodcont.2017.03.027

Noonan SC, Savage GP (1999) Oxalate content of foods and its effect on humans. Asia Pac J Clin Nutr 8:64–67. PMID: 24393738

Nørgaard JV, Malla N, Dionisio G, Madsen CK, Pettersson D, Lærke HN et al (2019) Exogenous xylanase or protease for pigs fed barley cultivars with high or low enzyme inhibitors. Anim Feed Sci Technol 248:59–66. https://doi.org/10.1016/j.anifeedsci.2018.12.005

Nwosu JN (2010) Effect of soaking, blanching and cooking on the anti-nutritional properties of asparagus bean (Vigna Sesquipedis) flour Obizoba C, J, Atii JV 1991. Effects of soaking, sprouting,fermentation and cooking on nutrient composition and some antinutritional factors of sorghum (Guinesia) seeds. Plant Foods Hum Nutr 41:203–212

Okafor PN, Okorie CN (2006) Determination of Cyanide, Nitrate and Nitrite, contents of poultry feeds of four top animal feed companies in Nigeria. J Food Technol 4:225–228. https://medwelljournals.com/abstract/?doi=jftech.2006.225.228

Oke OL (1979) Some aspects of the role of cyanogenic glucosides in nutrition. World Rev Nutr Diet 33:70–103. https://doi.org/10.1159/000402550

Oke O (1980) Toxicity of cyanogenic glycosides. Food Chem 6:97–109. https://doi.org/10.1016/0308-8146(80)90025-4

Olayemi FO (2010) A review on some causes of male infertility. Afr J Biotechnol 9:2834–2842. https://doi.org/10.5897/AJB2010.000-3110

Oleszek W (2002) Chromatographic determination of plant saponins. J Chromatogr 967:147–162. https://doi.org/10.1016/s0021-9673(01)01556-4

Olomu JM (1995) Non-ruminant animal nutrition: principles and practice. Jachem Publication, Benin City, p 320

Osuntokun BO (1972) The effects of malnutrition on the development of cognitive functions of the nervous system in childhood. Trop Geogr Med 24:295–310

Otlewski J, Jelen F, Zakrzewska M, Oleksy A (2005) The many faces of protease–protein inhibitor interaction. EMBO J 24:1303–1310. https://doi.org/10.1038/sj.emboj.7600611

Patel PAS, Alagundagi SC, Salakinkop SR (2013) The anti-nutritional factors in forages – a review. Curr Biotica 6:516–526

Patra AK, Saxena J (2010) A new perspective on the use of plant secondary metabolites to inhibit methanogenesis in the rumen. Phytochem 71:198–1222. https://doi.org/10.1016/j.phytochem.

Patton CS, Legendre AM, Gompf RE, Walker MA (1985) Heart failure caused by gossypol poisoning in two dogs. J Am Vet Med Assoc 187:625–627. PMID: 4086371

Petroski W, Minich DM (2020) Is there such a thing as anti-nutrients? A narrative review of perceived problematic plant compounds. Nutrients 12:2929. https://doi.org/10.3390/nu12102929

Phytolab (2007) Phytochemicals Polyphenols–Tannic Acid. Retrieved July 6, 2007 from http://www.Phytochemicalsinfolphytochemicals/tannic-acid.phD

Raboy V (2000) Low-phytic-acid grains. Food Nutr Bull 21:423–427. https://doi.org/10.1177/156482650002100416

Raes K, Knockaert D, Struijs K, Van Camp J (2014) Role of processing on bioaccessibility of minerals: influence of localization of minerals and antinutritional factors in the plant. Trends Food Sci Technol 37:32–41. https://doi.org/10.1016/j.tifs.2014.02.002

Rahman MM, Kawamura O (2011) Oxalate accumulation in forage plants: some agronomic, climatic and genetic aspects. Asian Aust J Anim Sci 24:439–448. https://doi.org/10.5713/ajas.2011.10208

Rahman KSM, Pasirayi G, Auger V, Ali Z (2009) Development of a simple and low cost microbioreactor for high-throughput bioprocessing. Biotechnol Lett 31:209–214. https://doi.org/10.1007/s10529-008-9853-8

Rahman MM, Abdullah RB, Wan Khadijah WE (2013) A review of oxalate poisoning in domestic animals: tolerance and performance aspects. J Anim Physiol Anim Nutr 97:605–614. https://doi.org/10.1111/j.1439-0396.2012.01309.x

Ramteke R, Doneria R, Gendley MK (2019) Antinutritional factors in feed and fodder used for livestock and poultry feeding. Acta Sci Nutr Health 3:39–48

Randel RD, Chase CC Jr, Wyse SJ (1992) Effects of gossypol and cottonseed products on reproduction of mammals. J Anim Sci 70:1628–1638. https://doi.org/10.2527/1992.7051628x

Rao P, Desothale YG (1998) Tannin content of pulses varietal difference and effects of germination and cooking. J Sci Food Agric 33:1013–1015

Redondo LM, Fernandez Miyakawa ME, Fortunato R, Salvat A, Chacana P (2013a) Eficacia de aditivos alimentarios basados en extractos vegetales para disminuir la excreción de *Salmonella enteritidis* en pollitos BB. II Seminario Internacional de Salmonelosis aviar. Medellin, Colombia

Redondo LM, Redondo EA, Delgado F, La Sala L, Pereyra A, Garbaccio S et al (2013b) Control of Clostridium perfringens necrotic enteritis by tannins added to the diet. In: Proceedings of the 8th international conference on the molecular biology and pathogenesis of the Clostridia (ClostPath 8), vol 5, Palm Cove

Reed JD (1995) Nutritional toxicology of tannins and related polyphenols in forage legumes. J Anim Sci 73:1516–1528. https://doi.org/10.2527/1995.7351516x
Reed JD, Solar H, Wood Ward A (1990) Foder tree and straw diets for sheep: intake, growth, digestibility and the effect of phenolics on nitrogen utilization. Anim Feed Sci Technol 30:39–50. https://hdl.handle.net/10568/28206
Robinson EH (1991) Improvement of cottonseed meal protein with supplemental lysine in feeds for channel catfish. J Appl Aquac 1:1–14. https://doi.org/10.1300/J028v01n02_01
Robson S (2007) Prussic acid poisoning in Livestock. www.dpi.nsw.gov.au/ Prime facts
Roeder E (1995) Medicinal plants in China containing pyrrolizidine alkaloids. Pharmazie 50:83–98. PMID: 7700976
Rosling H (1987) Cassava toxicity and food security: a review of health effects of cyanide exposure from cassava and of ways to prevent these effects. Uppsala. UNICEF/African Household Food Security Programme, 40p. https://agris.fao.org/agris-search/search.do?recordID=SE881037988
Ryden P, Selvendran RR (1993) In: Macrae R, Robinson RK, Sadler MJ (eds) Encyclopedia of food science, food technology, and nutrition. Academic, San Diego, pp 3582–3587
Saito K, Horie M, Hoshino Y et al (1990) High performance liquid chromatographic determination of glycoalkaloids in potato products. J Chromatogr 508:141–147. https://doi.org/10.1016/S0021-9673(00)91247-0
Salas CE, Dittz D, Torres MJ (2018) Plant proteolytic enzymes: their role as natural pharmacophores. In: Biotechnological applications of plant proteolytic enzymes. Springer, Cham, pp 107–127
Sallau AB, Mada SB, Ibrahim S, Ibrahim U (2012) Effect of boiling, simmering and blanching on the anti-nutritional content of Moringa oleifera leaves. Int J Food Nutr Saf 2:1–6. Corpus ID: 56219382
Salunkhe DK, Chavan JK, Kadam SS (1990) Dietary tannins: consequences and remedies. CRC Press, Boca Raton
Samtiya M, Rotimi EA, Dhewa T (2020) Plant food anti-nutritional factors and their reduction strategies: an overview. Food Prod Process Nutr. https://doi.org/10.1186/s43014-020-0020-5
Sarangthem K, Singh TN (2013) Fermentation decreases the anti-nutritional content in bamboo shoots. Int J Curr Microbiol Appl Sci 2:361–369. Corpus ID: 62835452
Sauvion N, Nardon C, Febvay G, Gatehouse AM, Rahbe Y (2004) Binding of the insecticidal lectin concanavalin a in pea aphid, Acyrthosiphon pisum (Harris) and induced effects on the structure of midgut epithelial cells. J Insect Physiol 50:1137–1150
Schneider IC, Ames ML, Rasmussen MA, Reilly PJ (2002) Fermentation of cottonseed and other feedstuffs in cattle rumen fluid. J Agric Food Chem 50:2267–2273
Seeman P (1974) Ultrastructure of membrane lesions in immune lysis, osmotic lysis and drug-induced lysis. Fed Proc 33:2116–2124. PMID: 4607618
Selle PH, Ravindran V, Caldwell RA, Bryden WL (2000) Phytate and phytase: consequences for protein utilization. Nutr Res Rev 13:255–278. https://doi.org/10.1079/095442200108729098
Serrano J, Puupponen-Pimiä R, Dauer A, Aura AM, Saura-Calixto F (2009) Tannins: current knowledge of food sources, intake, bioavailability and biological effects. Mol Nutr Food Res 53:S310–S329. https://doi.org/10.1002/mnfr.200900039
Sharma NK, Lodhi GN, Ichhponani JS (1978) Comparative feeding value of expeller processed undecorticated and decorticated cottonseed cakes for growing chicks. J Agric Sci 91:531–541. https://doi.org/10.1017/S0021859600059906
Shi J, Arunasalam K, Yeung D, Kakuda Y, Mittal G (2004 Spring) Jiang Y (2004) Saponins from edible legumes: chemistry, processing, and health benefits. J Med Food. 7:67–78. https://doi.org/10.1089/109662004322984734
Simee W (2011) Isolation and determination of antinutritional compounds from root to shells of peanut (Arachis Hypogaea). J Disper Sci Technol 28:341–347
Singh A, Tiwana US, Tiwana MS, Puri KP (2000) Effect of application method of level of nitrogen fertilizer on nitrate content in oat fodder. Indian J Anim Nutr 17:315–319

Singh A, Gupta S, Kaur R, Gupta HR (2017) Process optimization for antinutrient minimization of millets. Asian J Dairy Food Res 36:322–326. https://doi.org/10.18805/ajdfr.DR-1215
Skutches CL, Herman DL, Smith FH (1973) Effect of intravenous gossypol injection on iron utilization in swine. J Nutr 103:851–855. https://doi.org/10.1093/jn/103.6.851
Smeriglio A, Barreca D, Bellocco E, Trombetta D (2017) Proanthocyanidins and hydrolysable tannins: occurrence, dietary intake and pharmacological effects. Br J Pharmacol 174:1244–1262. https://doi.org/10.1111/bph.13630
Soetan KO (2008) Pharmacological and other beneficial effects of antinutriional factors in plants. –a review. Afr J Biotechnol 7:4713–4721. https://doi.org/10.5897/AJB08.024
Soetan KO, Oyewole OE (2009) The need for adequate processing to reduce the anti-nutritional factors in plants used as human foods and animal feeds: a review. Afr J Food Sci 3:223–232
Soto-Blanco B (2008) Gossipol effatores antinutricionais da soja. In: Spinosa HS, Górniak SL, Neto JP (eds) Toxicologia aplicada à medicina veterinária. Manole, Barueri, pp 531–545
Southon S, Johnson IT, Gee JM, Prince KR (1988) The effect of Gypsophylla saponins in the diet on mineral status and plasma cholesterol concentration in the rat. Br J Nutr 59:49–55
Sparg SG, Light ME, Van Staden J (2004) Biological activities and distribution of plant saponins. J Ethnopharmacol 94:219–243. https://doi.org/10.1016/j.jep.2004.05.016
Stegelmeier BL, Davis TZ, Clayton MJ (2020) Neurotoxic plants that poison livestock. Vet Clin N Am Food Anim Pract 36:673–688
Su B, Chen X (2020) Current status and potential of Moringa Oleifera leaf as an alternative protein source for animal feeds, review article. Front Vet Sci. https://doi.org/10.3389/fvets.2020.00053
Suryawanshi SN, Bhandarkar AG, Bhagwat SS (1993) Pathological investigations of acute cottonseed cake toxicity in rabbit and chicken. Livestock Advisor 18:30–32
Thakur A, Sharma V, Thakur A (2019) An overview of anti-nutritional factors in food. Int J Chem Stud 7:2472–2479
Timotheo CA, Lauer CM (2018) Toxicity of vegetable tannin extract from Acacia mearnsii in Saccharomyces cerevisiae. Int J Environ Sci Technol 15:659–664. https://agris.fao.org/agris-search/search.do?recordID=US201800213215
Udedibie ASI, Anyaegbu BC, Onyeckwu GC, Egbuokporo OC (2004) Effects of feeding different levels of fermented and unfermented cassava tuber meals on the performance of broilers. Niger J Anim Prod 31:211–219. https://doi.org/10.51791/njap.v31i2.1816
Uhegbu FO, Onwuchekwa CC, Emeka O, Iweala EJ, Kanu I (2009) Effect of processing methods on nutritive and antinutritive properties of seeds of *Brachystegia eurycoma* and *Detarium microcarpum* from Nigeria. Pak J Nutr 8:316–320. https://doi.org/10.3923/pjn.2009.316.320
Umaru HA, Adamu R, Dahiru D, Nadro MS (2007) Levels of antinutritional factors in some wild edible fruits of northern Nigeria. Afr J Biotechnol 6:1935–1938. https://doi.org/10.5897/AJB2007.000-2294
Van Der Ven C, Matser AM, Van den Berg RW (2005) Inactivation of soybean trypsin inhibitors and lipoxygenase by high-pressure processing. J Agric Food Chem 53:1087–1092. https://doi.org/10.1021/jf048577d
Van Leeuwen PV, Jansman AJM, Wiebenga J, Koninkx JFJG, Mouwen JMVM (1995) Dietary effects of faba bean (Vicia faba L.) tannins on the morphology and function of the small intestinal mucosa of weaned pigs. Br J Nutr 73:31–39. https://doi.org/10.1079/BJN19950006
Vaz Patto MC, Amarowicz R, Aryee AN, Boye JI, Chung HJ, Martín-Cabrejas MA et al (2015) Achievements and challenges in improving the nutritional quality of food legumes. Crit Rev Plant Sci 34:105–143. https://doi.org/10.1080/07352689.2014.897907
Vennesland B, Castric PA, Conn EE, Solomonson LP, Volini M, Westley J (1982) Cyanide metabolism. Fed Proc 41:2639–2648
Waghorn G (2008) Beneficial and detrimental effects of dietary condensed tannins for sustainable sheep and goat production—Progress and challenges. Anim Feed Sci Technol 147:116–139. https://doi.org/10.1016/j.anifeedsci.2007.09.013
Waldroup PW, Goodner TO (1973) Tolerance levels of free gossypol in layer diets as influenced by iron: gossypol ratios. Poult Sci 52:20–28. https://doi.org/10.3382/ps.0520020

Walter HL, Fanny L, Charles C, Christian R (2002) Minerals and phytic acid interaction: is it a real problem for human nutrition. Int J Food Sci Technol 37:727–739. https://doi.org/10.1046/j.1365-2621.2002.00618.x

WHO, World Health Organization (2010) Toxicological evaluation of certain food additives and naturally occurring toxicants, WHO Food Additive Series: 30. World Health Organization, Geneva. 9241660325_eng.pdf (10.55Mb)

Willard ST, Neuendorff DA, Lewis AW, Randel RD (1995) Effects of free gossypol in the diet of pregnant and postpartum Brahman cows on calf development and cow performance. J Anim Sci 73:496–507. https://doi.org/10.2527/1995.732496x

Woyengo TA, Nyachoti CM (2011) Supplementation of phytase and carbohydrases to diets for poultry: a review. Can J Anim Sci 91:177–192. https://doi.org/10.4141/cjas10081

Woyengo TA, Nyachoti CM (2013) Review: Antinutritional effects of phytic acid in diets for pigs and poultry – current knowledge and directions for future research. Can J Anim Sci 93:9–21. https://doi.org/10.4141/cjas2012-017

Yasmin A, Zeb A, Khalil AW, Paracha GMUD, Khattak AB (2008) Effect of processing on anti-nutritional factors of red kidney bean (*Phaseolus vulgaris*) grains. Food Bioproc Technol 1:415–419. https://doi.org/10.1007/s11947-008-0125-3

Yilkal T (2015) Important anti-nutritional substances and inherent toxicants of feeds. Food Sci Qual Manag 36:40–47. Corpus ID: 56333400

Zagrobelny M, Bak S, Rasmussen AV, Jorgensen B, Naumann CM, Moller BL (2004) Cyanogenic glucosides and plant insect interactions. Phytochemistry 65:293–306. https://doi.org/10.1016/j.phytochem.2003.10.016

Zhang WJ, Xu ZR, Pan XL, Yan XH, Wang YB (2007) Advances in gossypol toxicity and processing effects of whole cottonseed in dairy cows feeding. Livest Sci 111:1–9. https://doi.org/10.1016/j.livsci.2007.03.006

Chapter 5
Genetic Engineering Tools and Techniques in Livestock Production

H. B. Ranjitha, Madhu Ramesh, Subhasmita Behera, Dhanesh ValiyaValappil, Suresh H. Basagoudanavar, and Anjum Sherasiya

Abstract Biotechnology is an applied branch of biology that harnesses living systems or organisms to make useful products for the benefit of mankind. Animal biotechnology utilizes biotechnological techniques to alter the animal genome for the development of pharmaceutical, agricultural or industrial products. The application of biotechnology, in particular, the use of advanced genetic engineering tools holds great promise for improvement in animal health, which in turn can help improve the economy.

In this chapter, we review the application of biotechnology tools to enhance animal health by improving the ability to detect, treat and prevent diseases. The major techniques reported are artificial insemination, embryo transfer, animal cloning, sexing, marker-assisted selection that assist the animal breeding programs, allowing to produce healthier offsprings and thus improve animal production. The growing population of the world and exponential increase of the demand for the supply of livestock products need sophisticated techniques to boost animal production through genetic improvement by the use of advanced genetic engineering tools. The reproductive animal biotechnology techniques would bring about superior genetic improvement of productivity and developing desired healthy livestock. In addition, cloning and transgenesis help to develop bio factory and boost reproduction exponentially with limited time. Further, development of genetically modified animals enhances production by incorporating favourable traits. Also, the molecular techniques like polymerase chain reaction help in early diagnosis of diseases and genome editors are harnessed for improving the animal health for better production. We discuss on such biotechnology tools and techniques that are focused on improving the animal health and production.

H. B. Ranjitha · S. Behera · D. ValiyaValappil · S. H. Basagoudanavar
ICAR-Indian Veterinary Research Institute, Bengaluru, Karnataka, India

M. Ramesh
JNCASR, Jakkur, Bengaluru, Karnataka, India

A. Sherasiya (✉)
Veterinary World, Wankaner, Gujarat, India

V. K. Yata et al. (eds.), *Sustainable Agriculture Reviews 57*, Sustainable Agriculture Reviews 57, https://doi.org/10.1007/978-3-031-07496-7_5

Keywords Recombinant DNA technology · Expression systems · RT-PCR · Microarray · Recombinant vaccines · Genome editors · Insemination · Genetically modified animals · Embryo transfer · Sexing

5.1 Introduction

The livestock plays a vital role in developing economy of a country. In India, the livestock sector contributes about more than one-third of the agricultural gross domestic product (GDP) i.e., 25.6% of total agriculture GDP and 4.11% to overall GDP and provide employment to about 8.8% of the Indian population. Also, mixed farming system is mostly followed in India to make use of the resources efficiently by utilizing the output of one enterprise as the input of another (Patil et al. 2019). Advancements in the field of biotechnology have been successfully employed in enhancing the productivity of livestock. Apart from helping to alleviate poverty and hunger, biotechnology holds great promise for potential transformation in the veterinary field of treatment and significant improvement in the diagnosis of livestock diseases (Wheeler 2013). A major application of animal biotechnology is engineering biological systems by the usage of molecular biology techniques to alter the animal genome for the development of pharmaceutical, agricultural or industrial products (Davide 2010).

Biotechnology has a significant role in improving the health of animal by increasing the precision of disease diagnosis, disease control, and treatment. The polymerase chain reaction (PCR) based diagnostics are widely used in developing countries to allow early diagnosis of diseases. Also, vaccination is the best method to prevent contracting the diseases and to prevent transmission of disease. A number of different vaccines produced using biotechnology have several benefits including decreasing the risk of reversion to virulence of live vaccines, induction of passive immunization by antibodies, long-term storage of vaccines for proper immunization, and differentiation of infected from vaccinated animals (Babiuk 2002; Shams 2005; Henderson 2005). The application of molecular techniques and computational bioinformatics to diagnose various diseases, gene therapy, and *in vivo* delivery systems contribute significantly to control animal diseases and improve animal health, thereby stimulating both food production and livestock trade (Mohammed et al. 2016). The modern genetic engineering tools- genome editors like zinc finger nucleases (ZFNs), transcription activator-like effector endonucleases (TALENs), and the clustered regulatory interspaced short palindromic repeats (CRISPR) associated protein9 (CRISPR/Cas9) are very useful in the production of transgenic animals, creation of animal models for disease study, specific cell line production and genetic diseases treatment (Petersen 2017). Application of biotechnology in the field of animal nutrition has made tremendous improvement in animal production by the formulation of improvised feeds with increased digestibility of nutrients, silage inoculants, supplementation of amino acids, mycotoxins diagnosis, removal of anti-nutrients and toxins through enzyme treatment, supplementation of probiotics, prebiotics to promote gut growth and use of antibodies, hormones to improve health (Bonneau and Laarveld 1999).

The most widely applied animal biotechnology tool is perhaps artificial insemination (AI) to improve animal reproduction, genetics, and breeding. Artificial insemination has evolved as a powerful tool in genetic improvement for productivity and developing desired livestock with the evolvement of the various technologies, such as the methods of cryopreservation of semen and sperm sexing. Embryo transfer also provides the same opportunities (Foote 2010). Also, marker-assisted selection can also be used for the genetic improvement of livestock (Falconer and Mackay 1996). The development of genetically modified animals (GMAs) is being carried out for enhanced production by incorporating favourable traits to animals and birds like higher growth, reproduction, and disease resistance. In this regard, cloning and transgenesis created an opportunity to develop bio factory in the farm animals and produce pharmaceuticals on a large scale. Especially, cloning boosts reproduction exponentially in a limited time. These techniques help to incorporate the favourable genes in the immediate next generation bypassing the time-consuming breeding and selection strategies. This chapter has focused on such genetic engineering tools and techniques that emphasize the improvement of livestock production.

Role of Molecular Biotechnology in Animal Production and Health Modern biotechnology tools offer enormous prospects for the production of diagnostics and vaccines (Borroto 2009). The use of vaccine is the best method to prevent contracting the diseases and to prevent transmission of disease. Several different vaccines produced using biotechnology have several benefits. Also, the molecular techniques help in early diagnosis of diseases and modification of the genome for the animal health improvement. Different applications of biotechnology tools and techniques in several fields of livestock production and health are presented in Fig. 5.1.

5.2 Molecular Gene Cloning

The advancement in the knowledge of recombinant DNA (rDNA) technology, especially in the field of cloning and expression of the gene of interest in the particular host has brought about tremendous development in the discovery of vaccines, diagnostic antigens, and other rDNA products in human and animal healthcare. Molecular gene cloning is the process of making multiple copies of a gene of interest by inserting it into a vector and transforming into a suitable host capable of producing multiple copies.

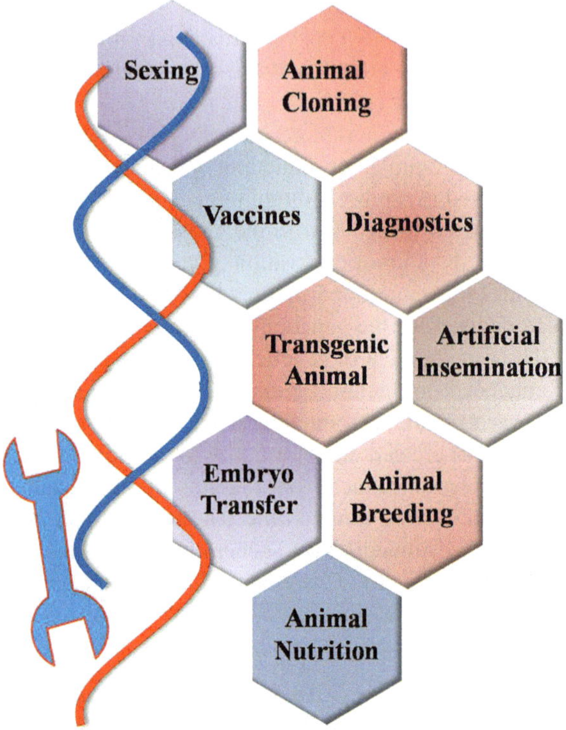

Fig. 5.1 An illustration that depicts the applications of biotechnological tools to improve the livestock production in various fields such as vaccines, diagnostics, nutritional biology and reproductive biotechnology

5.2.1 Steps in Gene Cloning

The molecular gene cloning basically consists of four major steps. The cloning starts with insertion of the gene of interest into the vector to form recombinant DNA (rDNA). Secondly the rDNA consist of gene of interest is introduced into host cells for the multiplication. The multiplied copies of rDNA were isolated and purified from host cells (Kfoury 2007). The gene of interest can be amplified by polymerase chain reaction and/or can be prepared using restriction endonuclease enzymes (RE). There are several types of vectors that can carry the insert based on its size into the host for making multiple copies. Plasmids, bacteriophages, yeast artificial chromosomes, bacterial and mammalian artificial chromosomes are most commonly used cloning vectors. The same restriction endonuclease enzyme is used to cut the vector as used for the insert. The insert and vector are ligated to each other by DNA ligase to form rDNA and transformed into suitable host by Ca++ ion treatment/ heat shock method or electroporation. Marker genes such as antibiotic resistance gene present in the vector are used for the selection of recombinant colonies.

5.2.2 Gateway Cloning Technology

It is an advanced cloning method based on the site-specific recombination system, commercialized by Invitrogen. It exploits the principle used by the phage lambda to integrates/excise its DNA from bacterial chromosomes. This method requires two sets of recombination sites- gateway 'att' sites and two enzymes- BP clonase and LR clonase. This quick technique allows easy transfer and multiple insertion of the inserts into any expression vectors without affecting the reading frame by >99% cloning efficiency (Katzen 2007).

5.2.3 Golden Gate Cloning

It is an another advanced method that allows directional cloning of multiple insertions using Type IIs restriction enzymes such as *BsaI*, *BsmBI*, and *BbsI* and T4 DNA ligase (Engler et al. 2008).

5.2.4 Expression Systems

Generally, prokaryotic, eukaryotic, and insect systems are used as an expression system/host for the expression of rDNA products.

(a) **Prokaryotic system**
E. coli is commonly used as a prokaryotic expression system as it offers several advantages including simple media requirements, less time for bacterial multiplication, and a high level of protein expression. However, the problems like the sequence of the foreign gene with the presence of introns, termination signals, and codon bias will affect the expression. Also, limitations of *E. coli* as a host include- post-translation modifications, and degradation of recombinant protein (Brown 2016).

(b) **Eukaryotic system**
The protein from microbial eukaryote- yeast (*Saccharomyces cerevisiae* and *Pichia pastoris*) or filamentous fungus is more closely related to an animal protein with higher levels of expression. Chinese hamster ovary (CHO) and Baby hamster kidney (BHK) cells are the commonly used mammalian expression systems that are also capable of producing post-translation modified proteins. However, the cost of production is comparatively higher. Insect cell culture system is a widely used system for making recombinant proteins (Sudeep et al. 2005; Mena and Kamen 2011). In particular, baculoviruses have revolutionized the protein expression system. They cause fatal disease in insects mainly and thus characterized by narrow host range with the incapability to multiply in vertebrates (Tinsley and Harrap 1978). The baculovirus has a large DNA genome of approximately 130 kb size, can be easily

manipulated, and propagated in cell lines such as those obtained from *Spodoptera frugiperda* (SF) (Granados 1994). The strong polyhedrin promoter is used for the production of the protein of interest. The first commercially available insect cell produced veterinary vaccine is based on the E2 antigen of classical swine fever virus (CSFV) (Van Aarle 2003). The PCV2 (ORF2 protein) vaccine against multi-systemic wasting syndrome with ORF2 protein of the virus as the antigen is manufactured in insect cell culture system as Porcilis® PCV(Merck) and CircoFLEX® (B. Ingelheim) (Blanchard et al. 2003; Fachinger et al. 2008).

(c) **Transgenic plants**

Transgenic plant provides an alternative platform to express immunogenic foreign bacterial and viral pathogen proteins with transient (viral vectors) or stable by Agrobacterium T-DNA vectors or microprojectile bombardment) of foreign genes. They act as a vehicle to produce vaccines that stimulate both the humoral immune system with the production of secretory IgA (S-IgA) at mucosal surfaces (McGhee and Kiyono 1993) and mucosal immune responses upon consumption as food (Mason and Arntzen 1995). The immunogenic Foot-and-mouth disease virus (FMDV) structural protein VP1 was expressed in various plants such as alfalfa, *Arabidopsis thaliana*, and potato plants and has been shown to induce protective neutralizing antibody response (Wigdorovitz et al. 1999; Carrillo et al. 1998). Bovine rotavirus epitope, eBRV4 also expressed in alfalfa plants and evaluated in suckling mice for the production of passive protection (Wigdorovitz et al. 2004). Also, Bovine viral diarrhoea virus (BVDV), glycoprotein E2 has been expressed in alfalfa. Thus, transgenic plant-based immunization methods are an alternative form and less expensive tool compared to conventional vaccine production system.

5.3 Diagnostics

The molecular diagnostic tools have led to control several diseases, specifically by the development of monoclonal antibodies (mAbs), nucleotide sequencing, polymerase chain reaction and antigen production. The mAbs permit the analysis of any antigenic molecule (Kenneth et al. 1980) and also helpful for checking vaccine efficacy. There are several kits and simple reagents available for the diagnosis of the diseases (Ferris et al. 1988). For example, kits for FMD virus diagnosis based on 3AB protein identification are used in endemic countries. Also, rDNA technology helps in the generation of specific antigens which further is useful in differentiating vaccinated and infected animals by enzyme-linked immunosorbent assays (ELISA) (Robinson and McEvoy 1993).

5.3.1 Polymerase Chain Reaction

The PCR, a highly sensitive method has revolutionized biological research that enables the production or amplification of multiple copies of a gene of interest within a short duration. Kary Mullis, a biochemist discovered this technique and was awarded with Nobel Prize in 1993. DNA polymerases require 3′-OH to add nucleotides, so there is requirement of primer to initiate replication (Mullis and Faloona 1987).

Components of Polymerase Chain Reaction

The main component of polymerase chain reaction is DNA polymerase which amplifies the target DNA (deoxyribonucleic acid) using template (as low as 100 pg). The other main building blocks are deoxyribonucleotides (200 μM). Other essential components for selective amplification of target DNA are primers (0.5 μM) which flank the target gene. The whole reaction will be carried out in buffer supplemented with essential ions like magnesium (1.5 mM). The reaction is carried out in a thermocycler which ramp between various temperatures required for the reaction to occur.

Three Temperature-Dependent Steps

Denaturation – It is usually between 90 and 100 °C. It breaks the hydrogen bonds between the strands of the template DNA.

Annealing – It occurs between 30 and 65 °C that facilitates the binding of primers to the template.

Elongation – It happens at 60–75 °C for less than a minute per kilobase strand synthesis.

5.3.2 Quantitative Real-Time Polymerase Chain Reaction

Quantitative real-time polymerase chain reaction (qRT-PCR) is most widely used technique in the field of molecular diagnostics. It possesses high sensitivity, high specificity, with the advantage of simultaneous amplification and real-time quantification of the target DNA after each cycle i.e., it determines the initial number of copies of template DNA (Higuchi et al. 1992). It allows detection of single nucleotide polymorphisms, viral load, gene expression analysis, genotyping, allelic discrimination. There are two methods of nucleic acid quantification- relative method of quantification and absolute quantification. Absolute method involves quantification of target DNA molecules by comparison with DNA standards using a standard curve. Whereas, relative quantification determines target gene expression based on internal reference genes. RT-PCR protocols have been used to detect conserved regions of several notifiable viral pathogens such as FMDV, CSFV, Blue tongue virus, Influenza and Newcastle disease virus. For example, detection of conserved regions of FMDV- 3D and 5′UTR by RT-PCR have been developed for the diagnosis (Hoffmann et al. 2009). In addition, analysis of gene expression, pathogen detection, single-nucleotide polymorphism (SNP) analysis, detection of chromosome aberrations, cancer phenotyping, measuring RNA interference can be carried out (Kubista et al. 2006).

Detection Systems

DNA binding dyes like ethidium bromide, SYBR® green I and Evagreen, etc., are intercalating dyes which increase fluorescence upon binding to the dsDNA minor groove, and can be measured in the elongation phase of the cycle. The drawback of using dyes for the detection includes- detection of nonspecific products and primer-dimers. It is a cheaper method but analysis of the melting curve is necessary to check the specificity of amplification. To overcome the limitations of dyes, various probes have been developed which are described as under.

(a) **Primer probes- example, Scorpions**

It combines the action of both primer and the probe. The increase in the fluorescence emission occurs when the probe binds to the newly amplified DNA complementary region by DNA polymerase at each cycle. The drawback of DNA binding dyes of detection of primer–dimers and formation of non–specific products is absent.

(b) **Hydrolysis probes- example, TaqMan**

When the two fluorophores of the probe are close, the fluorescent signal is quenched. Due to the 5′-3′-exonuclease action of DNA polymerase in the extension phase, the bound hydrolysis probe is degraded and fluorescence is emitted from the reporter.

(c) **Hybridization probes- example, Molecular beacon probes**

The Beacon probe unfolds and binds during the annealing phase to the target, leading to fluorescence emission (Navarro et al. 2015).

5.3.3 Types of Polymerase Chain Reaction and Their Applications

The polymerase chain reaction is a highly specific and reliable method that is used for identification of SNP in different breeds, STRs, genetic mutation testing, clinical diagnostics, blood screening, parentage identification, forensics, etc. The use of primers to the conserved regions of the pathogen helps in their identification. For example the use of universal primers along with oligonucleotide probes for the sequences of the conserved 16s rRNA gene aid in recognition of particular bacterial species (Greisen et al. 1994). The degenerate primers identify orthologous genes of different organisms where genome information is not available.

Allele-specific polymerase chain reaction allows detection of SNP (Newton et al. 1989). Hot-start polymerase chain reaction reduces non-specific product amplification (Sharkey et al. 1994). Multiplex-polymerase chain reaction allows simultaneous amplification of many targets in one reaction by the use of multiple primers (Elnifro et al. 2000). Assembly polymerase chain reaction synthesizes long DNA sequences by the use of short overlapping oligonucleotides (Stemmer et al. 1995). Asymmetric polymerase chain reaction preferentially amplifies one strand of the DNA by using an excess of one of the primers (Innis et al. 1988). Quantitative PCR or RT-PCR does simultaneous amplification and real-time quantification of the target DNA (Higuchi et al. 1992). Touchdown or step-down polymerase chain reaction uses higher annealing temperature in early cycles to avoid non-specific amplification (Don et al. 1991).

5.4 Vaccines

Vaccination is the process of injecting antigen to elicit an active immune response against a particular disease. It is considered as an effective method for the prevention and control of diseases. Herd protection is more important in livestock rather than the individual. A vaccine is either made from killed microbes, live attenuated microorganisms, toxoids, or subunits. The limitations of conventional vaccines like a reversion to virulence, requirement of cold storage, short term protection, frequent new outbreaks with new strains, and high level of biosafety requirement affect the vaccine usage. However, the application of modern biotechnology with genetic engineering techniques like reverse vaccinology helps in the development of novel safer vaccines with long term protection. The techniques like isolation of pure antigen by using specific monoclonal antibody, production of antigen by molecular cloning of gene and peptides synthesis as vaccines; pathogen genome analyses, and understanding their mechanisms of pathogenesis resulting in antigen discovery have revolutionized the development of vaccines (Jorge and Dellagostin 2017).

5.4.1 Recombinant Subunit Vaccines

Subunit vaccines consist of an immunogenic component of the pathogen of either protein or glycoprotein that is capable of inducing a protective immune response. The gene cloning and recombinant DNA technology is a powerful tool to prepare a subunit vaccine. Some examples of recombinant subunit vaccines- VP1 protein of FMDV expressed in a heterologous system, E2 subunit vaccine of BVDV, F subunit protein of PPRV and the like. However, the smaller portion of the pathogen does not undergo proper folding, difficulty in the purification of intracellular expressed proteins and requirement of the other components of the pathogen to elicit proper immune response limit the usage of subunit vaccines.

5.4.2 Virus-Like Particles

Virus-like particle vaccines are particles with the absence of replicative genetic material but the presence of viral structural proteins that closely resemble native viruses which would elicit immune response. The absence of genome is also safer to handle without the requirement of high biosafety measures and cause no harm to animals or environment. For instance, the expression of P1-2A, a capsid precursor of FMDV expressed in insect cell culture expression system was found to induce better immune protection against the disease (Subramanian et al. 2012).

5.4.3 Peptide Vaccine

The peptides known to be highly immunogenic to trigger the desired immune response are of 20–30 amino acids length, being included to manufacture synthetic peptide vaccines. Since it is homogenous in preparation it could limit the chances of allergenic complications, since they possess low molecular weight, they require carriers like keyhole limpet hemocyanin (KLH) and adjuvants to improve the efficiency. The amino-terminal region of parvoviral protein VP2 against canine parvovirus is effective (Langeveld et al. 1994).

5.4.4 DNA Vaccines

These emerging vaccines involve the insertion of a gene of interest into a vector, along with suitable genetic materials such as promoters, a polyadenylation signal sequence, and a bacterial origin of replication. Thus obtained plasmid is injected into host cells via several methods including direct injection or electroporation or gene gun. The host cell transcribes and translates the antigenic protein that can induce both cellular and humoral immune responses. The major advantages of DNA vaccines are less expensive, temperature stable and can be transported easily. They stimulate immune response in the presence of maternal antibodies as well. Multiple antigenic genes can be combined that allow for the vaccine development against many pathogens. Also, DNA vaccines induce both cellular and humoral immune response and can be injected via several routes. DNA vaccines against West Nile virus are commercially available by Fort dodge animal health and were released to the market in December 2008 (Redding and Weiner 2009). In addition, efficacy of DNA vaccines can be increased by the expression of co-stimulatory molecules like cytokines, tumor necrosis factor and granulocyte-macrophage colony stimulating factor.

5.4.5 Viral Vectored Vaccines

Commonly used vectors for vaccine generation includes- adenovirus, herpesvirus, and poxviruses. Around 12 viral vectored veterinary vaccines are licensed (Draper and Heeney 2010). Recombinant adenovirus expressing the surface glycoproteins of Rift Valley fever virus (RVFV) showed protection in sheep, goats, and cattle (Warimwe et al. 2016). Viral vector based vaccines exhibit high efficiency of gene transduction and more specific delivery into the target cells. They effectively induce robust immune responses without necessarily requiring an adjuvant.

5.4.6 Marker Vaccine

This vaccine allows differentiation of infected animals from vaccinated animals. The immunogenic antigenic region present in the pathogen will be deleted in the vaccine antigen thus no antibodies would develop against the deleted region. Since the deleted region is present in the original virus, antibodies will be present in the infected animals but not in the vaccinated animals.

In O/HN/CHA/93 FMDV amino acid deletion of 93–143 has been rescued in BHK-21 cells. The mutated virus vaccine has been used as a negative marker vaccine. It protected the pigs against challenge, facilitating differentiation from infection using monoclonal antibody-based indirect blocking ELISA for the deleted portion (Li et al. 2014).

5.4.7 Self-Replicating Vaccine

These are self-amplifying viral RNA sequences that consist of genes encoding an antigen of interest and all necessary elements for RNA replication. It results in viral RNA replication and endogenous expression of the proteins within the host cell; therefore, they are regarded as self-adjuvants. The main advantage is the high amount of antigen expression and triggers a potent interferon response along with T cell-mediated immune responses in the host. Most self-replicating vaccines are based on positive-stranded RNA viruses. T7 or SP6 promoters can be used to produce *in-vitro* transcribed RNA and can be injected directly or cytomegalovirus (CMV) promoter helps in RNA polymerase II-mediated transcription in the target cell. The most widely used replicon vaccines are based on alphaviruses, flaviviruses, and picornaviruses. The DNA-based replicons are delivered to the target cell as plasmid DNA or by using a viral vector. They can be administered via different methods such as naked RNA, RNA launched from a DNA plasmid, RNA packaged in viral replicon particles (VRPs) (Hikke and Pijlman 2017). Some of the examples are Semliki Forest virus-based self-replicating vaccine expressing NS3 of

Bovine viral diarrhoea virus (Reddy et al. 1999) and CS-E2 of Classical swine fever virus (Sun et al. 2011).

These new generation vaccines that are developed by the application of advanced genetic engineering tools have come up with several advantages including stability, non-infectious nature, homogeneity, and cost-effectiveness in their manufacture (Balamurugan et al. 2006).

5.5 Microarray

Microarrays are microscope slides that contain an ordered arrangement of either DNA, RNA, protein or tissue samples. Based on the sample types, they are categorised into DNA microarray, RNA microarray, protein microarray and tissue microarray. The microarray experimentation technique consists of mainly three parts- fabrication of chip called Microarray production, carrying out the reaction which includes isolation of mRNA, processing of the mRNA and finally, hybridization to DNA chip. This is followed by analyzing the results. Initially, it is evolved from southern blotting technique. The first whole genome expression was made on *Saccharomyces cerevisiae* in 1997 and SARS-corona virus is the first virus detected by using microarray.

5.5.1 Applications of microarrays.

(a) To analyze the expression levels of genes in a sample i.e., upregulation or downregulation, or unchanged during several conditions such as a change in nutrition, temperature, or chemical environment. This type of study determines the function of particular genes. Pathogenesis of Bovine Spongiform Encephalopathy (BSE) study by microarray reveals that 182 differentially expressed genes were identified between healthy and infected samples (>2.0-fold change, $P < 0.01$) (Almeida et al. 2011).

(b) It allows specific localization of genes in a cell or tissue and their encoded proteins.

(c) It explains the mechanism of drug action as the drugs regulate the expression of several genes, therefore; identification of the genes help in understanding the mechanism, prediction of drug targets and toxicological properties. DNA microarray helps in development of therapeutics by identifying the presence of drug-resistance genes/plasmids in the pathogens.

(d) It allows diagnosis of diseases. For instance, identification of cancer cells by the gene expression profile; differentiate between cancer cell types. Also prediction of outcome of clinical diseases by the gene profile and to design individual therapies (Macgregor and Squire 2002).

(e) Microarrays can be used for both structural genomics and functional genomics study. Comparative Genome analysis (CGH) can identify gene amplification or deletion of gene that can cause tumor and also to detect the genetic abnormality (Ullmann 2008).
(f) The difference in single nucleotide due to SNP between genomic samples occurs at frequency of about 1 in 1000 bases in animals. It helps in identifying genes responsible for diseases, predicting environmental effects as well as responses to therapeutic agents on their expression (Park et al. 2018).

The drawback of microarray is that the results are difficult to interpret, non-reproducible, expensive, and not always quantitative enough. Animal Viruses Probe Dataset (AVPDS) is used for the identification and microarray based diagnosis of viruses. The first chip based identification of animal virus is FMDV. It is serotype specific and has 35–45 bp long of 155 oligonucleotide probes, designed from VP3-VP1-2A region of virus, labelled with Alexa-Fluor 546 dye (Baxi et al. 2006). Thus, microarray is an advanced rapid technology for the diagnosis of several infectious diseases and oncology. It saves time and would aid in early diagnosis of diseases during outbreaks and/or disease surveillance. In addition, it helps in understanding host-pathogen interactions and development of vaccines.

5.6 Marker Assisted Selection in Animal Breeding

"Phenotype" refers to the observable physical traits of an organism. The phenotype is determined by its genotype, as well as by environmental factors that influence the genes. Marker-Assisted Selection (MAS) is a process that uses a conventional breeding program by which DNA markers that are associated with desirable production traits/ phenotypic information using linkage strategies are used to increase the selection response in a breeding program (Falconer and Mackay 1996). The quantitative traits of the milk includes yield, fat percentage, protein yield, long life, growth rate, fatness, and feed intake are controlled by multiple genes present in several genetic loci; such loci are referred to as quantitative trait loci (QTL) (Beattie 1994; Haley 1995). The markers that do not affect the phenotype are chosen for QTL analysis such as SNPs, microsatellites, restriction fragment length polymorphisms (RFLPs), and transposons position (Casa et al. 2000; Vignal et al. 2002). Finally, the markers that are genetically linked to a QTL influencing the phenotype are identified and scored for the candidate genes that mainly code for transcription factors, and other signalling molecules are finally used for the selection of superior animals for the desirable traits. However, a large sample size is required for the analysis of QTL (Miles and Wayne 2008). The MAS would allow selection of animals for breeding with tolerance or resistance to environmental stresses, including diseases like listeriosis or salmonellosis, African trypanosomiasis, *Boophilus microplus* (Kerr et al. 1994), *Trichostrongylus colubriformis* and *Haemonchus contortus* (Gogolin-Ewens et al. 1990).

5.7 Genome Editors

Genome editors are DNA nucleases that modify the target gene by induction of double-strand breaks at any particular region of the genome. This technique allows deletion, addition or gene silencing to achieve desired characteristics. It can be designed to alter any region of DNA with a target rate around 10,000- to 100,000-fold. They consist of two main domains-cleavage domains and a DNA-binding domain. The more specific DNA nucleases with long recognition sites include- transcription activator-like endonucleases (TALENs), zinc finger nucleases (ZFNs), and the clustered regularly interspaced short palindromic repeats/Cas9 system (CRISPR/Cas9). They are very useful in the production of transgenic animals, creation of animal models for disease study, specific cell line production, genetic diseases treatment, etc. (Petersen 2017).

5.7.1 Zinc Finger Nucleases

ZFNs are class of genetic editors that facilitates gene editing by the creation of double strand breaks at specific position. It was first developed by Kim et al. 1996. There are two functional domains in ZFN, a site-specific DNA-binding domain and cleavage domain. ZFNs typically contain three to six individual zinc finger repeats, each recognizing 3 bp of DNA. And a DNA- non-specific cleavage domain consists of the nuclease domain -Fok I. Two ZFNs are used to target and cut the DNA as dimerization of FokI nuclease is necessary to cut and they are separated by 5–7 bp at which DNA cleavage occurs. The double-strand breaks are repaired by DNA-repair processes of cells, either by homologous recombination or non-homologous end-joining (NHEJ), resulting in gene deletions, integrations, or modifications. A ZFN with new sequence specificities can be designed to cleave any targets (Urnov et al. 2010). The major limitations of ZFN are poor targeting density and cytotoxicity due to high levels of off-target effects.

5.7.2 Transcription Activator-Like Effector Endonucleases

TALEN has a TAL effector DNA-binding domain and a cleavage domain (FokI endonuclease, which cleaves DNA strands). The DNA binding domain has repeats of conserved 33–34 amino acids but differing at the 12th and 13th positions. These two highly variable positions are called the Repeat Variable Diresidue (RVD) and help in specific nucleotide recognition. The appropriate combination of RVD with repeat segments helps in recognition of specific DNA sites. It edits the genomes by inducing break in double-strand (DSB) and are corrected by Non-homologous end joining (NHEJ) and Homology directed repair.

5.7.3 *Clustered Regulatory Interspaced Short Palindromic Repeats/CRISPR Associated Protein9*

The CRISPR-Cas9 uses Watson-Crick base pairing for binding to the target DNA and a Cas9 nuclease, guided by small RNAs known as gRNA for editing the genome. Basically, CRISPR-Cas is an adaptive immune system in prokaryotes. It has two distinct characteristics: the presence of nucleotide repeats and protospacers. These spacers are added into the bacteria by the previously infected viruses and are linked with a protospacer adjacent motif (PAM). They enable the bacteria to recognize the viruses and destroy them. Out of the three types (I–III) of CRISPR systems, the best characterized is Type II CRISPR system. It has nuclease Cas9, the crRNA that codes for guide RNAs which is complementary to the DNA of interest to be edited, and tracrRNA (trans-activating crRNA) that facilitates the processing of the crRNA (Ran et al. 2013). The insulin deficit novel diabetic pig models have been produced by CRISPR/Cas9 editing for the development of diabetes therapeutics by islets transplantation (Cho et al. 2018).

5.7.4 *Applications of Genome Editors*

Genome editors are used for production of disease-resistant animals by genetic knockout of the receptors that are utilized by the organisms for the entry into a cell or knock-in of natural resistance to infection genes. For example- genetic knockout of CD163 receptor (Whitworth et al. 2015) and SRCR5 domain from exon 7 (Burkard et al. 2017) in pigs for the development of resistance against Porcine Reproductive and Respiratory Syndrome Virus infection. Genetic knock-in of NRAMP1 gene by Cas9 nickase has developed resistance to mycobacterial tuberculosis in bovines (Gao et al. 2017).They are used to enhance meat production in livestock. For example, genetic knockout of the MSTN gene- a growth hormone negative regulator of the cattle like Belgian Blue and Piedmontese or Texel sheep that leads to increased skeletal muscle formation for better meat production (Luo et al. 2014).They are applicable in increasing the production and nutritional value of milk; a dominant allergen β-lactoglobulin (blg) gene, was knocked out in cattle by ZFNs (Yu et al. 2011) and improving protein composition by elevating casein levels. Certain individuals are allergic to animal products. Production of allergen-free animal products like knockout of allergenic components from egg white -ovalbumin and ovomucoid (Oishi et al. 2016). An injury due to the horns of the cattle is a common serious problem, so the creation of polled animals is the better option to improve safety of the animals. Carlson et al. (2016) developed polled phenotype Celtic mutation (Pc) by using TALENs in Holstein cattle. Genome editors are also used extensively for development of animal models for disease study as animal models and xenotransplantation helped in research and biomedicine. Some examples of gene editing applications in livestock improvement are provided in Table 5.1.

Table 5.1 Gene editing applications in livestock improvement

Animal	Gene	Type of mutation	Phylogenetic changes	References
Goat	*FGF5*	C to T base editing (non-sense mutation)	Longer cashmere wool	Li et al. (2019)
Pig	*MSTN*	11 bp deletion using CRISPR\Cpf1 system	Double muscle	Zou et al. (2019)
Pig	*CEP112 locus*	β-glucanase, Xylanase & Phytase genes inserted	Promote the digestion of nitrogen and phosphorus in formula feed	Li et al. (2020)
Cattle	*LGB locus*	9 bp deletion in B-lactoglobulin gene using TALEN and SCNT	Knock-out (elimination) of B-lactoglobulin (allergen) from Cow's milk	Wei et al. (2018)
Cattle	*POLLED locus*	Introgression of putative Pc POLLED allele	Hornlessness	Carlson et al. (2016)
Pig	*ALB locus*	Human albumin expressed at swine albumin locus	Transgenic pigs producing hurpan albumin	Peng et al. (2015)
Pig	*hf9*	Gene insertion by CRISPR/Cas9	Gene therapy for hemophilia B	Chen et al. (2021)
Sheep	*FGFS (dominant inhibitor)*	CRISPR/Cas9 gene disruption	Increase in wool length	Hu et al. (2017)

5.8 Role of Biotechnology in Animal Nutrition

The application of biotechnology in the livestock production would benefit the development of improvised feeds with increased digestibility of nutrients, silage inoculants, supplementation of amino acids, mycotoxins diagnosis, removal of toxins and anti-nutritional factors by the enzymatic treatment, inclusion of probiotics, prebiotics, hormones, disease-specific antibodies, to promote gut growth and health. The use of glucanase to improve the digestion of non-starch carbohydrates, phytase to reduce phosphorous content in manure, and to improve the digestibility of anti-nutritional phytate content are some of the applications of enzymes to improve digestion in livestock (Bonneau and Laarveld 1999). Chemical treatments such as formaldehyde and physical treatments like heat treatment and extrusion cooking help in efficient utilization of degradable protein in the rumen for the synthesis of microbial proteins.

5.9 Artificial Insemination

It is one of the assisted reproduction technologies, wherein, the sperm from genetically superior male is introduced to the female oviduct or uterus to produce genetically superior offspring. Lazzaro Spallanzani reported the first successful AI in dogs in the year 1780. This is commonly practiced for the breeding of intensively kept domestic animals especially for breeding dairy cattle (Juneyid et al. 2017), sheep, goat, turkey, pig, and chicken improvement programs. It facilitates the recording of pedigrees with accuracy, reduces the chances of transmission of venereal disease, lessens the need for maintenance of breeding males, and minimizes the cost of introducing improved genetics. Artificial insemination has evolved as a powerful tool in developing desired livestock with the advancement of the various technologies, such as the methods of cryopreservation of semen and sperm sexing (Foote 2010). However, artificial insemination technology is mostly followed in dairy cattle than other livestock species. Difficulty in heat detection in beef cattle kept on ranches, complex surgical method of insemination in small ruminants and inability to successfully cryopreserve boar semen have limited the use of artificial insemination in them (Robinson and McEvoy 1993).

Advantages of Artificial Insemination

(a) Efficient usage of the semen
Dilution of the semen collected from a single superior male can be used to inseminate several females, thus increasing the efficiency of the usage of semen and the stress involved in the natural mating can be reduced as well. Easy transport of semen as compared to the male animals transport from one geographical location to another also allows the insemination simultaneously of several females of different locations. Besides, the stored semen can be used even after the end of natural reproductive lives and the death of the male (Abraham and Pal 2014).

(b) Increased productivity of the animals
Artificial insemination promotes animal productivity, especially milk yields with the well-defined breeding strategy of selected males and females. Also, offspring from artificial insemination sires produce significantly more and increases income.

(c) Reduced transmission of diseases
It avoids the spread of infectious or contagious diseases from a carrier animal to the other by direct contact or shares the same environment in an animal housing facility. The spread of exposure to infectious genital diseases such as granular vaginitis, trichomoniasis, navel ill, dourine, brucellosis, and coital exanthema through sexual contact is also prevented by artificial insemination.

(d) Conservation of rare or endangered animals
The semen can be collected from wild species and stored for the propagation and conservation of endangered animals.

(e) Genetic improvement and increased production
The usage of semen from selected males of superior genetic merit increases the selection intensity and productivity of the animals; also it helps in decreasing inbreeding due to an increase in the genetic pool.

The drawback of artificial insemination is that some of the carrier male animals shed the virus in the semen without showing any signs of disease affecting the individuals and the use of certain breeds for the breeding program may cause loss of genetic variation (Patel et al. 2017).

5.10 Cloning and Transgenic Animals in Livestock Production

Cloning and transgenesis are two major techniques used in generating genetically modified animals (GMAs) for enhanced production by incorporating the favorable traits to animals and birds like higher growth, reproduction, and disease resistance. These techniques help to incorporate the favourable genes in the immediate next generation bypassing the time-consuming breeding and selection strategies. Cloning and transgenesis created an opportunity to develop a bio factory in the farm of animals and produce pharmaceuticals on a large scale. Especially, cloning boosts reproduction exponentially with limited time (Heyman 2005). There are real concerns with cloning, like the survival of the offsprings and their production ability. There are many ethical concerns raised all over the world over the last decades that have suppressed the implication of genetically modified animals in real farm use. The safety of the consumption of such products is a concern and researchers all over the community are putting efforts to evaluate the safety of the GMAs. The growing population of the world and an exponential increase in the demand for the supply of livestock products need advanced techniques to boost production. Provided, GMAs are approved for safety, they will be an invaluable tool to meet the rising demand in the future and a promising route to combat the global hunger (Ibtisham et al. 2017).

5.10.1 Animal Cloning

Clone refers to 'identical' and animal cloning is the process of development of an identical individual from the donor without fertilization and reproduction. Cloning assists the breeding to a greater extent wherein the identical animal of superior quality is produced. The animal clone can be generated in two ways. One is by splitting of the embryonic mass and other by transfer of nuclear material from the donor cells to enucleated oocyte. This way, the identical animal can be produced without fertilization.

5.10.1.1 Types of Animal Cloning

(a) **Embryo splitting**
The simple method of developing the clones is splitting the early embryonic blastomere into two or more smaller embryos. In the early stage, embryo cells are exactly similar and differentiate into an identical organism. The derived embryos are viable when they are split in a very early stage of an embryo of four cells to the eight-cell stage. At a later stage embryo splitting mostly fails to develop into independent blastocysts. This led to looking for the alternative technique wherein the cells in the later stage can be utilized for clone development. The initial studies in amphibians

and mice show that the nucleus from the later stage is capable of imparting viable development of embryos. These findings lead to a new technique of nuclear transfer to embryos.

(b) **Somatic cell nuclear transferprocess**

The nucleus from the adult cells is capable of activating the oocyte and can be developed into a viable embryo. This concept triggered animal science researchers to think about somatic cell nuclear transfer of oocytes to generate embryos without fertilization. The oocytes are available in plenty which are collected and the nucleus was removed. This is called enucleation. Adult somatic cells were used to extract the nucleus and the whole nucleus is transferred into an enucleated oocyte. This results in the activation of oocytes similar to fertilization and differentiates into the embryo. These embryos are cultured in media and transferred into the womb of the surrogate recipient for the development of offspring. Using this technique, the first somatic cell origin clone of lamb was produced named -dolly. After this, there were successful clones in many other domestic animals like cattle, goats, buffalo, and many others. Many epigenetic and post-translational modifications regulate embryonic development. During cloning procedures, the altered conditions elicit the altered epigenetic modifications largely mediated by methylation and histone modifications. This altered regulation yields a large number of abnormal offsprings. These abnormalities are carried mainly through the culture of embryos in the early stages which were overcome by improved media. Still, there is a need to evolve better protocols to avoid environmental aberrations in the clones (Colman 1999). The schematics of reproductive cloning of embryo from a somatic cell and an egg by somatic cell nuclear transfer method is depicted in Fig. 5.2.

5.10.1.2 Applications of Cloning

Cloning is a very helpful tool in producing the identical individual *in vitro* and propagation. This technique also helps to manipulate the genome and transfer it to the next generation. Cloning has the following application that assists in elevating livestock production. It helps to generate genetically identical individuals that assist in epigenetic research. The technique fastens the breeding by propagating the best stocks by cloning. It helps to incorporate desirable qualities like high production and disease resistance in the livestock population. Cloned embryo can be surrogated in case of infertile individuals and dams suffering from diseases. Also, cloning is applicable in cell-based therapeutics, tissue and organ transplantation procedures. It will also be used in generating animals of disease models which assist in fastening research through transgenic animal production and production of biomedical and pharmaceutical products. Furthermore, it will be helpful in conservation of endangered species (Smith et al. 2000).

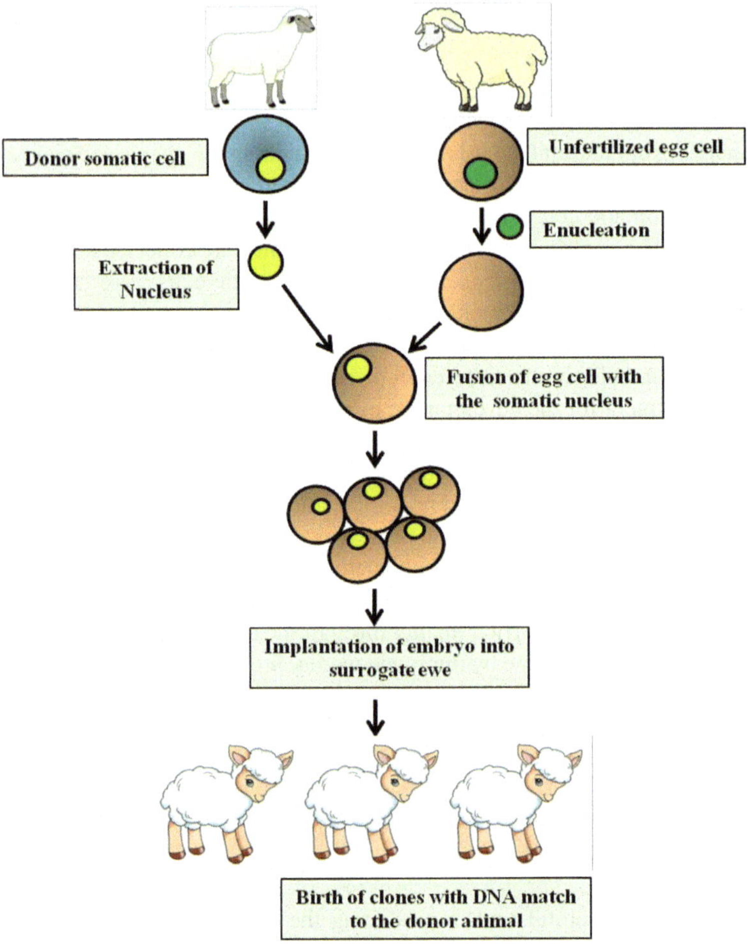

Fig. 5.2 SCNT involves the fusion of nucleus from the donor somatic cell with the enucleated unfertilized egg cell. Later, the blastocyst stage fused cell is transplanted into a surrogate mother to produce the animal clones

5.10.1.3 Limitations of Animal Cloning

Although animal cloning gained importance due to various applications, it has a few limitations as well. The success rate of clone generation is very poor. The offsprings generated shows structural and functional abnormalities. The low success in the reproductive traits in the subsequent generations limits the use of cloning in large scale breeding programs. Animal health and welfare concerns with ethical disputes are also one more concern. However, SCNT is an ineffective procedure in which only less than 10% of the embryos transferred to the recipients would result in the birth of viable offspring (Gurdon and Melton 2008; Wilmut et al. 2002; Yang et al. 2007). Whereas in pigs, only 3%–5% successful conception can be observed (Petersen et al. 2008).

5.10.2 Transgenic Livestock.

A transgenic animal carries an external gene of interest (transgene) added to the germline by biotechnology tools. There is a need to be stably incorporated into the offsprings and passed on to the next generation. Transgenic animals have extensive application in biopharmaceutical industries to produce products like insulin, omega-3 fatty acidand other pharmaceuticals. Many biotechnology tools and techniques were utilized to develop transgenic livestock. Increase in the production of milk by the incorporation of genetic material from high-yielding Holsteins into Guzerat cows could be possible through transgenesis (Wheeler et al. 2010; Wheeler 2013). The first transgenic animal was a supermouse generated by incorporating the growth hormone gene.

5.10.2.1 Methods of Transgenesis

Transgenic animals were developed by stably incorporating the foreign gene into the oocyte or early-stage embryos. This was achieved by various techniques discussed below.

(a) **DNA microinjection**
This was one of the first techniques adapted in transgenesis. In this, the DNA construct was directly injected into the pronucleus of the zygote for stable incorporation. Fine micro-needles and syringe were used and zygote patched in micropipette was injected with stereo-microscope assistance. The DNA gets directly integrated into the genomic DNA (Wahnschaffe et al. 2006). It is a proven method for introduction of transgenes (synthesis of mAbs in milk) of greater than 10 kb (Gavin et al. 2018).

(b) **Transposons**
The DNA injection to the pronucleus is rarely permanently incorporated into the nuclear genome. To overcome the limitation, the gene construct is complimented with transposons which integrate to genomic DNA with a high percentage (Ding et al. 2005).

(c) **Retroviral vector-mediated transfection**
Viral vectors capable of infecting the primordial cells are used to carry the transgene and successfully incorporated into the embryos.

(d) **Electroporation**
The technique in which the pulse of the electric field was applied to create transient pores in the membrane and transfer of DNA constructs (Potter and Heller 2018).

5.10.2.2 Applications of Transgenic Animals

In biomedical research, transgenic animals are extensively used to understand the pathophysiology of particular genes associated with diseases. They have advanced the research to a greater extent over the last few decades (Wheeler 2013). Toxicity studies were conducted with the sensitive animals generated by the transgenic technique. The pharmaceutical industry has explored transgenic animals for the bulk production of animal originating pharmaceutical products. Targeted production of pharmaceutical proteins, drugs, and bioactive products and product efficacy testing in transgenic animals are common applications. Transgenic animals are source for large scale production of the protein of interest for biotechnology applications like antibody production. The technique helped to generate animals capable of expression of insulin-like growth factor 1, epidermal growth factor, transforming growth factor beta, and lactoferrin in milk that improves the growth and survival of offspring (Grosvenor et al. 1993). Genetically engineered genes were added to animals to uplift the milk yield and meat production. Transgenic animals were developed to produce xenografts with human genes to reduce graft rejection in pigs. Production of disease resistance animals to reduce the mortality includes prion-free (Richt et al. 2007) and suppressed prion livestock (Golding et al. 2006) is also one of the important applications in improving production by combating diseases. Reproductive improvements like increase in the litter size by the incorporation of a mutated or engineered estrogen receptor (ESR) gene in pigs. The Boroola fecundity (*FecB*) gene in Merino sheep, an autosomal gene is found to increase the ovulation rate by approximately 1.5 ova (Piper et al. 1984) are classical examples. Many transgenic animals have been developed by inserting the genes which impart special growth and reproductive traits. These incorporations have enhanced the growth and production of animals which is believed to enhance production by manifold. Few examples of transgenic animals are listed below.

Growth hormone gene was inserted to increase the growth and size, enhancing the omega-3 fatty acid in fish known as Superfish. Some species, such as zebrafish, possess optically clear embryos and due to their rapid development, they are considered as important models for developmental and neuronal studies (Chitramuthu and Bennett 2018). Transgenic zebrafish overexpressing TP3 was resistant to *S. agalactiae* infection (Su et al. 2018). Pigs are incapable of digesting the phytate present in cereal grains. Hence transgenic Enviro pig with *E.coli* phytase gene was produced which produces phytase to digest the phytate. Transgenic cattle with lactoferrin and interferon genes. They have been utilized as bioreactors to produce these pharmaceuticals in their milk. Many transgenic sheep were generated by incorporating genes responsible for better wool production and growth in meat-based breeds. The transgenic goats were explored to produce bioactive products in the pharmaceutical industry. Transgenic goats secreting human plasminogen activator, antithrombin factors, silk protein in milk were produced. Many transgenic mice and rats were created by inserting human disease-causing genes into the lab animals. These animals help to advance biomedical research and drug discovery. There are three approved transgenic drugs for the use in humans (Woodfint et al. 2018)-antithrombin from transgenic goats, Ruconest (for treating hereditary angioedema) from transgenic rabbits) and 'farmaceutical' from transgenic chickens.

5.10.2.3 Limitations of Transgenesis

The transgenesis is useful technique to generate wide variety of animals with various capabilities. But this suffers from few limitations which need to be improvised in future. Sometimes there will be multiple copy insertions and overexpression of proteins. The new gene if inserted at essential gene results in the lethal effect.

This also results in transgenesis mediated gene silencing. Apart from main limitations, animal health and safety issues with transgenesis techniques, and need for robust protocols and ethically approved procedures to avoid the misuse of transgenesis technique (Houdebine 2002).

5.11 Embryo Transfer Technology

Embryo transfer (ET) involves the recovery of embryos of superior genotype prior to their implantation in uterus, from the best female (donor) and then transfer into recipient females, serving as surrogate mothers. It helps in achieving greater rate of conception, increase in the progeny number per year from the single superior genotype female. This method can be followed in every species of domestic animals, wildlife and exotic animals, including humans and non-human primates (Vijayalakshmy et al. 2019). The first successful ET in rabbit was made by Walter Heape in 1890 (Heape 1891).

5.11.1 Advantages

ET technology provides a rapid rate of improvement of the genetic quality of offspring at a relatively lower cost than purchasing a live animal. The number of offspring's per female can be increased (MOET-multiple ovulation embryo transfer technology) by ETT. It is easier and results in rapid exchange of genetic material between countries. Transport of live animals can be avoided and thereby reducing risks of transmission of disease. Also, it helps to conserve and expand rare genetic stock and wild animals. Technique is applicable to produce twins and to manipulate embryos for developing new breeding concepts like shortening of generation interval. The technique is valuable tool for expression of desired gene within a short duration of time.

5.11.2 Disadvantages

Although technique is impressive, the technique is costlier and requires advanced and complex manipulations. Also, it creates difficulty in parturition (large size fetus of donor in small recipient).

5.11.3 Steps of Embryo Transport Technology

(a) **Selection of donors**
The donor should be regular estrous cyclic cow or heifer. Usually, cow is preferable than heifer as embryo transfer is based on the milk yield and genetic superiority and age of the donor- between 3 and 10 years. They should be free from any diseases and conformational abnormalities. Donor should have superior production traits with good reproductive performance (a history of not more than two breeding). For dairy cows, a high cow index value that measures the genetic transmitting ability for milk, fat, and protein is the best indicator of good genetic potential. At least 60 days postpartum animals and previous calves with approximately 365 day intervals are chosen before transfer procedure begins. Also, there should not be any parturition difficulties or reproductive irregularities in donor animal.

(b) **Induction of super ovulation**
Multiple ovulations can be achieved by injecting exogenous hormones like PMSG – Pregnant Mare Serum gonadotrophins, FSH- Follicle Stimulating hormone, HMG – Human Menopausal Gonadotrophins in the early follicular phase of the oestrus cycle. Cows or heifers properly treated can release as many as ten or more viable eggs at one estrous. Application dose of FSH is 25–50 mg. FSH is administered at two doses per day as the half-life of FSH is only 2 h. Best time to start FSH administration is the 9th–14th days of the cycle. The PMSG is a glycoprotein having both FSH and LH effects. PMSG – having long half-life (about 5 days) given at dose rate of 2000–4000 IU of one single injection is sufficient to induce superovulation. Human menopausal gonadotrophin (hMG) obtained from urine of a menopausal woman was also showed effective induction effect. If corpus luteum (CL) is present on the ovary, prostaglandin injection given on the fourth day of the treatment schedule causes regression of CL and estrus induction approximately 48 h later. The embryos are recovered 8 days after insemination.

(c) **Recovery of embryo**
Embryo recovery is performed at morula stage and it is varied among species. In Cattle: 6–8 days, sheep and goat: 4–5 days, pig: 3–5 days and in buffalo: 150 h of post estrous (day 6 of the cycle). In bovine, non-surgical method is followed to collect embryos by injecting a local anaesthetic in between the vertebrae on the rump of the cow to reduce rectal contractions. Vulva of the donor is thoroughly cleansed so as to reduce contamination, then Foley's catheter is introduced via the cervix into

the uterine horn with embryo and are flushed with 250–300 ml flushing medium. The recovery rate is influenced by the embryo position in the uterus, flushing method, time of recovery, ovarian response and embryo viability. Surgical method is followed in other small farm animals for the recovery of embryo but it may cause adhesions.

(d) **Evaluation of embryo**

The morphological method and staining method are used to evaluate embryo viability after recovering the embryos.

Morphological Evaluation

The cell shape, colour, number and compactness, size of the perivitelline space, vesicles number and zona pellucida are examined by microscope. The ideal embryo is even, empty perivitelline space with regular diameter. According to their morphological appearance, embryos are classified into four groups (Vijayalakshmy et al. 2019).

(i) Excellent embryos: proper developmental stage with even perfect morphology, empty perivitelline space and regular diameter.
(ii) Good embryos: proper stage of development with minor morphological deviations.
(iii) Degenerated and/or retarded embryos: proper developmental stage with major deviations in morphology (degenerated embryos).
(iv) Unfertilized ova.

Staining Methods

The vital staining and fluorescence techniques are used to evaluate the embryo viability.

(e) **Selection of recipients**

Selection of recipients is based on the traits like normal physiological and health conditions. The animal should be in good reproductive status and lack of any reproductive disorders. The oestrus of donors and recipients should be synchronised within 24 h; otherwise pregnancy rates will be considerably lowered because highest conception is achieved when an embryo is transplanted to a uterine environment that most closely resembles the environment that embryo originated from.

(f) **Transfer of embryos**

It is important to transfer the quality embryo into the tip of the uterine horn bearing the corpus luteum without damaging the endometrium by the use of special catheters. Non-surgical method of transfer is followed in cattle. Following epidural anaesthesia, the embryo to be transferred is taken into a 0.25 ml straw and then placed into the AI gun, which is carefully passed through the cervix and into the uterus corresponding to the ovary that has a corpus luteum. The embryo should be disposed as deep into the uterine horn without much force. If twins are required, an embryo should be placed in both horns of the uterus of the recipient with presence of corpus luteum. The surgical method is followed in sheep, goat and pig, wherein;

abdomen is opened and embryo is placed into the tip of the uterine horn. Recently, laparoscopic method is followed to transfer the embryo. In pig, approximately 16–20 embryos are transferred to achieve a normal litter size.

5.11.4 Cryopreservation of Embryos

It is a crucial part of embryo transfer programme and this procedure allows transportation of embryos across worldwide. Mostly in bovines, 1.4 M glycerol as a cryoprotectant, 20 min equilibration period and 0.25 ml straws as embryo containers are used; it is then slowly cooled down to −35 °C and subsequent plunging into liquid nitrogen (−196 °C) (Vijayalakshmy et al. 2019).

5.12 Sperm Sexing and Embryo Sexing

Determination of the sex of embryo plays an important role in livestock breeding, reproduction and diagnosis of diseases. To meet the increasing demand of meat & milk, farmers always wish to produce young ones of desired sex which ultimately leads to economical benefit. Before the implantation of embryo its sex can be detected either by invasive or non –invasive method from the biopsy sample. Karyotyping is one of the embryo sexing methods, wherein some embryo cells are cultured in colchicine containing media, later staining is done for the visualization of the presence of 2X chromosomes in females and 1Y in males. The presence of barr body in females due to inactivation of the X chromosomes is another method of sexing. The Y-chromosome specific probes have also enabled male sex determination. The polymerase chain reaction (PCR) amplification of DNA sequence of Y chromosome and the visualization of the reaction products determines the presence of male sex (Wakchaure et al. 2015). In past, a variety of techniques have been used to separate the sperm bearing X and Y chromosome to get desired sex, which were based on principle of mass difference, size, swimming pattern, immunological structure and surface charges etc. Currently, flow cytometry is effective as this process constantly separates the semen with 90% of the desired sex (Bhalakiya et al. 2018).

5.13 Conclusion

The increase in the growing population would demand better quality food proteins from livestock, which essentially depends on biotechnology based techniques to get necessary traits. The use of advanced biotechnological tools such as diagnostics, vaccines to prevent the occurrences of the diseases, artificial insemination, embryo transfer, animal cloning, sexing, marker-assisted selection and transgenics would improve livestock production through genetic and animal health improvement. The new generation vaccines that are developed have come up with several advantages like stability, non-infectious nature, homogeneity as there exist batch to batch variation in the conventional system of antigen production and cost-effectiveness in the manufacture. The other biotechnology tools like cloning and transgenesis have created an opportunity to develop bio factory in the farm animals and produce pharmaceuticals on large scale. Especially, cloning boosts reproduction exponentially in limited time. Also, the embryo transfer helps to achieve a greater rate of conception increase in number of progeny per year from single female animal of superior genotype or to obtain desirable genes from superior female animals. The techniques such as application of polymerase chain reaction and bioinformatics to diagnose several diseases, gene therapy and delivery systems contribute significantly to control animal diseases and improve animal health thereby stimulating both food production and livestock trade. The role of biotechnology to increase productivity of animals by protection of amino acids, fats in the feed and modification of rumen microbes for better digestibility and improve animal health, also addition of vaccines or antibodies in feeds can protect the animals from diseases. The modern genetic engineering tools like genome editors such as ZFNs, TALENs and the CRISPR/Cas9 are very useful in the production of transgenic animals, creation of animal models for disease study, specific cell line production and genetic diseases treatment. Therefore, the advanced biotechnology tools would help in achieving quality food supply to the growing world population in a sustainable way. However, there are many ethical concerns raised all over the world over last decades that have suppressed the application of biotechnology and genetically modified animals in real farm use. The safety of the consumption and usage of such products is the concern and researchers all over the community are putting efforts to evaluate the safety. Growing population of the world and exponential increase of the demand for the supply of livestock products definitely need advanced techniques to boost production.

References

Abraham H, Pal SK (2014) Animal biotechnology options in improving livestock production in the horn of Africa. Int J Interdiscip Multidiscip Stud 1(3):1–8. ISSN: 2348-0343

Almeida LM, Basu U, Williams JL, Moore SS, Guan LL (2011) Microarray analysis in caudal medulla of cattle orally challenged with bovine spongiform encephalopathy. Genet Mol Res 10(4):3948–3962. https://doi.org/10.4238/2011.october.25.5

Babiuk LA (2002) Vaccination: a management tool in veterinary medicine. Vet J 164(3):188–201. https://doi.org/10.1053/tvjl.2001.0663

Balamurugan V, Sen A, Saravanan P, Singh RK (2006) Biotechnology in the production of recombinant vaccine or antigen for animal health. J Anim Vet Adv 5(6):487–495

Baxi MK, Baxi S, Clavijo A, Burton KM, Deregt D (2006) Microarray-based detection and typing of foot-and-mouth disease virus. Vet J 172(3):473–481. https://doi.org/10.1016/j.tvjl.2005.07.007

Beattie CW (1994) Livestock genome maps. Trends Genet 10(9):334–338. https://doi.org/10.1016/0168-9525(94)90037-X
Bhalakiya N, Haque N, Patel D, Chaudhari A, Patel G, Madhavatar M, Patel P, Hossain S, Kumar R (2018) Sperm sexing an its application in livestock sector. Int J Curr Microbiol App Sci 7:259–272. ISSN: 2319-7692
Blanchard P, Mahé D, Cariolet R, Keranflec'h A, Baudouard MA, Cordioli P, Albina E, Jestin A (2003) Protection of swine against post-weaning multisystemic wasting syndrome (PMWS) by porcine circovirus type 2 (PCV2) proteins. Vaccine 21(31):4565–4575. https://doi.org/10.1016/S0264-410X(03)00503-6
Bonneau M, Laarveld B (1999) Biotechnology in animal nutrition, physiology and health. Livest Prod Sci 59(2–3):223–241. https://doi.org/10.1016/S0301-6226(99)00029-9
Borroto CG (2009) Biotechnology and its application to veterinary science. Compendium of technical items presented to the International Committee or to Regional Commissions of the OIE, 2008, pp 231–250. ISBN: 9789290447382
Brown TA (2016) Gene cloning and DNA analysis: an introduction. Wiley, Hoboken
Burkard C, Lillico SG, Reid E, Jackson B, Mileham AJ, Ait-Ali T, Whitelaw CBA, Archibald AL (2017) Precision engineering for PRRSV resistance in pigs: macrophages from genome edited pigs lacking CD163 SRCR5 domain are fully resistant to both PRRSV genotypes while maintaining biological function. PLoS Pathogens 13(2):e1006206. https://doi.org/10.1371/journal.ppat.1006206
Carlson DF, Lancto CA, Zang B, Kim ES, Walton M, Oldeschulte D, Seabury C, Sonstegard TS, Fahrenkrug SC (2016) Production of hornless dairy cattle from genome-edited cell lines. Nat Biotechnol 34(5):479–481. https://doi.org/10.1038/nbt.3560
Carrillo C, Wigdorovitz A, Oliveros JC, Zamorano PI, Sadir AM, Gomez N, Salinas J, Escribano JM, Borca MV (1998) Protective immune response to foot-and-mouth disease virus with VP1 expressed in transgenic plants. J Virol 72(2):1688–1690. https://doi.org/10.1128/JVI.72.2.1688-1690.1998
Casa AM, Brouwer C, Nagel A, Wang L, Zhang Q, Kresovich S, Wessler SR (2000) The MITE family Heartbreaker (Hbr): molecular markers in maize. Proc Natl Acad Sci 97(18):10083–10089. https://doi.org/10.1073/pnas.97.18.10083
Chen J, An B, Yu B, Peng X, Yuan H, Yang Q, Chen X, Yu T, Wang L, Zhang X, Wang H (2021) CRISPR/Cas9-mediated knockin of human factor IX into swine factor IX locus effectively alleviates bleeding in hemophilia B pigs. Haematologica 106(3):829. https://doi.org/10.3324/haematol.2019.224063
Chitramuthu BP, Bennett HP (2018) Application of zebrafish and knockdown technology to define progranulin neuronal function. In: Progranulin. Humana Press, New York, pp 207–231. https://doi.org/10.1007/978-1-4939-8559-3_15
Cho B, Kim SJ, Lee EJ, Ahn SM, Lee JS, Ji DY, Lee K, Kang JT (2018) Generation of insulin-deficient piglets by disrupting INS gene using CRISPR/Cas9 system. Transgenic Res 27(3):289–300. https://doi.org/10.1007/s11248-018-0074-1
Colman A (1999) Somatic cell nuclear transfer in mammals: progress and applications. Cloning 1(4):185–200. https://doi.org/10.1089/15204559950019825
Davide E (2010) What is animal biotechnology? www.bio.org
Ding S, Wu X, Li G, Han M, Zhuang Y, Xu T (2005) Efficient transposition of the piggyBac (PB) transposon in mammalian cells and mice. Cell 122(3):473–483. https://doi.org/10.1016/j.cell.2005.07.013
Don RH, Cox PT, Wainwright BJ, Baker K, Mattick JS (1991) 'Touchdown' PCR to circumvent spurious priming during gene amplification. Nucleic Acids Res 19(14):4008. https://doi.org/10.1093/nar/19.14.4008
Draper SJ, Heeney JL (2010) Viruses as vaccine vectors for infectious diseases and cancer. Nat Rev Microbiol 8(1):62–73. https://doi.org/10.1038/nrmicro2240
Elnifro EM, Ashshi AM, Cooper RJ, Klapper PE (2000) Multiplex PCR: optimization and application in diagnostic virology. Clin Microbiol Rev 13(4):559–570. https://doi.org/10.1128/CMR.13.4.559

Engler C, Kandzia R, Marillonnet S (2008) A one pot, one step, precision cloning method with high throughput capability. PloS one 3(11):e3647. https://doi.org/10.1371/journal.pone.0003647

Fachinger V, Bischoff R, Jedidia SB, Saalmüller A, Elbers K (2008) The effect of vaccination against porcine circovirus type 2 in pigs suffering from porcine respiratory disease complex. Vaccine 26(11):1488–1499. https://doi.org/10.1016/j.vaccine.2007.11.053

Falconer DS, Mackay TFC (1996) Introduction to quantitative genetics. Longman Group, Essex

Ferris NP, Powell H, Donaldson AI (1988) Use of pre-coated immunoplates and freeze-dried reagents for the diagnosis of foot-and-mouth disease and swine vesicular disease by enzyme-linked immunosorbent assay (ELISA). J Virol Methods 19(3–4):197–206. https://doi.org/10.1016/0166-0934(88)90014-6

Foote RH (2010) The history of artificial insemination: selected notes and notables. J Anim Sci 80:1–10

Gao Y, Wu H, Wang Y, Liu X, Chen L, Li Q, Cui C, Liu X, Zhang J, Zhang Y (2017) Single Cas9 nickase induced generation of NRAMP1 knockin cattle with reduced off-target effects. Genome Biol 18(1):1–15. https://doi.org/10.1186/s13059-016-1144-4

Gavin W, Blash S, Buzzell N, Pollock D, Chen L, Hawkins N, Howe J, Miner K, Pollock J, Porter C, Schofield M (2018) Generation of transgenic goats by pronuclear microinjection: a retrospective analysis of a commercial operation (1995–2012). Transgenic Res 27(1):115–122. https://doi.org/10.1007/s11248-017-0050-1

Gogolin-Ewens KJ, Meeusen ENT, Scott PC, Adams TE, Brandon MR (1990) Genetic selection for disease resistance and traits of economic importance in animal production [gene mapping, immune mechanisms, linkage analysis, marker assisted trait selection; RFLP (restriction fragment length polymorphism), QTL (quantitative trait loci), VNTR]. Revue Scientifiqueet Technique de l'OIE (France) (1990). https://doi.org/10.20506/rst.9.3.519

Golding MC, Long CR, Carmell MA, Hannon GJ, Westhusin ME (2006) Suppression of prion protein in livestock by RNA interference. Proc Natl Acad Sci USA 103:5285–5290. https://doi.org/10.1073/pnas.0600813103

Granados RR, Boyce Thompson Inst for Plant Res Inc (1994) Trichoplusiani cell line which supports replication of baculoviruses. U.S. Patent 5,300,435. Patent Number: US-5300435-A5,300,435

Greisen K, Loeffelholz M, Purohit A, Leong D (1994) PCR primers and probes for the 16S rRNA gene of most species of pathogenic bacteria, including bacteria found in cerebrospinal fluid. J Clin Microbiol 32(2):335–351. https://doi.org/10.1128/jcm.32.2.335-351.1994

Grosvenor CE, Picciano MF, Baumrucker CR (1993) Hormones and growth factors in milk. Endocr Rev 14(6):710–728. https://doi.org/10.1210/edrv-14-6-710

Gurdon JB, Melton DA (2008) Nuclear reprogramming in cells. Science 322(5909):1811–1815. https://doi.org/10.1126/science.1160810

Haley CS (1995) Livestock QTLs—bringing home the bacon? Trends Genet 11(12):488–492. https://doi.org/10.1016/S0168-9525(00)89158-1

Heape W (1891) III. Preliminary note on the transplantation and growth of mammalian ova within a uterine foster-mother. Proc R Soc Lond 48(292–295):457–458. https://doi.org/10.1098/rspl.1890.0053

Henderson LM (2005) Overview of marker vaccine and differential diagnostic test technology. Biologicals 33(4):203–209. https://doi.org/10.1016/j.biologicals.2005.08.006

Heyman Y (2005) Nuclear transfer: a new tool for reproductive biotechnology in cattle. Reprod Nutr Dev 45(3):353–361. https://doi.org/10.1051/rnd:2005026

Higuchi R, Dollinger G, Walsh PS, Griffith R (1992) Simultaneous amplification and detection of specific DNA sequences. Bio/Technology 10(4):413–417. https://doi.org/10.1038/nbt0492-413

Hikke MC, Pijlman GP (2017) Veterinary replicon vaccines. Ann Rev Anim Biosci 5:89–109. https://doi.org/10.1146/annurev-animal-031716-032328

Hoffmann B, Beer M, Reid SM, Mertens P, Oura CA, van Rijn PA, Slomka MJ, Banks J, Brown IH, Alexander DJ, King DP (2009) A review of RT-PCR technologies used in veterinary virology and disease control: sensitive and specific diagnosis of five livestock diseases notifiable to the World Organisation for Animal Health. Vet Microbiol 139(1–2):1–23. https://doi.org/10.1016/j.vetmic.2009.04.034

Houdebine LM (2002) The methods to generate transgenic animals and to control transgene expression. J Biotechnol 98(2–3):145–160. https://doi.org/10.1016/S0168-1656(02)00129-3
Hu R, Fan ZY, Wang BY, Deng SL, Zhang XS, Zhang JL, Han HB, Lian ZX (2017) RAPID COMMUNICATION: generation of FGF5 knockout sheep via the CRISPR/Cas9 system. J Anim Sci 95(5):2019–2024. https://doi.org/10.2527/jas.2017.1503
Ibtisham F, Qadir MF, Xiao M, An L (2017) Animal cloning applications and issues. Russ J Genet 53(9):965–971. https://doi.org/10.1134/S102279541709006X
Innis MA, Myambo KB, Gelfand DH, Brow MA (1988) DNA sequencing with Thermus aquaticus DNA polymerase and direct sequencing of polymerase chain reaction-amplified DNA. Proc Natl Acad Sci USA 85(24):9436–9440. https://doi.org/10.1073/pnas.85.24.9436
Jorge S, Dellagostin OA (2017) The development of veterinary vaccines: a review of traditional methods and modern biotechnology approaches. Biotechnol Res Innov 1(1):6–13. https://doi.org/10.1016/j.biori.2017.10.001
Juneyid R, Hassen A, Kemal J, Welay K (2017) Assessment on problems associated with artificial insemination service in dairy cattle in Tullo district, West Hararghe, Ethiopia. Ethiopian Veterinary Journal 21(2):62–74. https://doi.org/10.4314/evj.v21i2.5
Katzen F (2007) Gateway® recombinational cloning: a biological operating system. Expert Opin Drug Discovery 2(4):571–589. https://doi.org/10.1517/17460441.2.4.571
Kenneth HF, McKearn TJ, Behtol KB (1980) Monocloal antibodies: a new dimension in biological analysis. Plenum Press, New York. https://doi.org/10.1007/978-1-4615-7505-4
Kerr RJ, Frisch JE, Kinghorn BP (1994) Evidence for a major gene for tick resistance in cattle. In: Proceedings, 5th World Congress, University of Guelph, Guelph, Ontario, Canada, 7–12 August, 1994. Volume 20. Poultry breeding; avian biotechnology; behaviour genetics; reproductive biotechnology; immunogenetics and disease resistance genetics; breeding in the tropics and extreme environments, pp 265–268. International Committee for World Congresses on Genetics Applied to Livestock Production
Kfoury C (2007) Therapeutic cloning: promises and issues. McGill J Med 10:112. PMID: 18523539
Kim YG, Cha J, Chandrasegaran S (1996) Hybrid restriction enzymes: zinc finger fusions to Fok I cleavage domain. Proc Natl Acad Sci 93(3):1156–1160. https://doi.org/10.1073/pnas.93.3.1156
Kubista M, Andrade JM, Bengtsson M, Forootan A, Jonák J, Lind K, Sindelka R, Sjöback R, Sjögreen B, Strömbom L, Ståhlberg A (2006) The real-time polymerase chain reaction. Mol Asp Med 27(2–3):95–125. https://doi.org/10.1016/j.mam.2005.12.007
Langeveld JP, Casal JI, Osterhaus AD, Cortes E, De Swart R, Vela C, Dalsgaard K, Puijk WC, Schaaper WM, Meloen RH (1994) First peptide vaccine providing protection against viral infection in the target animal: studies of canine parvovirus in dogs. J Virol 68(7):4506–4513. https://doi.org/10.1128/jvi.68.7.4506-4513.1994
Li P, Lu Z, Bai X, Li D, Sun P, Bao H, Fu Y, Cao Y, Chen Y, Xie B, Yin H (2014) Evaluation of a 3A-truncated foot-and-mouth disease virus in pigs for its potential as a marker vaccine. Vet Res 45(1):51. https://doi.org/10.1186/1297-9716-45-51
Li X, Hao F, Hu X, Wang H, Dai B, Wang X, Liang H, Cang M, Liu D (2019) Generation of Tβ4 knock-in Cashmere goat using CRISPR/Cas9. Int J Biol Sci 15(8):1743. https://doi.org/10.7150/ijbs.34820
Li G, Zhang X, Wang H, Mo J, Zhong C, Shi J, Zhou R, Li Z, Yang H, Wu Z, Liu D (2020) CRISPR/Cas9-mediated integration of large transgene into pig CEP112 locus. G3 Genes Genomes Genetics 10(2):467–473. https://doi.org/10.1534/g3.119.400810
Luo J, Song Z, Yu S, Cui D, Wang B, Ding F, Li S, Dai Y, Li N (2014) Efficient generation of myostatin (MSTN) biallelic mutations in cattle using zinc finger nucleases. PloS one 9(4):e95225. https://doi.org/10.1371/journal.pone.0095225
Macgregor PF, Squire JA (2002) Application of microarrays to the analysis of gene expression in cancer. Clin Chem 48(8):1170–1177. https://doi.org/10.1093/clinchem/48.8.1170
Mason HS, Arntzen CJ (1995) Transgenic plants as vaccine production systems. Trends Biotechnol 13(9):388–392. https://doi.org/10.1016/S0167-7799(00)88986-6

McGhee JR, Kiyono H (1993) New perspectives in vaccine development: mucosal immunity to infections. Infect Agents Dis 2(2):55–73. PMID: 8162356

Mena JA, Kamen AA (2011) Insect cell technology is a versatile and robust vaccine manufacturing platform. Expert Rev Vaccines 10(7):1063–1081. https://doi.org/10.1586/erv.11.24

Miles CM, Wayne M (2008) Quantitative trait locus (QTL) analysis. Nature Education 1(1):208

Mohammed BR, Malang SK, Mailafia S, Agbede RIS (2016) Application of biotechnology towards diagnosis and treatment in veterinary medicine in Africa: potentials and future developments. J Biotechnol Biomater 6(245):2. https://doi.org/10.4172/2155-952X.1000245

Mullis KB, Faloona FA (1987) Specific synthesis of DNA in vitro via a polymerase-catalyzed chain reaction. Methods Enzymol 155:335–350. https://doi.org/10.1016/0076-6879(87)55023-6

Navarro E, Serrano-Heras G, Castaño MJ, Solera JJCCA (2015) Real-time PCR detection chemistry. Clin Chim Acta 439:231–250. https://doi.org/10.1016/j.cca.2014.10.017

Newton CR, Graham A, Heptinstall LE, Powell SJ, Summers C, Kalsheker N, Smith JC, Markham AF (1989) Analysis of any point mutation in DNA. The amplification refractory mutation system (ARMS). Nucleic Acids Res 17(7):2503–2516. https://doi.org/10.1093/nar/17.7.2503

Oishi I, Yoshii K, Miyahara D, Kagami H, Tagami T (2016) Targeted mutagenesis in chicken using CRISPR/Cas9 system. Sci Rep 6:23980. https://doi.org/10.1038/srep23980

Park JH, Jung YL, Lee K, Lee C, Bhagwan B, Park HG (2018) DNA microarray-based technologies to genotype single nucleotide polymorphisms. In: Chang HN (ed) Emerging areas in bioengineering. Wiley, Weinheim. https://doi.org/10.1002/9783527803293.ch30

Patel G, Haque N, Madhavatar M, Kumar Chaudhari A, Kumar Patel D, Bhalakiya N, Jamnesha N, Rajesh Kumar PP (2017) Artificial insemination: a tool to improve livestock productivity. P-ISSN: 2349-8234

Patil SS, Hiremath GM, Ram J, Koppalkar BG (2019) Growth performance of livestock population in India, Karnataka and North-Eastern region of Karnataka. Econ Aff 64(4):783–787. https://doi.org/10.30954/0424-2513.4.2019.13

Peng J, Wang Y, Jiang J, Zhou X, Song L, Wang L, Ding C, Qin J, Liu L, Wang W, Liu J (2015) Production of human albumin in pigs through CRISPR/Cas9-mediated knockin of human cDNA into swine albumin locus in the zygotes. Sci Rep 5(1):1–6. https://doi.org/10.1038/srep16705

Petersen B (2017) Basics of genome editing technology and its application in livestock species. Reprod Domest Anim 52:4–13. https://doi.org/10.1111/rda.13012

Petersen B, Lucas-Hahn A, Oropeza M, Hornen N, Lemme E, Hassel P, Queisser AL, Niemann H (2008) Development and validation of a highly efficient protocol of porcine somatic cloning using preovulatory embryo transfer in peripubertal gilts. Cloning Stem Cells 10(3):355–362. https://doi.org/10.1089/clo.2008.0026

Piper LR, Bindon BM, Davis GH (1984) Single gene inheritance of the high litter size of the Booroola Merino. In: Land RB, Robinson DW (eds) Genetics of reproduction in sheep. Elsevier, Kent

Potter H, Heller R (2018) Transfection by electroporation. Curr Protoc Mol Biol 121(1):9–3. https://doi.org/10.1002/cpmb.48

Ran FA, Hsu PD, Wright J, Agarwala V, Scott DA, Zhang F (2013) Genome engineering using the CRISPR-Cas9 system. Nat Protoc 8(11):2281–2308. https://doi.org/10.1038/nprot.2013.143

Redding L, Weiner DB (2009) DNA vaccines in veterinary use. Expert Rev Vaccines 8(9):1251–1276. https://doi.org/10.1586/erv.09.77

Reddy JR, Kwang J, Varthakavi V, Lechtenberg KF, Minocha HC (1999) Semiliki forest virus vector carrying the bovine viral diarrhea virus NS3 (p80) cDNA induced immune responses in mice and expressed BVDV protein in mammalian cells. Comp Immunol Microbiol Infect Dis 22(4):231–246. https://doi.org/10.1016/S0147-9571(99)00014-4

Richt JA, Kasinathan P, Hamir AN, Castilla J, Sathiyaseelan T, Vargas F, Sathiyaseelan J, Wu H, Matsushita H, Koster J, Kato S (2007) Production of cattle lacking prion protein. Nat Biotechnol 25(1):132–138. https://doi.org/10.1038/nbt1271

Robinson JJ, McEvoy TG (1993) Biotechnology—the possibilities. Anim Sci 57(3):335–352. https://doi.org/10.1017/S1357729800042661
Shams H (2005) Recent developments in veterinary vaccinology. Vet J 170(3):289–299. https://doi.org/10.1016/j.tvjl.2004.07.004
Sharkey DJ, Scalice ER, Christy KG, Atwood SM, Daiss JL (1994) Antibodies as thermolabile switches: high temperature triggering for the polymerase chain reaction. Bio/Technology 12(5):506–509. https://doi.org/10.1038/nbt0594-506
Smith LC, Bordignon V, Babkine M, Fecteau G, Keefer C (2000) Benefits and problems with cloning animals. Canadian Vet J 41(12):919. PMID: 11143925
Stemmer WP, Crameri A, Ha KD, Brennan TM, Heyneker HL (1995) Single-step assembly of a gene and entire plasmid from large numbers of oligo deoxyribonucleotides. Gene 164(1):49–54. https://doi.org/10.1016/0378-1119(95)00511-4
Su BC, Lai YW, Chen JY, Pan CY (2018) Transgenic expression of tilapia piscidin 3 (TP3) in zebrafish confers resistance to Streptococcus agalactiae. Fish Shellfish Immunol 74:235–241. https://doi.org/10.1016/j.fsi.2018.01.001
Subramanian BM, Madhanmohan M, Sriraman R, Reddy RC, Yuvaraj S, Manikumar K, Rajalakshmi S, Nagendrakumar SB, Rana SK, Srinivasan VA (2012) Development of foot-and-mouth disease virus (FMDV) serotype O virus-like-particles (VLPs) vaccine and evaluation of its potency. Antivir Res 96(3):288–295. https://doi.org/10.1016/j.antiviral.2012.09.019
Sudeep AB, Mourya DT, Mishra AC (2005) Insect cell culture in research: Indian scenario. Indian J Med Res 121(6):725. ISSN 0971-5916
Sun Y, Li HY, Tian DY, Han QY, Zhang X, Li N, Qiu HJ (2011) A novel alphavirus replicon-vectored vaccine delivered by adenovirus induces sterile immunity against classical swine fever. Vaccine 29(46):8364–8372. https://doi.org/10.1016/j.vaccine.2011.08.085
Tinsley TW, Harrap KA (1978) Viruses of invertebrates. In: Newly characterized Protist and invertebrate viruses. Springer, Boston, pp 1–101. https://doi.org/10.1007/978-1-4684-2724-0_1
Ullmann R (2008) Array comparative genomic hybridization in pathology. In: Molecular pathology of lung diseases. Springer, New York, pp 97–107. https://doi.org/10.1007/978-0-387-72430-0_10
Urnov FD, Rebar EJ, Holmes MC, Zhang HS, Gregory PD (2010) Genome editing with engineered zinc finger nucleases. Nat Rev Genet 11(9):636–646. https://doi.org/10.1038/nrg2842
Van Aarle P (2003) Suitability of an E2 subunit vaccine of classical swine fever in combination with the E (rns)-marker-test for eradication through vaccination. Dev Biol 114:193–200
Vignal A, Milan D, SanCristobal M, Eggen A (2002) A review on SNP and other types of molecular markers and their use in animal genetics. Genet Select Evol 34(3):275–305. https://doi.org/10.1051/gse:2002009
Vijayalakshmy K, Virmani M, Verma R, Manimegalai J, Choudhiry UL (2019) Micromanipulation of embryos: an overview. P-ISSN:2349–6800
Wahnschaffe U, Kielhorn J, Bitsch A, Mangelsdorf I (2006) Transgenic animal Mutagenicity assays. World Health Organization
Wakchaure R, Ganguly S, Kumar P, Mahajan T (2015) Methods for embryos sexing and their applications in animal breeding: a review. Octa J Biosci 3(2):47–49
Warimwe GM, Gesharisha J, Carr BV, Otieno S, Otingah K, Wright D, Charleston B, Okoth E, Elena LG, Lorenzo G, Ayman EB (2016) Chimpanzee adenovirus vaccine provides multispecies protection against Rift Valley fever. Sci Rep 6(1):1–7. https://doi.org/10.1038/srep20617
Wei J, Wagner S, Maclean P, Brophy B, Cole S, Smolenski G, Carlson DF, Fahrenkrug SC, Wells DN, Laible G (2018) Cattle with a precise, zygote-mediated deletion safely eliminate the major milk allergen beta-lactoglobulin. Sci Rep 8(1):1–13. https://doi.org/10.1038/s41598-018-25654-8
Wheeler MB (2013) Transgenic animals in agriculture. Nature Educ Knowl 4(11):1–9
Wheeler MB, Monaco E, Bionaz M, Tanaka T (2010) The role of existing and emerging biotechnologies for livestock production: toward holism. Acta Sci Vet 38(Supl 2):s463–s484. ISSN 1679-9216

Whitworth KM, Rowland RR, Ewen CL, Trible BR, Kerrigan MA, Cino-Ozuna AG, Samuel MS, Lightner JE, McLaren DG, Mileham AJ, Wells KD (2015) Gene-edited pigs are protected from porcine reproductive and respiratory syndrome virus. Nature Biotechnol 34(1):20. https://doi.org/10.1038/nbt.3434

Wigdorovitz A, Carrillo C, Santos MJD, Trono K, Peralta A, Gómez MC, Ríos RD, Franzone PM, Sadir AM, Escribano JM, Borca MV (1999) Induction of a protective antibody response to foot and mouth disease virus in mice following oral or parenteral immunization with alfalfa transgenic plants expressing the viral structural protein VP1. Virology 255(2):347–353. https://doi.org/10.1006/viro.1998.9590

Wigdorovitz A, Mozgovoj M, Santos MJD, Parreno V, Gomez C, Perez-Filgueira DM, Trono KG, Ríos RD, Franzone PM, Fernandez F, Carrillo C (2004) Protective lactogenic immunity conferred by an edible peptide vaccine to bovine rotavirus produced in transgenic plants. J Gen Virol 85(7):1825–1832. https://doi.org/10.1099/vir.0.19659-0

Wilmut I, Beaujean N, De Sousa PA, Dinnyes A, King TJ, Paterson LA, Wells DN, Young LE (2002) Somatic cell nuclear transfer. Nature 419(6907):583–587. https://doi.org/10.1038/nature01079

Woodfint RM, Hamlin E, Lee K (2018) Avian bioreactor systems: a review. Mol Biotechnol 60(12):975–983. https://doi.org/10.1007/s12033-018-0128-x

Yang X, Smith SL, Tian XC, Lewin HA, Renard JP, Wakayama T (2007) Nuclear reprogramming of cloned embryos and its implications for therapeutic cloning. Nature Genet 39(3):295–302. https://doi.org/10.1038/ng1973

Yu S, Luo J, Song Z, Ding F, Dai Y, Li N (2011) Highly efficient modification of beta-lactoglobulin (BLG) gene via zinc-finger nucleases in cattle. Cell Res 21(11):1638–1640. https://doi.org/10.1038/cr.2011.153

Zou YL, Li ZY, Zou YJ, Hao HY, Hu JX, Ning L, Li QY (2019) Generation of pigs with a Belgian Blue mutation in MSTN using CRISPR/Cpf1-assisted ssODN-mediated homologous recombination. J Integr Agric 18(6):1329–1336. https://doi.org/10.1016/S2095-3119(19)62694-8

Chapter 6
Mineral Ions in Regulation of Hypothalamic-Pituitary-Ovarian Axis

Mustafa Hassan Jan, Harpreet Singh, and Shikha Kapil

Abstract Reproductive capabilities of females in all species have been known to be affected by their nutritional status. This is true particularly of malnutrition, for example, mineral imbalance that influences almost all facets of female reproduction, right from the onset of puberty to ovarian activity, ovulation, estrus exhibition, fertility, and conception rate. Mineral deficiencies lead to disturbances along hypothalamic-pituitary-ovarian axis, causing reproductive failure. Evidence gathered from *in vitro* studies and laboratory animal models suggests involvement of mineral ions in regulation of upstream pathways involved in synthesis and release of gonadotropin-releasing hormone inside hypothalamic nuclei as well as downstream pathways involved in its action on pituitary gonadotrophs. In contract, mineral availability of specific reproductive tissues during various physiological states and their actions in these tissues are poorly defined in any livestock species. Declaration of mineral deficiency of individual animals based on concentration in blood is often misleading due to extreme dietary influences. Moreover, mineral deficiencies are mostly subclinical and by the time symptoms start appearing, the reproductive symptoms thereof get overlooked due to the overall deterioration of the animal's physiology. The present review offers a summary of biochemical, enzymatic, and endocrine actions of macromineral (calcium, phosphorus, and magnesium) and micromineral (copper, zinc, and manganese) ions along the reproductive axis. We hope to encapsulate research findings to understand the role of specific minerals in female reproduction and identification of biomarkers of impairment of reproductive axis in event of mineral malnutrition, to help us in the development of new nutritional and reproductive strategies for increasing fertility of livestock.

Keywords Minerals · Reproduction · Fertility · FSH/LH · GnRH · Steroidogenesis

M. H. Jan
ICAR-Central Institute for Research on Buffaloes, Sub-Campus, Nabha, Punjab, India

H. Singh (✉)
Lala Lajpat Rai University of Veterinary and Animal Sciences, Hisar, Haryana, India

S. Kapil
University Institute of Biotechnology, Chandigarh University, Mohali, Punjab, India

V. K. Yata et al. (eds.), *Sustainable Agriculture Reviews 57*, Sustainable Agriculture Reviews 57, https://doi.org/10.1007/978-3-031-07496-7_6

Abbreviations

Akt	serine/threonine kinase, also known as protein kinase B
GnRH	gonadotropin-releasing hormone
GPCR	G-protein-coupled receptors
HPO	hypothalamic-pituitary-ovarian axis
IGF1	insulin-like growth factor 1
LCAT	lecithin cholesterol acyltransferase
mTOR	mammalian target of rapamycin
PAL	peptidylglycine α-amidating lyase
PGF2α	prostaglandin F2α
PHM	peptidylglycine α-hydroxylating-monooxygenase
PIP2	phosphatidylinositol 4,5-bisphosphate
SHBG	sex hormone-binding globulin
SNARE	soluble N-ethylmaleimide-sensitive factor attachment protein receptors
SOD	superoxide dismutase

6.1 Introduction

Mineral ions are inorganic substances that serve as essential components of physicochemical processes vital to life. Minerals are distributed throughout all body tissues, having a structural role in some tissues and a regulatory role in others (Underwood and Suttle 1999). The major minerals such as calcium, magnesium, sodium, potassium, phosphorus, sulfur, and chlorine are present in the body in relatively larger amounts. Minerals like Ca and P are important components of bone and other tissues; K, Na and Cl play an important part in the maintenance of acid/base balance and membrane electric potential; Na helps to maintain osmotic pressure, whereas Ca has a role to play in nerve transmission. On the other hand, the trace minerals which include iron, copper, zinc, cobalt, molybdenum, manganese, iodine, and selenium, influence enzyme activity as metalloenzymes (Mn, Zn) or as cofactors (Co), are associated with endocrine hormones (I) and serve as stabilizers of secondary molecular structure. Productive and reproductive efficiency of the animal is the most important factor for the success of a dairy farm. Mineral deficiencies are associated with loss of reproductive performance and related production parameters of economic importance. The hypothalamic–pituitary–gonadal axis as a vital bodily system governs the production and release of gonadal steroid hormones, which further regulate species-specific patterns of sexual development and behavior. Presumably, altered hormonal milieu and ultimately disruption of reproductive functions occur due to disturbance of one or more minerals along this axis. In general, minerals especially, Ca, P and Mg (major elements), and Cu, Zn and Mn (trace elements) influence the reproductive process by their action on hypothalamo-pituitary-gonadal axis (Michaluk and Kochman 2007; Pradhan and Nakagoshi 2008; Dicken et al. 2010). Reproductive performance is, however, affected not only by the absolute concentrations of the major elements, but also by their relative proportions, most notably that of Ca and P (Steevens et al. 1971).

6.2 Hypothalamic-Pituitary-Ovarian Axis

Hypothalamic-pituitary-ovarian axis refers to endocrine coordination between the hypothalamus, pituitary, and the gonads (ovaries) responsible for initiating and regulating the cyclic changes in the female reproductive tract (Fig. 6.1). The gonadotropin-releasing hormone (GnRH), produced by the hypothalamus, regulates gonadotropin secretion by the anterior pituitary, which in turn regulates ovarian functions. Potent stimulators of GnRH include members of the kisspeptin family of peptides. It is believed that GnRH neurons express GPR54 (kisspeptin receptor, KISS1R) and kisspeptin neurons target the GnRH neuronal network to influence the release of GnRH. The hypophyseal portal system picks up GnRH at the median eminence and transports it to the anterior pituitary. The pulsatile release of follicle-stimulating and luteinizing hormones is in response to rhythmic pulses of GnRH reaching the pituitary gonadotrophs. The pulsating nature of gonadotropin release drives follicular maturation, secretion of estrogen during proestrus, and ovulation

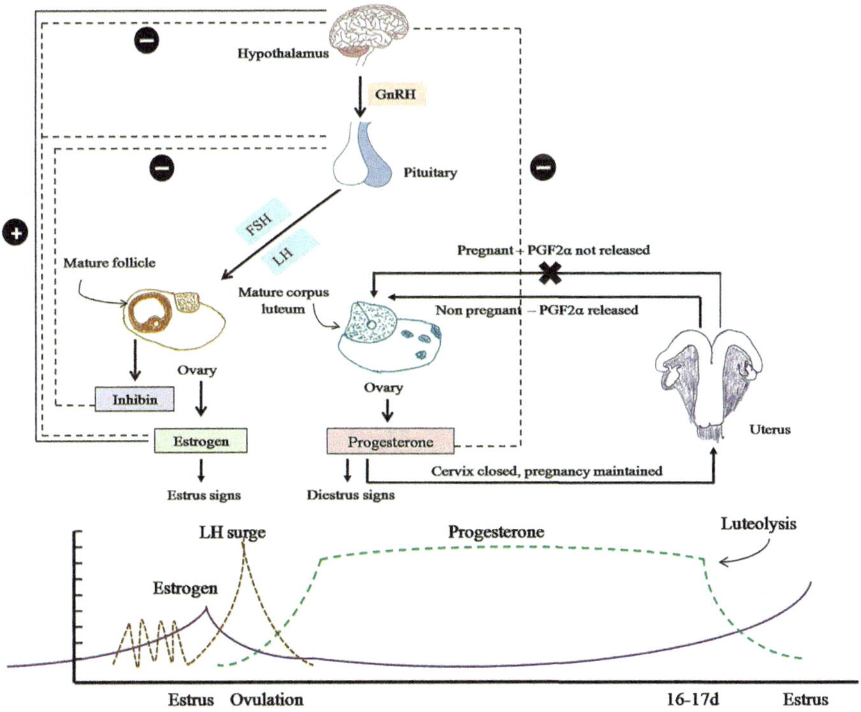

Fig. 6.1 The regulation of hypothalamic-pituitary-ovarian axis and the associated hormonal dynamics. Positive (+) and negative (−) feedback loops are indicated

and secretion of progesterone during diestrus. At the ovaries, the granulosa cells bind follicle-stimulating hormone and theca cells bind luteinizing hormone, to produce progestins, androgens, and estradiol – a model popularly known as two-cell–two-gonadotropin theory (Schiffer et al. 2019).

The neurosecretory mechanism responsible for GnRH secretion from the mammalian hypothalamus is under the influence of a complex interaction involving several excitatory and inhibitory signals. The control of GnRH release from the hypothalamus during the estrous cycle depends mainly upon the feedback mechanism from downstream molecules such as estrogen and progesterone (Kasa-Vubu et al. 1992; Evans et al. 1997; Harris et al. 1999; Richter et al. 2001). The largest healthy follicle in the cohort releases estradiol and inhibin which through a negative feedback mechanism prevent gonadotropin secretion. The lack of gonadotropin stimulation prevents the growth of other follicles in the cohort, whereas only the dominant follicle is selected to grow and destined to be the preovulatory follicle (Hillier 1994). Towards the end of follicular phase (especially during estrous), the peak estrogen production from preovulatory follicle induces GnRH surge from the hypothalamus through a positive feedback mechanism followed by a concomitant luteinizing hormone surge (Adams et al. 2008). This luteinizing hormone surge is responsible for final growth, maturation and ovulation of preovulatory follicle, and formation of corpus luteum. Following ovulation, the granulosa and theca cells of the ovulated follicle luteinize to produce progesterone. In non-pregnant animals, during the late-luteal phase, the luminal epithelium of the endometrium releases luteolytic signal (prostaglandin F2α). However, in a pregnant animal, embryo releases enough quantities of the maternal recognition of pregnancy signal (interferon tau) between days 12 and 32 of the estrous cycle (Bazer 2013). The interferon tau prevents the formation of the luteolytic signal, thereby leading to the persistence of corpus luteum. Evidence gathered from *in vitro* studies and laboratory animal models suggests involvement of mineral ions in several steps along the reproductive axis.

6.3 Macrominerals

Reproductive events are cyclic in nature so the nutrient requirement of tissues also varies across different physiological stages (Hurley and Doane 1989; Robinson 1990). Macro-minerals regulate various reproductive functions including biosynthesis, secretion, and function of hormones, and their requirement may vary independently according to various physiological stages such as puberty, pregnancy, lactation, and the postpartum period (Ahmed et al. 2000; Ali et al. 2010). A perusal of the scientific literature suggests that even though several macrominerals affect the reproductive health and fertility in farm animals, their precise roles in reproduction are not clearly defined. Among the macro-minerals, because of the role they play in the hypothalamic-pituitary-ovarian axis, Ca, P and Mg have been associated with the various reproductive processes (Table 6.1).

Table 6.1 Role of macrominerals along the hypothalamo-pituitary-ovarian axis

Macrominerals	Hypothalamus	Pituitary	Ovary	Other tissues
Calcium	GnRH release from nerve terminals hypophyseal portal vessels (Bourguignon et al. 1987; Goor et al. 2000)	Exocystosis of secretary vesicle of gonadotrophs (Martin 2003)	Steroidogenesis (Shemesh et al. 1984), ovulation (Espey 1970), follicular atresia (Lebedeva et al. 1998), luetolysis (Davis et al. 1987; Choudhary et al. 2005)	Uterine contractility
Phosphorus	Part of second messenger of hormones (IP3, PIP2, cAMP, cGMP, etc)			
Magnesium	GnRH release (Burrows and Barnea 1982a), circadian rhythm (Kruijver and Swaab 2002), cofactor of enzymes that require ATP (such as Protein kinases) (de Baaij et al. 2015)	Cofactor of enzymes that require ATP (such as Protein kinases) (de Baaij et al. 2015)	Enzyme involved in cholesterol esterification (Gueux et al. 1984)	Part of sex hormone binding globulin (Maggio et al. 2011), influences parathyroid hormone (PTH) secretion (Rodríguez-Ortiz et al. 2014)

6.3.1 Calcium

Calcium (Ca) is a major macromineral that affects reproduction. This element is involved in various reproductive processes and acts via the hypothalamic-pituitary-gonadal axis to regulate follicular development and atresia. An optimum Ca concentration ensures that reproductive organs are properly sensitized to various hormones and that the normal reproductive cycle is maintained. Blood concentration of Ca varies across the estrous cycle, being maximum at estrous in cattle (Burle et al. 1995) demonstrating the critical role it plays during the follicular phase especially in and around estrus. In contrast, subclinical hypocalcemia has been reported to cause delayed puberty and anestrus in cows (Dutta et al. 2001). The blood Ca homeostasis is under the influence of parathyroid hormone, 1,25-dihydroxyvitamin D_3 and calcitonin. The parathyroid hormone and 1,25-dihydroxyvitamin D_3, released in response to low Ca concentration, increase blood Ca by increasing bone resorption, decreasing renal excretion, and enhancing Ca absorption from the intestines. Calcitonin from the thyroid gland, on the other hand, decreases blood Ca by inhibiting reabsorption of Ca from bones and increasing urinary Ca excretion (Murray et al. 2008).

At the level of the hypothalamus, the neuronal circuitry responsible for the pulsatile secretion of GnRH from axonal terminals near median eminence is dependent on voltage-sensitive Ca^{2+} influx for its maintenance (Bourguignon et al. 1987; Goor et al. 2000). As GnRH concentrations increase beyond a threshold, isolated gonadotrophs have also been shown to respond by presenting dose-dependent intracellular Ca^{2+} signals (Leong and Thorner 1991; Stojilković et al. 1993; Tomić et al. 1996; Sánchez-Cárdenas and Hernández-Cruz 2010). The GnRH receptor (GnRHR) on the gonadotroph membrane is a member of the G-protein-coupled receptor (GPCR) family (Naor 2009). The binding of GnRH to the receptor, activates phospholipase C, which in turn hydrolyzes membrane phosphatidylinositol 4,5-bisphosphate (PIP2) into inositol 1,4,5-trisphosphate (IP_3) and diacylglycerol (DAG) (Naor 2009). IP_3 initiates Ca^{2+} release from the intracellular pool, the rise which is responsible for the fusion of gonadotropin-laden secretory vesicles with the plasma membrane to release hormones (Stojilković et al. 1991; Martin 2003). In addition, DAG-induced protein kinase-C activation also drives the influx of extracellular calcium into gonadotrophs through voltage-gated Ca^{2+} channels for sustained release of gonadotropins (Durán-Pastén and Fiordelisio 2013). Ca withdrawal either due to deficiency or due to presence of Ca blockers inhibits the release of GnRH, follicle-stimulating hormone and luteinizing hormone (Stojilković et al. 1988; Krsmanović et al. 1992; Dhanvantari and Wiebe 1994). It is believed that the detection of Ca by Ca-binding synaptotagmins is involved in both vesicle docking as well as synaptic vesicle fusion with the cell membrane by using SNARE (soluble *N*-ethylmaleimide-sensitive factor attachment protein receptors) proteins as effectors (Durán-Pastén and Fiordelisio 2013).

At the ovarian level, the level of Ca in the follicular fluid increases with the increase of follicular size in cattle (Wise 1987), suggesting a possible role in the gonadotropic regulation of ovarian steroidogenesis (Veldhius and Klase 1982; Carnegie and Tsang 1984). Ca influences cholesterol delivery to and utilisation by mitochondria, and stimulates conversion of pregnenolone to progesterone, which might further explain how this mineral affects ovarian stereogenesis (Shemesh et al. 1984). As the follicle develops from early follicle to the ovulatory stage, an increase in the mineral content (including Ca) of follicular fluid osmotically drives the movement of water from the blood to antrum (Kalmath and Ravindra 2007). Moreover, an increase in Ca concentration increases plasmin activity which in turn weakens the follicular wall and initiates the process of ovulation (Espey 1970). Calcium dependent endonucleases have also been recognized for causing follicular and luteal cell apoptosis.

6.3.2 *Phosphorus*

The importance of Phosphorus (P) for the normal functioning of all animal tissues is underlined by its involvement in the process of energy exchange at tissue level, besides its role in growth, lactation and reproduction (Little 1970). P is an essential component of phospholipid–dependent protein kinase and cAMP-dependent protein kinase, and is, therefore, crucial in mediating hormone action at the target tissues (Hurley and Doane 1989). The deficiency of P leads to disturbances along

HPO-axis, resulting in dysregulation of the normal reproductive rhythm including ovulation (Bhaskaran and Abdulla Khan 1981; Ullah et al. 2010). P-deficiency is manifested as an alteration of ovarian activity resulting in delayed puberty, irregular cyclicity and cessation of behavioral estrous (Hurley and Doane 1989; Ahmed et al. 2010). As compared to anestrous cattle and buffaloes, significantly higher serum concentrations of P (total as well as free form) have been reported in normal cyclic animals (Dutta et al. 2001; Chaurasia et al. 2010). Similarly, in comparison to blood, higher concentrations of P have been reported in follicular fluid (Abd Ellah et al. 2010; Tabatabaei and Mamoei 2011). The concentration in follicular fluid tends to increase towards (Abd Ellah et al. 2010; Eissa 1996).

The importance of dietary calcium and phosphorus ratio (Ca:P) for reproductive performance has been stressed by Pugh et al. (1985), suggesting that a disturbed dietary Ca:P ratio with a resultant improper serum ratio of the ions has a blocking action on the pituitary gland and consequently on the ovarian function. Low Ca and inorganic P levels with resultant serum Ca:P imbalance might be responsible for anestrus status in heifers under poor managemental practices (Dunn and Moss 1992). 1α-hydroxylase (key enzyme in synthesis of 1,25-dihydroxyvitamin D_3) deficient [1α(OH)ase$^{-/-}$] female mice have been reported to be infertile and exhibit uterine hypoplasia and absence of corpora lutea which was attributed to the resulting hypocalcemia and hypophosphatemia (Panda et al. 2001). Infertility accompanied by decreased synthesis of sex steroids, defects in follicular and luteal development, decreased expression of ovarian angiogenic factors and hypoplasia of endometrium was observed in response to hypocalcemia and hypophosphatemia in 1,25-dihydroxyvitamin D_3-deficient female mice (Sun et al. 2010). The infertility and the associated defective reproductive phenotype in 1α-hydoxylase deficient female mice were reversed when serum Ca and P levels were restored by the rescue diet (Sun et al. 2010). Besides, reproductive problems such as low first service conception rates and silent heat in ewes (Mosaad and Derar 2009) and delayed puberty in buffalo heifers (Ahmed et al. 2010) have been related to abnormally wide Ca:P ratios.

6.3.3 Magnesium

Magnesium (Mg) has a role in reproduction (Stolkowski 1977), but a direct effect has not been demonstrated yet. Low Serum Mg levels have been reported during anestrous in cattle (Dutta et al. 2001) and buffaloes (Chaurasia et al. 2010). Mg on account of its association with ATP is involved in enzymatic reactions that require the transfer of a phosphate group such as protein kinases and ATPases involved in the transport of various ions (Weisinger and Bellorín-Font 1998; Romani 2007; de Baaij et al. 2015). Mg and ATP are essential substrates for adenyl cyclase, an enzyme required for the synthesis of cyclic adenosine monophosphate (cAMP) – an important second messenger for several reproductive hormones. Protein kinases and cAMP, on the other hand, are involved in hormone synthesis, storage and release at different sites in the hypothalamo-pituitary-ovarian axis (Poisner and Douglas 1968; Moriyama et al. 2000), and physiological response in target cells.

High Mg and low Ca in the perifusion medium increases activity of suprachiasmatic nucleus in brain tissue slices (Pan et al. 1992). The nucleus is involved in circadian rhythm and has receptors for estrogen and progesterone (Kruijver and Swaab 2002), and possibly be involved in the regulation of cyclicity. The release of luteinizing hormone releasing hormone (LH-RH) occurs in an Mg dependent manner (Burrows and Barnea 1982a). In an *in-vitro* assembly, Mg and ATP act jointly to facilitate the release of LH-RH in hypothalamic granules, though Mg alone can also release LH-RH but to a lower magnitude (Burrows and Barnea 1982a). In humans, the risk of polycystic ovarian syndrome was 19 times greater in Mg deficient females than those with normal serum concentrations (Sharifi et al. 2012).

Acute magnesium deficiency is associated with diminished activity of lecithin cholesterol acyltransferase (LCAT), an enzyme involved in reverse cholesterol transport (Gueux et al. 1984). LCAT converts cholesterol to cholesterol esters, which are taken up by follicular cells for steroidogenesis (Cigliano et al. 2002). Mg deficiency can impair estrogen metabolism and deplete brain dopamine (Sircus 2011). In addition, Mg is an essential component of sex hormone-binding globulin (SHBG), a glycoprotein that transports the sex steroids in the blood and regulates their activity in target cells (Maggio et al. 2011). Mg status affects concentrations of cytochrome P450 (CYP450) enzymes involved in both vitamin D–activating (25-hydroxylase and 1α-hydroxylase) and deactivating enzymes (24-hydroxylase) (Dai et al. 2018). It has also been demonstrated that Mg can modulate parathyroid hormone (PTH) secretion (Rodríguez-Ortiz et al. 2014). Hence, Mg influences the absorption of Ca and P, and therefore, its imbalance may also hamper reproductive efficiency indirectly.

6.4 Micro/Trace Minerals

Trace minerals are essential for the normal functioning of the body as cofactors, as activators of enzymes, or as stabilizers of secondary molecular structures, even though they constitute a miniscule of the total mass of the organism (Rabiee et al. 2010). These minerals play vital roles in vitamin synthesis, hormone production, collagen formation, oxygen transport, energy production, and other physiological processes related to growth, reproduction and health (Ceko et al. 2016). However, some processes are prioritized over others under various conditions, e.g., reproduction or immune competence may be altered during subclinical deficient states without any effect on growth and feed intake. Most micronutrient deficiencies affect reproduction through analtered enzyme activity that disrupts energy and protein metabolism, hormone synthesis, and the integrity of rapidly dividing cells within the reproductive system (Table 6.2). Besides, microminerals serve as antioxidants and scavenge free radicals to protect cells from free radical-induced oxidative damage (Leung 1998; Spears and Weiss 2008).

Table 6.2 Role of microminerals along the hypothalamo-pituitary-ovarian axis

Microminerals	Hypothalamus	Pituitary	Ovary	Other tissues
Copper	Conversion of prohormone to active GnRH (Prigge et al. 2000), GnRH release (Burrows and Barnea 1982b)	Modulate action of GnRH on gonadotrophs (Michaluk and Kochman 2007), maintain activity of FSH & LH in blood (Georgievskii 1981)	Ovulatory follicle growth and development (Kendall et al. 2003)	Part of superoxide dismutase (SOD1) enzyme (Matzuk et al. 1998)
Zinc	Conversion of prohormone to active GnRH (Michaluk and Kochman 2007)	Part of steroid receptors (Spelsberg et al. 1989), influences Prolactin release (Brandao-Neto et al. 1995)	Steroidogenesis (Hurley and Doane 1989), cumulus cell expansion, completion of meiosis-I, ovulation (Tian and Diaz 2012)	Part of superoxide dismutase (SOD1) enzyme (Matzuk et al. 1998), mobilization of vitamin A from liver (Christian and West 1998)
Manganese	GnRH release (Lee et al. 2007), *Kiss-1* gene expression (Srivastava et al. 2016)	LH release (Pine et al. 2005)	Luteolysis (Sugino et al. 2000)	Part of superoxide dismutase (SOD2) enzyme (Kasahara et al. 2005)

6.4.1 Copper

The influence of copper (Cu) on both production and reproduction renders this trace mineral of utmost importance to the livestock industry. Cu related physiological disorders may be due to Cu deficiency in diet or due to presence of molybdenum (Mo) and/or sulfur (S) in diet which interferences with Cu bioavailability (Clarkson et al. 2019).

Cu deficiency appears to play significant roles in two key areas i.e. altered reproductive performance and immune suppression (Corah 1996). The consequences of Cu deficiency (high dietary Mo and S) in grazing ruminants may vary from no alteration in estrous behavior to delayed onset of puberty and depressed estrous, but severe copper-deficient cows may show anovulation and retardation of future estrous cycles leading to low fertility (Annenkov 1981; Ingraham et al. 1987; Phillippo et al. 1987; Corah and Ives 1991). The mean serum Cu level in anestrous cattle and buffaloes is reported to be lower than in cyclic animals (Dutta et al. 2001; Akhtar et al. 2009; Ahmed et al. 2010). Neutrophils and mononuclear cells from heifers fed low Cu diet prepartum exhibited reduced phagocytic activity (Torre et al. 1995; Torre et al. 1996). Cu is essential for the activity of superoxide dismutase

(SOD1) enzyme, the deficiency of which is associated with defects in ovarian folliculogenesis and dysregulation of luteal function leading to subfertility and infertility in mice (Matzuk et al. 1998; Noda et al. 2012).

Cu has specific effects on the reproductive axis at the level of the hypothalamus, pituitary and ovary. It maintains optimum fertility by affecting GnRH, FSH, LH and estrogen activity (Desai et al. 1982; Michaluk and Kochman 2007). Administration of Cu salt leads to ovulation in female rabbits (Suzuki et al. 1972; Tsou et al. 1977) and ewes (Murawski et al. 2006), through hypothalamic action. Divalent Cu ion is essential for the activity of peptidylglycine α-hydroxylating-monooxygenase (PHM), an enzyme involved in the first step of generation of C-terminal carboxamides of peptide hormones, the reaction necessary for activation of active GnRH peptide from the prohormone (Prigge et al. 2000) (Fig. 6.2). Cu complex variant of GnRH has been shown to interact with GnRH receptors with enhanced affinity, to bring about a more potent release of LH, and to modulate intracellular signaling by increasing cAMP accumulation in the gonadotrope cells, whereas the non-complexed GnRH utilizes IP3/DAG signaling pathway (Kochman et al. 2005; Michaluk and Kochman 2007; Gajewska et al. 2016).

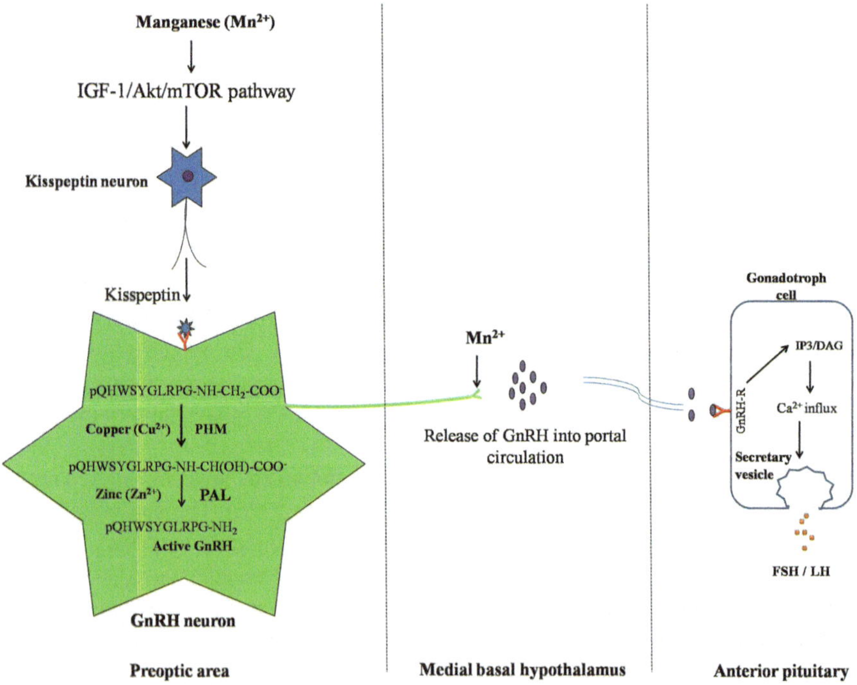

Fig. 6.2 The regulation of upstream and downstream pathways of hormone synthesis and release along hypothalamic-pituitary-ovarian axis by minetral ions and the associated hormonal dynamics. *IGF1* insulin-like growth factor 1, *Mn* manganese, *mTOR* mammalian target of rapamycin complex 1, *Akt* serine/threonine kinase, *PHM* peptidylglycine α-hydroxylating-monooxygenase, *PAL* peptidylglycine α-amidating lyase, *IP3* inositol 1,4,5-trisphosphate, *DAG* diacylglycerol

Cu-deficiency induced perturbation of the follicular growth and development may be due to disturbance in Cu-containing enzyme, lysyl oxidase, which promotes cross-linking of collagen and elastin, and stabilizes the extracellular matrix (Kendall et al. 2003). The basal lamina is constantly remodeled as the follicle matures (Rodgers et al. 1999), thereby requiring the expression of the lysyl oxidase in granulosa cells (Slee et al. 2001). Interestingly, a high concentration of Cu in follicular fluid is seen as an endocrine disruptor leading to polycystic ovary syndrome (Sun et al. 2019).

6.4.2 *Zinc*

Zinc (Zn) plays a pivotal role in reproduction as an essential component of enzymes in different biochemical pathways (Hurley and Doane 1989). It acts indirectly through the pituitary to influence the release of gonadotropic hormones (Dicken et al. 2010) and directly owing to its presence in Zn fingers (two in number) in steroid receptors (Spelsberg et al. 1989). The second step of the amidation reaction for activation of active GnRH from prohormone is carried out by peptidylglycine α-amidating lyase which requires divalent Zn as a cofactor (Michaluk and Kochman 2007) (Fig. 6.2). It has a physiological role in regulating pituitary prolactin secretion (Brandao-Neto et al. 1995; Lee and Kelleher 2016), and its deficiency is associated with hyperprolactinemia (Dicken et al. 2010), resulting in suppression of hypothalamo-pituitary-gonadal axis.

Zinc deficient animals have depressed serum vitamin A levels as it's deficiency hampers mobilization of vitamin A from the liver. Zn and vitamin A are necessary for the normal functioning of the germinal epithelium of the ovary, and their deficiency leads to deformed or atretic follicles resulting in failure of the ovarian function (Chhabra and Arora 1985). Lower mean serum Zn levels have been reported in buffaloes suffering from anestrous (Akhtar et al. 2009; Ahmed et al. 2010) than their cyclic counterparts.

Zn deficiency is associated with suboptimal steroid hormone concentrations i.e. estrogen and progesterone (Akhtar et al. 2009) which is attributed to the involvement of the ion in the process of steroidogenesis (Hurley and Doane 1989). Ovarian (follicular fluid and granulosa cell) zinc content is reduced during follicular atresia (Kaswan and Bedwal 1995; Bhardwaj and Sharma 2011; Mahavar 2011), indicating that the ion can serve as a marker for follicular health. Higher follicular fluid concentrations of Zn and Cu positively influence the outcome of IVF in the form of greater MII oocyte retrieval, and higher fertilization and cleavage rates (Sun et al. 2017). Zn is essential for cumulus cell expansion and completion of meiosis-I, and an acute Zn deficiency blocks follicle rupture during ovulation (Tian and Diaz 2012). Additionally, Zn supplementation during *in vitro* maturation (IVM) significantly increased the meiotic competence of bovine oocytes (Barros et al. 2018), and supplementation during *in vitro* embryo culture improved the cell number of inner cell mass (Woolridge et al. 2019).

6.4.3 Manganese

Manganese (Mn) is an essential component of bovine nutrition required for carbohydrate metabolism, bone growth, normal brain function, growth, reproduction, and a variety of enzymatic systems (Aschner and Aschner 2005). Corah (1996) suggested that Mn has far more influence on reproduction than actually realized. The effects of Mn deficiency on reproduction have been reviewed over the last few decades (Hidiroglou 1979; Pugh et al. 1985; Sharma 2006; Kumar et al. 2011). Poor follicular development with delayed ovulation, delayed postpartum estrous and delayed puberty in heifers, reduced intensity of estrous, and reduced conception rates are also associated with Mn deficiency (Bentley and Philips 1951; Bourne 1967).

Although the precise mechanism of the involvement of Mn in reproduction is unknown, evidence suggests its role in the activity of certain endocrine organs. Both pituitary and ovarian tissues, particularly Graafian follicle and corpus luteum, are rich in Mn content (Hidiroglou 1979). Mn is carried by transport proteins such as transferrin and divalent metal transporter-1 across the blood-brain barrier into the hypothalamus (Garcia et al. 2006; Aschner et al. 2008). Chronic administration of Mn at low doses has been reported to increase serum levels of LH, FSH and estradiol in female rats, possibly due to the hypothalamic action of the ion (Pine et al. 2005; Lee et al. 2006; Lee et al. 2007). Mn supplementation in female rats activates upstream genes, mainly the *Kiss-1* gene, which regulate GnRH release from hypothalamic nuclei (Srivastava et al. 2013). The *Kiss-1* gene is responsible for the synthesis of kisspeptins (Thompson et al. 2004; Caraty et al. 2007; Keen et al. 2008; Lehman et al. 2010). The Mn-stimulated kisspeptin synthesis is mediated by the IGF-1/Akt/mTOR pathway in the prepubertal female rat (Srivastava et al. 2016; Dees et al. 2017) (Fig. 6.2).

Ovarian Mn content is particularly responsive to Mn deficiency (Wilson 1966; Hurley and Doane 1989). Mn deficiency alters the synthesis of gonadal hormones such as estrogen and progesterone in the female (Keen and Zidenberg-Cheer 1990; Pradhan and Nakagoshi 2008), possibly through inhibition of cholesterol and cholesterol precursor synthesis (Doisy 1974). Mn is an essential component of mitochondrial superoxide dismutase (SOD2). SOD2 expression in corpus luteum has been reported to increase from mid-luteal to late-luteal phase in women (Sugino et al. 2000) and sheep (Al-Gubory et al. 2005), but does not alter in pregnant ewes between days 12 and 20 (Arianmanesh et al. 2011) suggesting a role in corpus luteum regression.

Mn supplementation shortens the postpartum anestrous period and increases conception rates in dairy cows (Krolak 1968; Ahmed et al. 2010), and is effective in suspending summer acyclicity in buffaloes (Ahmed et al. 2010). The serum concentrations of Mn are lower in delayed ovulating and anovulating heifers as compared to normal ovulating heifers (Das et al. 2009). This report confirms the earlier claims that delayed ovulation can be experimentally induced by dietary withdrawal of Mn from dairy cows (Rojas et al. 1965), thus underlining the crucial role Mn plays in the normal ovulatory process.

6.5 Conclusion

Dietary mineral elements in excess or otherwise affect the normal physiological processes of an animal. Imbalances in the mineral availability also affect the reproductive process leading to subnormal fertility, anestrous and ovulatory disturbances. The hypothalamo–pituitary–gonadal axis is the primary site of action of mineral ions. Derangements of mineral availablility end up altering the hormonal milieu along the hypothalamo–pituitary–gonadal axis and cause normal reproductive functions to fail. Evidence gathered from *in vitro* studies and laboratory animal models suggests involvement of mineral ions in regulation of upstream pathways involved in synthesis and release of gonadotropin-releasing hormone (GnRH) inside hypothalamic nuclei as well as downstream pathways involved in its action on pituitary gonadotrophs. However, advances in the understanding of such molecular mechanisms of action of mineral ions have not found their way into ruminant nutrition research and practice, leading to a widespread mineral imbalance in prized livestock. These imbalances not only affect the health of an individual animal or that of a herd, but tend to be a major hurdle in livestock rearing from an economical dimension as well.

Future research should focus on (1) finding better ways of supplementation that will increase bioavailability of minerals ions at specific sites along reproductive axis, (2) the establishment of a panel of biomarkers of reproductive functionality to be used in farm animals for early detection of specific mineral deficiency, (3) understanding how genetics and stages of (re)production of an animal affect the dynamics of mineral requirements as well as their utilisation; (4) understanding the effect of regional as well as seasonal variations on specific and general mineral requirements of an animal instead of a one-size-fits-all approach; and (5) extension of molecular research from laboratory animal models to ruminant nutrition practice.

Acknowledgments We would like to thank Dr. S.S. Dahiya (Director, ICAR-CIRB) for valuable suggestions and comments.

References

Abd Ellah MR, Hussein HA, Derar DR (2010) Ovarian follicular fluid constituents in relation to stage of estrus cycle and size of the follicle in Buffalo. Vet World 3(6):263–267

Adams GP, Jaiswal R, Singh J, Malhi P (2008) Progress in understanding ovarian follicular dynamics in cattle. Theriogenology 69:72–80. https://doi.org/10.1016/j.theriogenology.2007.09.026

Ahmed MM, Siham AK, Barri MES (2000) Macromineral profile in the plasma of Nubian goats as affected by the physiological state. Small Rum Res 38:249–254. https://doi.org/10.1016/S0921-4488(00)00166-8

Ahmed WM, Bashandy MM, Ibrahim AK, Shalaby SIA, El-Moez SIA, El-Moghazy FM, Ibrahim SRE (2010) Investigations on delayed puberty in Egyptian buffalo-heifers with emphasis on clinicopathological changes and treatment using GnRH (Receptal®). Glob Vet 4:78–85

Akhtar MS, Farooq AA, Mushtaq M (2009) Serum concentrations of copper, iron, zinc and selenium in cyclic and anoestrusNili-Ravi buffaloes kept under farm conditions. Pak Vet J 29:47–48

Al-Gubory KH, Ceballos-Picot I, Nicole A, Bolifraud P, Germain G, Michaud M, Mayeur C, Blachier F (2005) Changes in activities of superoxide dismutase, nitric oxide synthase, glutathione-dependent enzymes and the incidence of apoptosis in sheep corpus luteum during the estrous cycle. Biochim. Biophys. Acta. Gen Subj 1725:348–357. https://doi.org/10.1016/j.bbagen.2005.06.018

Ali F, Lodhi LA, Qureshi ZI, Younis M (2010) Serum macromineral levels in estrual, fertile, subfertile and pregnant mares kept under two different managemental conditions. Pak Vet J 30:87–90

Annenkov BN (1981) Mineral feeding of cattle. In: Georgievskii VI, Annenkov BN, Samokbin VT (eds) Mineral nutrition of animals. Butterworths, London, p 285

Arianmanesh M, McIntosh RH, Lea RG, Fowler PA, Al-Gubory KH (2011) Ovine corpus luteum proteins, with functions including oxidative stress and lipid metabolism, show complex alterations during implantation. J Endocrinol 210:47–58. https://doi.org/10.1530/JOE-10-0336

Aschner JL, Aschner M (2005) Nutritional aspects of manganese homeostasis. Mol Asp Med 26:353–362. https://doi.org/10.1016/j.mam.2005.07.003

Aschner M, dos Santos AP, Erikson KM, Zheng W (2008) Manganese transport into the brain:putative mechanisms. In: Metal ions in biology and medicine. Proceedings of the 10th international symposium on metal ions in biology and medicine. Corsica, France, pp 695–700

Barros RG, Lodde V, Dieci C, Fraciosi F, Luciano AM (2018) Study on the effects of zinc supplementation during in vitro embryo production technologies in cattle. Int J Health Anim Sci Food Safety 5(1s). https://doi.org/10.13130/2283-3927/10021

Bazer FW (2013) Pregnancy recognition signaling mechanisms in ruminants and pigs. J Anim Sci Biotechnol 4:23. https://doi.org/10.1186/2049-1891-4-23

Bentley OG, Philips PH (1951) The effect of low manganese rations upon dairy cattle. J Dairy Sci 34:396–403. https://doi.org/10.3168/jds.S0022-0302(51)91727-4

Bhardwaj JK, Sharma RK (2011) Changes in trace elements during follicular atresia in goat (*Capra hircus*) ovary. Biol Trace Elem Res 140:291–298. https://doi.org/10.1007/s12011-010-8700-7

Bhaskaran R, Abdulla Khan CK (1981) Role of blood serum inorganic phosphorus in post-parturient anoestrus cows. Livest Adv 9:138–143

Bourguignon JP, Gerard A, Debougnoux G, Rose J, Franchimont P (1987) Pulsatile release of gonadotropin releasing hormone (GnRH) from the rat hypothalamus in vitro:calcium and glucose dependency and inhibition by superactive GnRH analogs. Endocrinology 121:993–999. https://doi.org/10.1210/endo-121-3-993

Bourne F (1967) Manganese:its relationship to infertility in the bovine. Feed Forum 2:33

Brandao-Neto J, Madureira G, Mendonca BB, Bloise W, Castro AV (1995) Endocrine interaction between zinc and prolactin. An interpretative review. Biol Trace Elem Res 49:139–149. https://doi.org/10.1007/BF02788963

Burle PM, Mangle NS, Kothekar MD, Kalorey DR (1995) Blood biochemical profiles during various reproductive states of Sahiwal and Jersey X Sahiwal cattle. Livest Advisor 20:13–20

Burrows GH, Barnea A (1982a) Comparison of the effects of ATP, Mg^{2+}, and MgATP on the release of luteinizing hormone-releasing hormone from isolated hypothalamic granules. J Neurochem 38:569–573. https://doi.org/10.1111/j.1471-4159.1982.tb08664.x

Burrows GH, Barnea A (1982b) Copper stimulates the release of luteinizing hormone releasing hormone from isolated hypothalamic granules. Endocrinology 110:1456–1458. https://doi.org/10.1210/endo-110-4-1456

Caraty A, Smith JT, Lomet D, Ben Said S, Morrissey A, Cognie J, Doughton B, Baril G, Briant C, Clarke IJ (2007) Kisspeptin synchronizes preovulatory surges in cyclical ewes and causes ovulation in seasonally acyclic ewes. Endocrinology 148:5258–5267. https://doi.org/10.1210/en.2007-0554

Carnegie JA, Tsang BK (1984) The calcium-calmodulin system: participation in the regulation of steroidogenesis at different stages of granulosa cell differentiation. Biol Reprod 30:515–522. https://doi.org/10.1095/biolreprod30.2.515

Ceko MJ, O'Leary S, Harris HH, Hummitzsch K, Rodgers RJ (2016) Trace elements in ovaries:measurement and physiology. Biol Reprod 94:1–14. https://doi.org/10.1095/biolreprod.115.137240

Chaurasia R, Kushwaha HS, Chaurasia D, Gendley MK, Santra AK (2010) Comparative studies of certain macro minerals during various reproductive states in buffaloes. Buffalo Bull 29:291–298

Chhabra A, Arora SP (1985) Effect of Zn deficiency on serum vitamin A level, tissue enzymes and histological alterations in goats. Livest Prod Sci 12:69–77. https://doi.org/10.1016/0301-6226(85)90041-7

Choudhary E, Sen A, Inskeep EK, Flores JA (2005) Developmental sensitivity of the bovine corpus luteum to prostaglandin F2α (PGF2α) and endothelin-1 (ET-1):is ET-1 a mediator of the luteolytic actions of PGF2α or a tonic inhibitor of progesterone secretion? Biol Reprod 72:633–642. https://doi.org/10.1095/biolreprod.104.034736

Christian P, West KP Jr (1998) Interactions between zinc and vitamin A: an update. Am J Clin Nutr 68:435S–441S. https://doi.org/10.1093/ajcn/68.2.435S

Cigliano L, Balestrieri M, Spagnuolo MS, Dale B, Abrescia P (2002) Lecithin-cholesterol acyltransferase activity during maturation of human preovulatory follicles with different concentrations of ascorbate, a-tocopherol and nitrotyrosine. Reprod Fertil Dev 14:15–21. https://doi.org/10.1071/RD01044

Clarkson AH, Paine S, Martín-Tereso J, Kendall NR (2019) Copper physiology in ruminants: trafficking of systemic copper, adaptations to variation in nutritional supply and thiomolybdate challenge. Nutr Res Rev 33:43–49. https://doi.org/10.1017/S0954422419000180

Corah L (1996) Trace mineral requirements of grazing cattle. Anim Feed Sci Technol 59:61–70. https://doi.org/10.1016/0377-8401(95)00887-X

Corah LR, Ives S (1991) Effect of essential trace minerals on reproduction in beef cattle. Vet Clin North Am Food Anim Pract 7:41–57. https://doi.org/10.1016/S0749-0720(15)30809-4

Dai Q, Zhu X, Manson JE, Song Y, Li X, Franke AA, Costello RB, Rosanoff A, Nian H, Fan L, Murff H (2018) Magnesium status and supplementation influence vitamin D status and metabolism: results from a randomized trial. Am J Clin Nutr 108:1249–1258. https://doi.org/10.1093/ajcn/nqy274

Das JM, Dutta P, Deka KC, Biswas RK, Sarmah BC, Dhali A (2009) Comparative study on serum macro and micro mineral profiles during oestrus in repeat breeding crossbred cattle with impaired and normal ovulation. Livest Res Rural Dev 21:Article 72

Davis JS, Weakland LL, Weiland DA, Farese RV, West LA (1987) Prostaglandin F2 alpha stimulates phosphatidylinositol 4, 5-bisphosphate hydrolysis and mobilizes intracellular Ca^{2+} in bovine luteal cells. Proc Natl Acad Sci 84:3728–3732. https://doi.org/10.1073/pnas.84.11.3728

de Baaij JH, Hoenderop JG, Bindels RJ (2015) Magnesium in man–implication for health and disease. Physiol Rev 95:1–46. https://doi.org/10.1152/physrev.00012.2014

Dees WL, Hiney JK, Srivastava VK (2017) Influences of manganese on pubertal development. J Endocrinol 235:R33–R42. https://doi.org/10.1530/JOE-17-0237

Desai MC, Thakkar TP, Dharshana RA, Janakiraman K (1982) Serum copper and iron in Surti buffalo in relation to reproduction and gonadotropins. Indian J Anim Sci 52:443–444

Dhanvantari SA, Wiebe JP (1994) Suppression of follicle-stimulating hormone by the gonadal-and neurosteroid 3 alpha-hydroxy-4-pregnen-20-one involves actions at the level of the gonadotrope membrane/calcium channel. Endocrinology 134:371–376. https://doi.org/10.1210/en.134.1.371

Dicken C, Menke M, Neal-Perry G (2010) The hypothalamic-pituitary-ovarian axis. In: Santoro N, Neal-Perry G (eds) Amenorrhea, contemporary endocrinology series. Humana Press, Totowa, pp 1–19. https://doi.org/10.1007/978-1-60327-864-5_1

Doisy EA (1974) Effects of deficiency in manganese upon plasma levels of clotting proteins and cholesterol in man. In: Hoeskstra WG, Suttie JW, Ganther HE, Mertz W (eds) Trace element metabolism in animals-2. University Park Press, Baltimore, pp 668–670

Dunn TG, Moss GE (1992) Effects of nutrient deficiencies and excesses on reproductive efficiency of livestock. J Anim Sci 70:1580–1593

Durán-Pastén ML, Fiordelisio T (2013) GnRH-induced Ca2+ signaling patterns and gonadotropin secretion in pituitary gonadotrophs. Functional adaptations to both ordinary and extraordinary physiological demands. Front Endocrinol 4:127

Dutta A, Baruah B, Sharma BC, Baruah KK, Goswami RN (2001) Serum macromineral profiles in cyclic and anoestrus local heifers in lower Brahmaputra valley of Assam. Indian J Anim Res 35:44–46

Eissa HM (1996) Concentrations of steroids and biochemical constituents in follicular fluid of buffalo cows during different stages of the oestrous cycle. Br Vet J 152:573–581. https://doi.org/10.1016/S0007-1935(96)80009-1

Espey LL (1970) Effect of various substances on tensile strength of sow ovarian follicles. Am J Phys 219:230–233

Evans NP, Dahl GE, Padmanabhan V, Thrun LA, Karsch FJ (1997) Estradiol requirements for induction and maintenance of the gonadotropin-releasing hormone surge: implication for neuroendocrine processing of the estradiol signal. Endocrinology 138:5408–5414. https://doi.org/10.1210/endo.138.12.5558

Gajewska A, Zielinska-Gorska M, Wolinska-Witort E, Siawrys G, Baran M, Kotarba G, Biernacka K (2016) Intracellular mechanisms involved in copper-gonadotropin-releasing hormone (Cu-GnRH) complex-induced cAMP/PKA signaling in female rat anterior pituitary cells in vitro. Brain Res Bull 120:75–82. https://doi.org/10.1016/j.brainresbull.2015.11.002

Garcia SJ, Gellein K, Syversen T, Aschner M (2006) A manganese-enhanced diet alters brain metals and transporters in the developing rat. Toxicol Sci 92:516–525. https://doi.org/10.1093/toxsci/kfl017

Georgievskii VI (1981) The physiological role of microelements. In: Georgievskii VI, Annenkov BN, Samokhin VT (eds) Mineral nutrition of animals. Butterworths, London, p 171

Goor FV, Krsmanovic LZ, Catt KJ, Stojilkovic SS (2000) Autocrine regulation of calcium influx and gonadotropin-releasing hormone secretion in hypothalamic neurons. Biochem Cell Biol 78:359–370. https://doi.org/10.1139/o00-058

Gueux E, Rayssiguier Y, Piot MC, Alcindor L (1984) Reduction of plasma lecithin—cholesterol acyltransferase activity by acute magnesium deficiency in the rat. J Nutr 114:1479–1483. https://doi.org/10.1093/jn/114.8.1479

Harris TG, Dye S, Robinson JE, Skinner DC, Evans NP (1999) Progesterone can block transmission of the estradiol-induced signal for luteinizing hormone surge generation during a specifi c period of time immediately after activation of the gonadotropin releasing hormone surge-generating system. Endocrinology 140:827–834. https://doi.org/10.1210/endo.140.2.6490

Hidiroglou M (1979) Trace element deficiencies and fertility in ruminants:a review. J Dairy Sci 62:1195–1206. https://doi.org/10.3168/jds.S0022-0302(79)83400-1

Hillier SG (1994) Current concepts of the roles of follicle stimulating hormone and luteinizing hormone in folliculogenesis. Hum Reprod 9:188–191. https://doi.org/10.1093/oxfordjournals.humrep.a138480

Hurley WL, Doane RM (1989) Recent developments in the roles of vitamins and minerals in reproduction. J Dairy Sci 72:784–804. https://doi.org/10.3168/jds.S0022-0302(89)79170-0

Ingraham RH, Kappel LC, Morgan EB, Srikandakumar A (1987) Correction of subnormal fertility with copper and magnesium supplementation. J Dairy Sci 70:167–180. https://doi.org/10.3168/jds.S0022-0302(87)79991-3

Kalmath GP, Ravindra JP (2007) Mineral profiles of ovarian antral follicular fluid in buffaloes during follicular development. Indian J Anim Res 41:87–93

Kasahara E, Lin LR, Ho YS, Reddy VN (2005) Manganese superoxide dismutase protects against oxidation-induced apoptosis in mouse retinal pigment epithelium:implications for age-related macular degeneration. Investig Ophthalmol Vis Sci 46:3426–3434. https://doi.org/10.1167/iovs.05-0344

Kasa-Vubu J, Dahl GE, Evans NP, Thrun LA, Moenter SM, Padmanabhan V, Karsch FJ (1992) Progesterone blocks the estradiol-induced gonadotropin discharge in the ewe by inhibiting the surge of gonadotropin-releasing hormone. Endocrinology 131:208–212. https://doi.org/10.1210/endo.131.1.1611998

Kaswan S, Bedwal RS (1995) Light and electron microscopic changes in the ovary of zinc deficient BALB/c mice. Indian J Exp Biol 33:469–479

Keen CL, Zidenberg-Cheer S (1990) Manganese. In: Brown ML (ed) Present knowledge in nutrition. International Life Science Institute Nutrition Foundation, Washington, DC, pp 279–268

Keen KL, Wegner FH, Bloom SR, Ghatei MA, Terasawa E (2008) An increase in kisspeptin-54 release occurs with the pubertal increase in luteinizing hormone-releasing hormone-1 release in the stalk-median eminence of female rhesus monkeys in vivo. Endocrinology 149:4151–4157. https://doi.org/10.1210/en.2008-0231

Kendall NR, Marsters P, Scaramuzzi RJ, Campbell BK (2003) Expression of lysyl oxidase and effect of copper chloride and ammonium tetrathiomolybdate on bovine ovarian follicle granulosa cells cultured in serum-free media. Reproduction 125:657–665

Kochman K, Blitek A, Kaczmarek M, Gajewska A, Siawrys G, Counis R, Ziecik AJ (2005) Different signaling in pig anterior pituitary cells by GnRH and its complexes with copper and nickel. Neuroendocrinol Lett 26:377–382

Krolak M (1968) Effect of manganese added to diet on cattle fertility and manganese content of hair. Pol Arch Wet 11:293–304

Krsmanović LZ, Stojilković SS, Merelli F, Dufour SM, Virmani MA, Catt KJ (1992) Calcium signaling and episodic secretion of gonadotropin-releasing hormone in hypothalamic neurons. Proc Natl Acad Sci 89:8462–8466. https://doi.org/10.1073/pnas.89.18.8462

Kruijver FP, Swaab DF (2002) Sex hormone receptors are present in the human suprachiasmatic nucleus. Neuroendocrinology 75:296–305. https://doi.org/10.1159/000057339

Kumar S, Pandey AK, AbdulRazzaque WA, Dwivedi DK (2011) Importance of micro minerals in reproductive performance of livestock. Vet World 4:230–233

Lebedeva IY, Denisenko VY, Lebedev VA, Kuzmina TI (1998) Prolactin in follicular fluid and intracellular store calcium in follicular cells are related to morphological signs of ovarian follicle atresia in cows:work in progress. Theriogenology 49:509–519. https://doi.org/10.1016/S0093-691X(98)00002-8

Lee S, Kelleher SL (2016) Molecular regulation of lactation: the complex and requisite roles for zinc. Arch Biochem Biophys 611:86–92. https://doi.org/10.1016/j.abb.2016.04.002

Lee B, Pine M, Johnson L, Rettori V, Hiney JK, Les Dees W (2006) Manganese acts centrally to activate reproductive hormone secretion and pubertal development in male rats. Reprod Toxicol 22:580–585. https://doi.org/10.1016/j.reprotox.2006.03.011

Lee B, Hiney JK, Pine MD, Srivastava VK, Dees WL (2007) Manganese stimulates luteinizing hormone releasing hormone secretion in prepubertal female rats:hypothalamic site and mechanism of action. J Physiol 578:765–772. https://doi.org/10.1113/jphysiol.2006.123083

Lehman MN, Merkley CM, Coolen LM, Goodman RL (2010) Anatomy of the kisspeptin neural network in mammals. Brain Res 1364:90–102. https://doi.org/10.1016/j.brainres.2010.09.020

Leong DA, Thorner MO (1991) A potential code of luteinizing hormone-releasing hormone-induced calcium ion responses in the regulation of luteinizing hormone secretion among individual gonadotropes. J Biol Chem 266:9016–9022. http://intl.jbc.org/cgi/content/abstract/266/14/9016

Leung FY (1998) Trace elements that act as antioxidants in parenteral micronutrition. J Nutr Biochem 9:304–307. https://doi.org/10.1016/S0955-2863(98)00018-7

Little DA (1970) Factors of importance in the phosphorus nutrition of beef cattle in northern Australia. Aust Vet J 46:241–248. https://doi.org/10.1111/j.1751-0813.1970.tb15767.x

Maggio M, Ceda GP, Lauretani F, Bandinelli S, Corsi AM, Giallauria F, Guralnik JM, Zuliani G, Cattabiani C, Parrino S, Ablondi F (2011) SHBG, sex hormones, and inflammatory markers in older women. J Clin Endocrinol Metab 96:1053–1059. https://doi.org/10.1210/jc.2010-1902

Mahavar D (2011) Changes in trace elements during follicular atresia in rat. Int Referred Res J 2:88–89. https://doi.org/10.1007/s12011-010-8700-7

Martin TF (2003) Tuning exocytosis for speed:fast and slow modes. Biochim Biophys Acta (BBA)-Mol Cell Res 1641:157–165. https://doi.org/10.1016/S0167-4889(03)00093-4

Matzuk MM, Dionne L, Guo Q, Kumar TR, Lebovitz RM (1998) Ovarian function in superoxide dismutase 1 and 2 knockout mice. Endocrinology 139:4008–4011. https://doi.org/10.1210/endo.139.9.6289

Michaluk A, Kochman K (2007) Involvement of copper in female reproduction. Reprod Biol 7:193–205

Moriyama Y, Hayashi M, Yamada H, Yatsushiro S, Ishio S, Yamamoto A (2000) Synaptic-like microvesicles, synaptic vesicle counterparts in endocrine cells, are involved in a novel regulatory mechanism for the synthesis and secretion of hormones. J Exp Biol 203:117–125

Mosaad GM, Derar DR (2009) Effect of dietary energy and phosphorus on nutrients digestibility, blood constituents, and ovarian structures in ewes. Vet World 2:456–461

Murawski M, Bydlon G, Sawicka-Kapusta K, Wierzchos E, Zakrzewska M, Wlodarczyk S, Molik E, Zieba D (2006) The effect of long term exposure to copper on physiological condition and reproduction of sheep. Reprod Biol 6:201–206

Murray RD, Horsfield JE, McCormick WD, Williams HJ, Ward D (2008) Historical and current perspectives on the treatment, control and pathogenesis of milk fever in dairy cattle. Vet Rec 163:561–565. https://doi.org/10.1136/vr.163.19.561

Naor Z (2009) Signaling by G-protein-coupled receptor (GPCR): studies on the GnRH receptor. Front Neuroendocrinol 30:10–29. https://doi.org/10.1016/j.yfrne.2008.07.001

Noda Y, Ota K, Shirasawa T, Shimizu T (2012) Copper/zinc superoxide dismutase insufficiency impairs progesterone secretion and fertility in female mice. Biol Reprod 86:1–8. https://doi.org/10.1095/biolreprod.111.092999

Pan JT, Li CS, Ka-Choi T, Jing-Ying L (1992) Low calcium/high magnesium medium increases activities of hypothalamic arcuate and suprachiasmatic neurons in brain tissue slices. Neurosci Lett 144:157–160. https://doi.org/10.1016/0304-3940(92)90739-T

Panda DK, Miao D, Tremblay ML, Sirois J, Farookhi R, Hendy GN, Goltzman D (2001) Targeted ablation of the 25-hydroxyvitamin D 1α-hydroxylase enzyme:evidence for skeletal, reproductive, and immune dysfunction. Proc Natl Acad Sci 98:7498–7503. https://doi.org/10.1073/pnas.131029498

Phillippo M, Humphries WR, Atkinson T, Henderson GD, Garthwaite PH (1987) The effect of dietary molybdenum and iron on copper status, puberty, fertility and oestrous cycles in cattle. J Agric Sci 109:321–336. https://doi.org/10.1017/S0021859600080758

Pine M, Lee B, Dearth R, Hiney JK, Dees WL (2005) Manganese acts centrally to stimulate luteinizing hormone secretion:a potential influence on female pubertal development. Toxicol Sci 85:880–885. https://doi.org/10.1093/toxsci/kfi134

Poisner AM, Douglas WW (1968) A possible mechanism of release of posterior pituitary hormones involving adenosine triphosphate and an adenosine triphosphatase in the neurosecretory granules. Mol Pharmacol 4:531–540

Pradhan R, Nakagoshi N (2008) Reproductive disorders in cattle due to nutritional status. J Int Dev Coop 14:45–66

Prigge ST, Mains RE, Eipper BA, Amzel LM (2000) New insights into copper monooxygenases and peptide amidation:structure, mechanism and function. Cell Mol Life Sci 57:1236–1259. https://doi.org/10.1007/PL00000763

Pugh DG, Elmore RG, Hembree TR (1985) A review of the relationship between mineral nutrition and reproduction in cattle. Bovine Pract 20:10

Rabiee AR, Lean IJ, Stevenson MA, Socha MT (2010) Effects of feeding organic trace minerals on milk production and reproductive performance in lactating dairy cows:a meta-analysis. J Dairy Sci 93:4239–4251. https://doi.org/10.3168/jds.2010-3058

Richter TA, Robinson JE, Evans NP (2001) Progesterone treatment that either blocks or augments the estradiol-induced gonadotropin-releasing hormone surge is associated with different patterns of hypothalamic neural activation. Neuroendocrinology 73:378–386. https://doi.org/10.1159/000054656

Robinson JJ (1990) Nutrition in the reproduction of farm animals. Nutr Res Rev 3:253–276. https://doi.org/10.1079/NRR19900015

Rodgers, R.J., Van Wezel, I.L., Irving-Rodgers, H.F., Lavranos., T.C., Irvine, C.M., Krupa, M., 1999. Roles of extracellular matrix in follicular development. J Reprod Fertil Suppl 54, 343–352

Rodríguez-Ortiz ME, Canalejo A, Herencia C, Martínez-Moreno JM, Peralta-Ramírez A, Perez-Martinez P, Navarro-González JF, Rodríguez M, Peter M, Gundlach K, Steppan S (2014) Magnesium modulates parathyroid hormone secretion and upregulates parathyroid receptor expression at moderately low calcium concentration. Nephrol Dial Transpl 29:282–289. https://doi.org/10.1093/ndt/gft400

Rojas MA, Dyer IA, Cassatt WA (1965) Manganese deficiency in the bovine. J Anim Sci 24:664–667. https://doi.org/10.2527/jas1965.243664x

Romani AM (2007) Magnesium homeostasis in mammalian cells. Front Biosci 12:308–331. https://doi.org/10.2741/2066

Sánchez-Cárdenas C, Hernández-Cruz A (2010) GnRH-induced [Ca2+] i-signalling patterns in mouse gonadotrophs recorded from acute pituitary slices in vitro. Neuroendocrinology 91:239–255. https://doi.org/10.1159/000274493

Schiffer L, Barnard L, Baranowski ES, Gilligan LC, Taylor AE, Arlt W, Shackleton CH, Storbeck KH (2019) Human steroid biosynthesis, metabolism and excretion are differentially reflected by serum and urine steroid metabolomes: a comprehensive review. J Steroid Biochem Mol Biol 194:105439. https://doi.org/10.1016/j.jsbmb.2019.105439

Sharifi F, Mazloomi S, Hajihosseini R, Mazloomzadeh S (2012) Serum magnesium concentrations in polycystic ovary syndrome and its association with insulin resistance. Gynecol Endocrinol 28:7–11

Sharma MC (2006) Role of minerals in reproductive management with special reference to health and production,in:Compendium of national training course on reproductive health management of farm animals. Ministry of Agriculture, New Delhi, pp 14–25

Shemesh M, Hansel W, Strauss JF (1984) Calcium-dependent, cyclic nucleotide-independent steroidogenesis in the bovine placenta. Proc Natl Acad Sci 81:6403–6407. https://doi.org/10.1073/pnas.81.20.6403

Sircus M (2011) Transdermal magnesium therapy:a new modality for the maintenance of health. iUniverse, Bloomington

Slee RB, Hillier SG, Largue P, Harlow CR, Miele G, Clinton M (2001) Differentiation-dependent expression of connective tissue growth factor and lysyl oxidase messenger ribonucleic acids in rat granulosa cells. Endocrinology 142:1082–1089. https://doi.org/10.1210/endo.142.3.7990

Spears JW, Weiss WP (2008) Role of antioxidants and trace elements in health and immunity of transition dairy cows. Vet J 176:70–76. https://doi.org/10.1095/biolreprod40.1.54

Spelsberg TC, Rories C, Rejman JJ, Goldberger A, Fink K, Lau CK, Colvard DS, Wiseman G (1989) Steroid action on gene expression:possible roles of regulatory genes and nuclear acceptor sites. Biol Reprod 40:54–69. https://doi.org/10.1093/toxsci/kft195

Srivastava VK, Hiney JK, Dees WL (2013) Early life manganese exposure upregulates tumor-associated genes in the hypothalamus of female rats:relationship to manganese-induced precocious puberty. Toxicol Sci 136:373–381

Srivastava VK, Hiney JK, Dees WL (2016) Manganese-stimulated kisspeptin is mediated by the IGF-1/Akt/mammalian target of rapamycin pathway in the prepubertal female rat. Endocrinology 157:3233–3241. https://doi.org/10.1210/en.2016-1090

Steevens BJ, Bush LJ, Stout JD, Williams EI (1971) Effects of varying amounts of calcium and phosphorus in rations for dairy cows. J Dairy Sci 54:655–661. https://doi.org/10.3168/jds.S0022-0302(71)85902-7

Stojilković SS, Izumi SI, Catt KJ (1988) Participation of voltage-sensitive calcium channels in pituitary hormone release. J Biol Chem 263:13054–13061

Stojilković SS, Iida T, Merelli F, Torsello A, Krsmanović LZ, Catt KJ (1991) Interactions between calcium and protein kinase C in the control of signaling and secretion in pituitary gonadotrophs. J Biol Chem 266:10377–10384

Stojilković SS, Kukuljan M, Tomić M, Rojas E, Catt KJ (1993) Mechanism of agonist-induced [Ca2+] i oscillations in pituitary gonadotrophs. J Biol Chem 268:7713–7720

Stolkowski J (1977) Magnesium in animal and human reproduction. Rev Can Biol 36:135–177

Sugino N, Takiguchi S, Kashida S, Karube A, Nakamura Y, Kato H (2000) Superoxide dismutase expression in the human corpus luteum during the menstrual cycle and in early pregnancy. Mol Hum Reprod 6:19–25. https://doi.org/10.1093/molehr/6.1.19

Sun W, Xie H, Ji J, Zhou X, Goltzman D, Miao D (2010) Defective female reproductive function in 1, 25 (OH) 2D-deficient mice results from indirect effect mediated by extracellular calcium and/or phosphorus. Am J Physiol-Endocrinol Metab 299:E928–E935. https://doi.org/10.1152/ajpendo.00378.2010

Sun Y, Lin Y, Niu M, Kang Y, Du S, Zheng B (2017) Follicular fluid concentrations of zinc and copper are positively associated with in vitro fertilization outcomes. Int J Clin Exp Med 10:3547–3553

Sun Y, Wang W, Guo Y, Zheng B, Li H, Chen J, Zhang W (2019) High copper levels in follicular fluid affect follicle development in polycystic ovary syndrome patients:population-based and in vitro studies. Toxicol Appl Pharm 365:101–111. https://doi.org/10.1016/j.taap.2019.01.008

Suzuki M, Tanemoto Y, Takahashi K (1972) The effect of copper salts on ovulation, especially on hypothalamic ovulatory hormone releasing factor. Tohoku J Exp Med 108:9–18. https://doi.org/10.1620/tjem.108.9

Tabatabaei S, Mamoei M (2011) Biochemical composition of blood plasma and follicular fluid in relation to follicular size in buffalo. Comp Clin Pathol 20:441–445. https://doi.org/10.1007/s00580-010-1014-5

Thompson EL, Patterson M, Murphy KG, Smith KL, Dhillo WS, Todd JF, Ghatei MA, Bloom SR (2004) Central and peripheral administration of kisspeptin-10 stimulates the hypothalamic-pituitary-gonadal axis. J Neuroendocrinol 16:850–858. https://doi.org/10.1111/j.1365-2826.2004.01240.x

Tian X, Diaz FJ (2012) Zinc depletion causes multiple defects in ovarian function during the peri-ovulatory period in mice. Endocrinology 153:873–886. https://doi.org/10.1210/en.2011-1599

Tomić M, Jobin RM, Vergara LA, Stojilkovic SS (1996) Expression of purinergic receptor channels and their role in calcium signaling and hormone release in pituitary Gonadotrophs – integration of P2 channels in plasma membrane and endoplasmic reticulum derived calcium oscillations. J Biol Chem 271:21200–21208. https://doi.org/10.1074/jbc.271.35.21200

Torre PM, Harmon RJ, Sordillo LM, Boissonneault GA, Hemken RW, Trammell DS, Clark TW (1995) Modulation of bovine mononuclear cell proliferation and cytokine production by dietary copper insufficiency. J Nutr Immunol 3:3–20. https://doi.org/10.1300/J053v03n04_02

Torre PM, Harmon RJ, Hemken RW, Clark TW, Trammell DS, Smith BA (1996) Mild dietary copper insufficiency depresses blood neutrophil function in dairy cattle. J Nutr Immunol 4:3–24. https://doi.org/10.1300/J053v04n03_02

Tsou RC, Dailey RA, McLanahan CS, Parent AD, Tindall GT, Neill JD (1977) Luteinizing hormone releasing hormone (LHRH) levels in pituitary stalk plasma during the preovulatory gonadotropin surge of rabbits. Endocrinology 101:534–539. https://doi.org/10.1210/endo-101-2-534

Ullah N, Anwar M, Andrabi SM, Ansari MS, Murtaza S, Ali Q, Asif M (2010) Effect of mineral supplementation on post partum ovarian activity in Nili-Ravi buffaloes (*Bubalus bubalis*). Pak J Zool 43:195–200

Underwood EJ, Suttle NF (1999) The mineral nutrition of livestock. CABI Publishing, Wallingford

Veldhuis JD, Klase PA (1982) Mechanisms by which calcium ions regulate the steroidogenic actions of luteinizing hormone in isolated ovarian cells in vitro. Endocrinology 111:1–6. https://doi.org/10.1210/endo-111-1-1

Weisinger JR, Bellorín-Font E (1998) Magnesium and phosphorus. Lancet 352:391–396. https://doi.org/10.1016/S0140-6736(97)10535-9

Wilson JG (1966) Bovine functional infertility in Devon and Cornwall: response to manganese therapy. Vet Rec 79:562–566

Wise T (1987) Biochemical analysis of bovine follicular fluid:albumin, total protein, lysosomal enzymes, ions, steroids and ascorbic acid content in relation to follicular size, rank, atresia classification and day of estrous cycle. J Anim Sci 64:1153–1169. https://doi.org/10.2527/jas1987.6441153x

Wooldridge LK, Nardi ME, Ealy AD (2019) Zinc supplementation during in vitro embryo culture increases inner cell mass and total cell numbers in bovine blastocysts. J Anim Sci 97:4946–4950. https://doi.org/10.1093/jas/skz351

Chapter 7
Molecular Insights of Compromised Female Reproduction in Ruminants Under Metabolic and Nutritional Stress

S. Nandi, S. K. Tripathi, P. S. P. Gupta, and S. Mondal

Abstract The global increase in population and increased demand for livestock have made stress responsive decline in fertility (Metabolic stress and unbalanced diet feeding condition) a major challenge for the livestock industry. Metabolic stressors negatively affect the growth and development of ovarian cells and lead to reduced infertility in animals. Fertility in animals mainly categorized into two part one inherent property and second reproductive ability of animals which totally depends upon the feeding behaviour of animals. Stress negatively affects reproductive hormones. Reduced LH concentration associated with reduced estradiol secretion causing reduced fertility in animals by declined ovarian activity. Stress impaired folliculogenesis and ovulation process in animals. Metabolic stressors significantly altered the reproductive health of ruminants by alteration in biochemical composition of serum which directly and/or indirectly reflect in follicular, oviduct and uterine fluidresulting hampered reproductive health of ruminants. Relationship between diet and reproduction, and their interaction has long been understood to have important implications for reproductive success. The high level of urea (protein metabolites) alters glutamine metabolism and probably the function of the TCA cycle.

Non-enzymatic and enzymatic antioxidants, enzymatic cleansing is more effective. Antioxidants, like vitamins A, C and E and superoxide dismutase, glutathione peroxidase, catalase, glutathione S-transferase, peroxiredoxin and thioredoxin help cells from damaging effect of reactive oxygen species against the metabolic stress.

Keywords Antioxidant · Metabolic stress · Amelioration · Oocytes · Granulosa cells · Ruminants

---

S. Nandi (✉) · S. K. Tripathi · P. S. P. Gupta · S. Mondal
ICAR-National Institute of Animal Nutrition and Physiology, Bangalore, India

V. K. Yata et al. (eds.), *Sustainable Agriculture Reviews 57*, Sustainable Agriculture Reviews 57, https://doi.org/10.1007/978-3-031-07496-7_7

7.1 Introduction

The inability to achieve pregnancy is known as infertility (Adams et al. 2012). Anovulation is a potential cause of infertility in mammalian females. Attributed to poor control of the hypothalamic-pituitary-gonadal axis, endocrine hormone imbalances commonly lead to ovulatory disorders (Fig. 7.1), which account for infertility. Identifying what affects the ovulation process and recognizing what could happen with the complicated folliculogenesis or atresia process would encourage novel solutions to improve female fertility (Abedal-Majed and Cupp 2019). In addition, a deeper understanding of the biochemical processes that regulate the production of ovarian follicles and contribute to the development of a matured egg within a follicle would enable animals and women to have much more optimized assisted reproductive technologies (Allan et al. 1998; Abedal-Majed and Cupp 2019).

A stressful situation or stressor is a phenomenon outside the body that serves to decimate the body system (Lucy 2019). It is possible to quantify a stress condition and apply it uniformly across species. A strain is the reaction of the individual (the intensity of the displacement) to stress. Strong health is an unambiguous objective for farmers which are necessary for optimising animal production and well-being (Starkey et al. 1995; Pearson et al. 1997; Wilkanowska and Kokoszyn'ski 2015). Attentiveness in the care of animals has boosted over the past few decades. Wellbeing

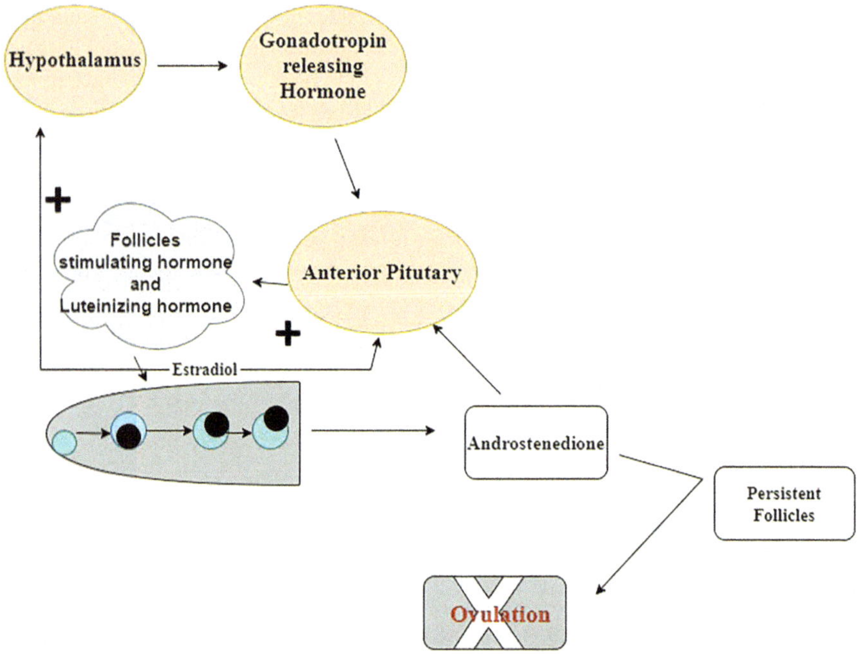

Fig. 7.1 Conceptual model of how androstenedione affects the normal hypothalamic–anterior pituitary–ovarian axis resulting in anovulation

has become the focus of many studies performed on cattle, pigs, sheep and poultry (Carlsson et al. 2007; Popescu et al. 2009; Carenzi and Verga 2009). In addition to making it possible to advance the conceptuality of the animals behaviourism, these experiments often provide a realistic view of how mammals interpret and how biological influences can impact their welfare globally (Carenzi and Verga 2009). According to Eler et al. (2002) with several components, reproduction is a complex process, some of which have been used as reproductive efficiency metrics. However, rather than by the successfulness of gestation, associated features can't be calculated easily in mammals (Eler et al. 2002). When management and nutritional conditions are optimum, most animals will reproduce, on the other hand, folks with thesuperior genetic characteristics will produce offspring under less favourable conditions (Morris 1980).

Several attributes, like physiological, surroundings environs and food intake are hypothesized to upset the reproductive profitability of animals (Wilkanowska and Kokoszyn'ski 2015). Health and dietary aspects are the most important, in relationships with their strong stimuluson reproductive capacity of livestock/animals (Smith and Somade 1994; Smith and Akinbamijo 2000; Bindari et al. 2013). In addition, more than any other, nutritional variables are readily manipulated in order to ensure beneficial results (Smith and Somade 1994; Smith and Akinbamijo 2000).

This chapter is an attempt to cover in details the effect of metabolic and nutritional stress on the reproduction of livestock, in mammals. The chapter also addresses available mitigation methods aimed at preserving livestockreproduction, in addition to these factors.

7.2 Nutrition and Reproductive Efficiency

Nutrition plays vital role in the sustaining animal reproductive performance. While genetic improvement has increased animal production, high producing animals (HPA) have also experienced a decrease in fertility at the same time. One of the main problems in HPA is reduced fertility. A network of genetic, environmental and managerial factors primarily influence the fertility of HPA, and their complex interactions make it more difficult to ascertain the exact reason for this decline (Ibtisham et al. 2017, 2018). Relationship between diet and reproduction, and their interaction has long been understood to have important implications for reproductive success (Nandi et al. 2015, 2016). In preserving the body condition and reproductive ability of dairy animals, nutrition plays a crucial role (Hoedemaker et al. 2009; Tripathi et al. 2016a, b).

Energy is particularly the key nutrient required by animals and any kind of discrepancy in the energy consumption has a deleterious effect on dairy cow reproductive activity. Energy deficiencies delay the initiation of estrus and decreases fertility in animal, as growth and follicular maturation are unsatisfactory during the Negative Energy Balance (NEB), ensuing in feeble estrus sign and diminish the likelihood of a large proportion of animal having initiated oestrus cycles for further breeding.

During the early postpartum phase, high-yielding animals (ruminants) face metabolic stress. This lead to negative energy balance (NEB) caused by the loss of energy through the processing of milk which cannot be substituted by the intake of energy (Rukkwamsuk et al. 1999; Walters et al. 2002; Ardema et al. 2013; Nandi et al. 2016). It is assumed that the negative energy balance would affect the fertility of ruminants (Walters et al. 2002; Britt 1992; Butler 2003; Tripathi et al. 2016a, b; Nandi et al. 2016, 2017). Significant body fat mobilisation raises the concentration of free fatty acid {Non-esterified fatty acid (NEFA) andbeta-hydroxy-butyrate (BHB)} in blood, follicular (Leroy et al. 2005, 2008; Nandi et al. 2016; Farman et al. 2015, 2018), oviduct and uterine fluid (Tripathi et al. 2016a, b) during times of metabolic stress.

Related to the vitality needed for upkeep and lactation, the energy costs of synthesizing and secreting hormones, ovulation and nurturing an early fertilized embryo are presumably modest. On the other hand, the biochemical manifestations connected with NEB prejudice the re-establishment of oocyte growth and production, and successfully pregnancy achievement as well as retention in dairy cattle. These results in severe economic loss to the dairy industry due to sluggish uterine involution, abridged reproductive performance, and calving durations, detrimental impact on productiveness, improved drug costs, reduced production (milk), reduced veal crops and early amortization of highly helpful livestock. Short or long term shortage cause result in under feeding. Animals with unbalanced feeding behaviour are stressed and attempt toendure metabolic revision to the stress [lipid mobilization and non-esterified fatty acid (NEFA) release] (D'Occhioet al. 2019). Obesity linked through insulin resistance, and the body storage (adipose mass) is less responsive in controlling effects of insulin. Therefore, obesity impair the brutality of metabolic ailments by the establishment of vicious response that lead to heightened lipolysis and amplified non-esterified fatty acid and beta-hydroxy-butyrate concentrations in circulation. Even though the effect of fatness prompted insulin resistance is not well-known in animals, few study demonstrating that this may be the case and disproportionate body fat is a well-known encouraging threat for metabolic related abnormalities in animals.

Dietary status is a key factor that influences the ability of an animal to produce (O'Callaghan et al. 2000; Tripathi et al. 2016a, b). Embryo viability has been related to nutritional status and is a vital factor affecting proficiency in assisted reproductive technologies (ART) (Webb et al. 2004). Feed intake and subsequent nutritional prominence may affect animal's fertility (Bridges et al. 2012). The perceptible measure of metabolic health is the body condition score (BCS) and continuous variation in it is used to analysethe state of animals (nutritional and physical) mostly for production (Berry et al. 2007). Body condition score (BCS) is correlated both phenotypically and genetically with reproductive success (Kubovicova et al. 2013). Found that high body condition score and/or adipose animal showed considerably elevated concentration of glucose on the day of birth than with ideal circumstance. Moreover, amount of glucose (sugar) substantially declined next 12 days. Most possible reason might be because of lower appetite in animals after parturition resulting high negative energy balance. Farman et al. (2018) also reported low blood glucose concentration in metabolic stressed ewes.

Disproportionate lipolysis during the early lactation associated with numerous reproductive abnormalities along with placental retention, and endometritis and reduced mRNA expression of immune; growth and development related gene in the endometrium. The effect of fertility and lipid metabolism in bovine are studied strongly, moreover, exact mechanisms responsible for the negative effect of lipolysis on animals fertility must be solved. Before and after pregnancy (transition period) and the commencing of lactation, white adipose tissue is the chief organ accountable for fat deployment and subsequently for alteration in non-esterified fatty acid concentration prompted by the negative energy balance in animals. Nutrition play vital role in determining the reproductive efficiency of animals. Unbalanced feeding significantly alter the biochemical environment of follicular fluid and/or uterine fluid, these alteration in biochemical environment reduce reproductive performance of animals by altering hormonal concentration along with changes in cell growth and development.

7.3 Metabolic Stressors and Oocyte Development and Molecular Approach

Significant but confined range of multifarious molecular structures in the ovary known as the ovarian follicles. Initially, they are formed in the form of precursor follicles residing single layer of flattened cells adjacent to the dormant oocyte. Basal cells enclose each precursor follicles, which separates the follicle throughout follicular development from the majority of the ovarian cytoplasm (Baddela et al. 2020). For months or years, most primordial follicles remain in a quiet phase before further development begins (Hirshfield 1991). Follicles with a single granulosa cells (GCs) layer are called major follicles, as well as those with numerous levels of granulosa cells layer are called secondary follicles.

It is argued that the growth of primary to secondary follicles is hormone (follicles stimulating hormone (FSH) and luteinizing hormone (LH)) responsive process. In secondary follicles, additional layers such as interna and externa cellsbegin to appear on the periphery of the basement membrane and make a significant contribution in follicular growth (Orisaka et al. 2009; Baddela et al. 2020). Gonadotropin-dependent progressive proceedings from secondary to ovulatory follicles occur as the granulosa cells triggered by stimulators like follicles stimulating hormone. Follicles stimulating hormone prompts growth and biosynthesis of hormones (steroid) in granulosa cells in combination with insulin-like growth factor 1 (IGF-1) and promotes follicular antrum development (Baddela et al. 2020). Follicular growth influenced mostly by on the animal's healthiness and vitality status, in addition to timely endocrine stimulation. Folliculogenesis has been shown to be significantly impaired during pre and post-partum condition (Tripathi et al. 2016a, b) unbalanced feeding (High protein diet feeding) (Aardema et al. 2013; Leory et al. 2008; Nandi et al. 2018). According Valckx et al. (2014) and Nandi et al. (2016) raised on esterified-fatty acid concentrations possibly prejudice fertility, by fluctuating physiology and sinking oocyte development ability. Treatment with elevated concentration of stearic acid (SA) tended to decrease glucose absorption and depletion of mural granulosa cells (decreased mRNA expression of Slc2a1 and GAPDH gene) compared to physiological and/or high non esterified-fatty acid treatment (Baddela et al. 2020). Long-term exposures to murine follicles to elevated NEFA concentrations (720 mM) has been known to decrease follicular development, with substantial outcome triggered by elevated stearic acid (280 mM, Valckx et al. 2014).

Higher serum on esterified-fatty acid levels are linked with a decrease glucose concentration in follicular fluid (Roth 2017; Leroy et al. 2015; De Bie et al. 2017). This decrease in the amount of glucose can have a major impact on the production of oocytes because glucose is the main source of energy and responsible for the vital biological components for maturation, together with pyruvate, adenosine triphosphate (ATP), and nicotinamide adenine dinucleotide phosphate (NADPH) and glutathione that neutralize the reactive oxygen species (Sutton-McDowall et al. 2010). There is bidirectional coordination between the oocytes and the corresponding cell types. Gonadotropins stimulate meiotic oocyte commencement via triggering of the signaling pathway (EGFR) in granulosa cells (Richani 2014). Growth factors such asGrowth differentiation factor 9 (GDF9), Bone morphogenetic protein 15(BMP15), and Fibroblast growth factor 8 (FGF8B) also regulate the role of granulosa cells (Emori and Sugiura 2014). Nutritional factors may influence reproduction separately or in combination, at the level of the hormones production, oocyte growth and embryo uterus interaction (Bilodeau-Goeseels and Kastelic 2003). Oocytes resulting from follicles subjected to stress (environmental and metabolic stressors) have reduced reproductive fitness (Nabenishi et al. 2012; Gebremedhn et al. 2020), accompanied by reduced fertilization ability and blastocyst stage development and establishment of pregnancy (Roth and Hansen 2005; Gebremedhn et al. 2020). The association between both the oocyte and the biochemical environment is affected by exposure to stress. Communication between follicle and the oocyte is predominantly befell by paracrine and autocrine molecules via gap junction and/or by discharge of molecules (Bosco et al. 2011; Gebremedhn et al. 2020).

The recent developments and characterization of vesicles derivedfrom different types of cells represent eminent cell communication mechanism (Yáñez-Mó et al. 2015). In transporting biomolecules like mRNAs, miRNAs, and proteins, EVs significantly indicate the physiological status of the origin of cells (Valadi et al. 2007a, b; Silva and Melo 2015; Gebremedhn et al. 2020). EV is characterized by the existence of unique membrane proteins, along with clusters of differentiation −9, 63 and 81 (CD9, CD63, CD81), tetraspanins and additional proteins such as: ALIX and TSG101 (Kumar et al. 2015). EVs have been reported to be present in bovine (Hung et al. 2015; Hailay et al. 2019a, b), equine and human follicular fluids and are necessary for transporting biomolecules (RNA, miRNA) during growth and development of follicles (da Silveira et al. 2012; Santonocito et al. 2014). During early lactation, high producing animals underwent a state of conditions known as negative energy balance. This in fact, outcomes in fluctuations in the amount of different biomolecules in the microenvironment of the body and follicular fluid (Farman et al. 2018) and oviduct/uterine fluid (Tripathi et al. 2016a, b) which lead to disrupted fertility. In cell-to-cell contact, extracellular vesicles play a critical role and hold a large number of biomolecules that can be supplied to act in other cells (Valadi et al. 2007a, b). Thousands of exosome-mediated molecules, including around 9769 proteins, 3408 mRNAs, 2838 miRNAs, and 1116 lipids, have been reported (Keerthikumar et al. 2016). Animals with negative energy status in the displayed exclusive expression of eight miRNAs, five of which were located on chromosome 21 (bta-miR-431, bta-miR-370, bta-miR-136, bta-miR-376e and bta-miR-411c-3p) (Hailay et al. 2019a, b). A strong correlation between metabolic stress and the release of Extracellular vesicles-coupled miRNAs was revealed by hierarchical clustering of differentially expressed Extracellular vesicles (EVs)-coupled miRNAs between cows with divergent metabolic states (Hailay et al. 2019a, b).

MiR-21, believed to be active in follicular development. High expression of MiR-21 in mice related with better subsistence and ovulation (Christenson 2010). In regulating maternal-to-embryonic transfer and early growth, increase expression of miR-21 was involved (Mondou et al. 2012). Thus in the research analysis, the down-regulation of EV-coupled miR-21 due to metabolic stress can indicate a corresponding reduction in follicular cell expression, that could hinder follicular growth. Additionally, accelerated miR-20b transcription in bovine CCs (cumulus cells) high maturation rate and production of progesterone by pointing INHBA, MAPK1, PTGS2, PTX3, and EGFR55 (Andreas et al. 2021; Hailay et al. 2019a, b). The molecular analysis of the package of EVs with a possible influence on the development of oocytes and embryos responsible for the development of bio- markers for the implementation of systems for addressing infertility. The components of Extracellular vesicles (EVs) produce from granulosa cells and oviductal epithelial cells in response to stress are defined (Gebremedhn et al. 2020). By upsetting steroid synthesis, cell growth, and cell apoptosis essential for follicles growth, fatty acids have been described to amend granulosa cells functions (Elis et al. 2015; Tripathi et al. 2016a, b; Nandi et al. 2018; Sharma et al. 2019). The induction of extreme morphological changes in granulosa cells is one of the anticipated effects of highernon-esterified fatty acid concentration. Yenuganti et al. (2016) and Sharma et al. (2019) conveyed the development of fizzlike cell structures in cultured bovine GCs on treatment with higher concentration non-esterified fatty acid concentration[(Oleic acid (OA), Palmitic acid (PA), and Stearic acid (SA)] in addition to the major impact on the expression of essential genes and hormone production (Nandi et al. 2018; Farman et al. 2015). Granulosa cells synthesise and synthesize sufficient quantities of estrogen after encouragement with follicle stimulating hormone and insulin growth factor −1 (IGF-1) under healthy conditions Adversarial morphological changes in granulosa cells occurred because of saturated fatty acids with an increase apoptosis rate in granulosa cells. In granulosa cells, decreased phosphorylation of Akt has been Increased expression of genes regulated by CD36 (fatty acid transporter), insulin growth factor −1 and follicle stimulating hormone reported in granulosa cells (Hailay et al. 2019a, b).

Ammonia is unfavorable for embryo development. Urea negatively affect the oocyte by need of amino acid, and overall amino acid loss and turnover at higher amount (Kowsar et al. 2018)."Exhaustion of essential amino acids (EAAs) (histidine, tryptophan, lysine, isoleucine and leucine), semi-essential amino acids (serine, arginine and glutamine), and non-essential amino acids (ornithine and aspartic acid)" was increased by elevated urea concentration (Kowsar et al. 2018). Hemming et al. (2012, 2013) observed that in incompetent oocytes the overall turnover and loss of amino acids was greater than their split counterparts. Dynamic genetic assortment combined with feeding high protein diets to upsurge milk production which was accompanying with declined fertility in animals. The high level of urea will alter glutamine metabolism and probably the function of the TCA cycle (Kowsar et al. 2018).

Elevated urea concentration is responsible for increase lysine turnover. Kowsar et al. (2018) reported that elevated urea concentration, significantly washed-out more lysine and methionine in cumulus oocyte cells complex (COCs) and DOs (Denuded oocytes), which is the most limiting amino acid in dairy animals during lactation (Schwab et al. 1992). Urea considerably declined the viability of epithelial cells in oviduct (Kowsar et al. 2016). The viability of cumulus cells cumulus oocyte cells showed a negative correlation with alanine turnover (Kowsar et al. 2018). Alanine has been recognized as a marker of apoptosis, with an increase in its concentration in apoptotic cells (Halama et al. 2013). This suggests a relationship between lower viability of cumulus oocyte cells and greater alanine appearance (Kowsar et al. 2018). Several studies reviewed about the impact of metabolic stressors [NEFA, BHB, and protein metabolites (ammonia and urea)] in oocyte maturation and development as mentioned in Table 7.1.

Animals with declined levels of insulin growth factor-1 in serum collected after parturition, showed endometritis than animals with normal insulin growth factor-1 concentrations. Mattiauda et al. (2017) reported, the higher transcript expression of insulin growth factor-1 in total mixed ration and high grazing animals than medium grazing animals and lower grazing animal) cows— animals with enhanced nutritional grade during lactation, for example, indicate a favorable uterine environment for the development of embryos. Progesterone stimulates the growth of the embryos and contributes to the secretion of interferon gamma (IFNÿ) by acting on the endometrium and resulting release of various embryotrophic factors. In fact it was shown that transcript expression of insulin growth factor-1 is predominantly controlled by the action of progesterone on the endometrial uterine.

Reproductive tract environment is not only affected by physiological processes of reinforcement, inflammation and infection but also by the metabolic condition of the animal. An earlier study found altered gene expression of insulin growth factor binding protein in cow oviducts with negative energy balance; however, the influence of changes in gene expression on embryo development was not analysed. During the dry and postpartum cycle, it seems that feed restriction generally influences global gene expression in the oviduct. Therefore, it's important to consider the ruminant's physiological condition in studies that further scrutinize the root causes for reproductive disorders. Fatty acids were also suggested as key midstream

Table 7.1 Studies of metabolic stress and nutritional stress in in-vitro fertilization (IVF)

Animal	Experimental findings	References
Bovine	Elevated non-esterified fatty acid (NEFA) concentration, responsible for epigenome alterations in matured oocyte or in embryo. Cell survivability, invulnerability, metabolic rate are associated with major impaired pathways.	Desmet et al. (2016)
Bovine	Low cell numbers in blastocyst, augmented apoptosisrate and altered gene expression (DNA Methyltransferase 3 Alpha, Insulin Like Growth Factor 2 Receptor and Solute Carrier Family 2 Member 1) resulted in maturation in elevated non-esterified fatty acid concentrations. Additionally the blastocysts showed declined consumption of oxygen, and glucose, higher lactate consumption and metabolism of amino acids.	Van Hoeck et al. (2011)
Cattle	By increasing the expression of endoplasmic reticulum stress marker genes: activating transcription factor 4 and heat shock protein family A (Hsp70) member 5, higher non esterified fatty acid concentrations at maturation provoke endoplasmic reticulum stress in cumulus..	Sutton-McDowall et al. (2016)
Mouse	Feeding high fat diet and exposure of oocyte to elevated lipid responsible for increased expression of activating transcription factor 4 and heat shock protein family A (Hsp70) member 5, in mouse COCs.	Wong et al. (2015)
Bovine	Expression of glutathione peroxidase-1 (reduced glutathione to oxidized glutathione) in oocyte, has been declined by non-esterified fatty acid supplementation.	Van Hoeck et al. (2015)
Ovine	Metabolic stressors repressed granulosa cells proliferation, increase apoptosis rate, declined hormones production rate, and reduced steroid-related gene expression. The expression ofapoptosis related geneBCL-2 and BAX were significantly increased in higher levels of metabolic stressors, ratio of BAX: BCL2 ratiowas significantly in higher elevated level of metabolic stressors.	Nandi et al. (2018)
Human	Unsaturated fatty acid significantly reduce the transcript expression of BAX and encourage BCL2 and transcript expression in GC	Valckx et al. (2014)
Bovine	Unsaturated fatty acid lower the transcript expression of genes Steroidogenic Acute Regulatory (STAR), Cytochrome P450 Family 19 Subfamily A Member 1 (CYP19A1), Follicle Stimulating Hormone Receptor (FSHR), Cytochrome P450 Family 11 Subfamily A Member 1(CYP11A1), Cyclin D2 (CCND2) and Proliferating Cell Nuclear Antigen (PCNA).	Sharma et al. (2019)
Ovine	Ammonia (250 μM and 150 μM) negatively affectgrowth and secretary activity of granulosa cells isolated from small and/or medium follicles and large follicles respectively.	Nandi et al. (2016)
Bovine	Supplementation of elevated concentration of ammonia and urea on bovine endometrium cells *In vitro* significantly reduce the mRNA expression of Insulin Like Growth Factor Binding Protein 1and Fibroblast Growth Factor 2 (FGF2). However, moderate concentration of ammonia and urea significantly increase the mRNA expression of Heat Shock Protein Family A (Hsp70) Member 1A (HSPA1A), Insulin Like Growth Factor Binding Protein 3 (IGFBP3) and Serine Protease Inhibitor-14(SERPINA14) genes.	Gunaretnam et al. (2013)

regulatory agencies of modifications in genome sequences detected in trophectoderm cells when elongation started. Consequently, sufficient concentrations of peroxisome proliferator-activated receptor gamma (PPARÿ) fatty acid ligands in the histotrophics result in increased transcription factor activity and subsequent alterations in cell environment necessary for conceptual elongation. Additionally, due to inadequate concentrations of peroxisome proliferator-activated receptor gamma fatty acid ligands or inconsistent fatty acid profile in the histotrophic slow down conceptus elongation which lead to a loss of pregnancy. These theory are endorsed by the significance of peroxisome proliferator-activated receptor gamma in trophoblast and placenta in mice, humans and sheep. For example, the genetic ablation of PPARÿ in mice resulted in embryonic lethality due to placental inefficiencies.

After parturition negative energy balance last to 70–84 days (10–12 weeks), during this period of time in animal fertility is reduced (Butler 2003). After examining the biochemical environment in the serum and follicular (Leroy et al. 2005; Nandi et al. 2013; Farman et al. 2015) oviduct and uterine fluid (Tripathi et al. 2016a, b) found that its administration of metabolic stressors (NEFA, BHB, ammonia and urea) to the cell culture decreases ovarian cells (granulosa cells, and oocyte) and endometrium growth and development. Antioxidant capacity decreased during stress in mammals (Ferreira et al. 2016). In addition, oocyte and embryo exposed to HS in vitro, elevated concentration of reactive oxygen species (ROS) impairs embryo development (Sakatani et al. 2004; Roth 2017). In addition, Van Hoeck et al. (2013) and Nandi et al. (2018) reported that metabolic stressors induced oxidative stress is responsible for reducing growth and development of oocyte in animals suffering from negative energy balance. Furthermore, animals displaying signs of negative energy balance and imbalanced feeding-(loss of body weight, elevated NEFA and BHB, ammonia and urea levels)-have oxidative stress (Tripathi et al. 2016a, b), which can be explained by decreased concentration of antioxidants (β-carotene, vitamins C and E, superoxide dismutase, and glutathione peroxidase) in serum and in follicular fluid. After pregnancy has been developed, ROS negatively affect the placenta and soon or later fetus survival, with consequences for the pathophysiology of abortion, impulsive membrane rupture and fatal disease.

Oocyte production affected by extrinsic oxidative stress parameters present in metabolically compromised mothers' follicular fluid, but also by intrinsic oxidative stress in the oocyte caused by exposure to extrinsic oxidative stress and/or altered metabolic factors. In the peritoneal fluid of endometriosis patients significant elevation in the concentration of oxidative stress markers, particularly lipid peroxides (LPO), was observed. Additionally, women with endometriosis reported that the peritoneal fluid contains low levels of antioxidant. In addition to the correlations described above, glutathione peroxidase (GPx) and total antioxidant status (TAS) levels in follicular fluid have been shown to be higher, achieving efficiently and effectively fertilized oocytes than non-fertilized oocytes in follicular fluid. In conclusion, metabolic stressors (NEFA, BHB, ammonia and urea) significantly alter the growth and development of oocyte by significant alteration at gene level. Metabolic stressors increased the reactive oxygen species concentration leads to reduced growth of cells; metabolic stressors significantly lowered the mRNA expression of growth related gene.

7.4 Strategies to Ameliorate the Negative Effect on Reproduction During Metabolic and Nutritional Stress

As described above, stress has a negative impact on the anti-oxidant scavenging activity of ruminants, oocytes and the subsequent embryos, so antioxidant supplementation responsible for alleviating the action against stress. Reactive oxygen species and the antioxidants were necessary for ovulation and a disproportion among them was responsible for the oocyte‘s inferior quality. While it needs both non-enzymatic and enzymatic antioxidants, enzymatic cleansing is more effective."Antioxidants, like vitamins A, C and E and superoxide dismutase, glutathione peroxidase, catalase, glutathione S-transferase, peroxiredoxin and thioredoxin helps cells from damaging effect of reactive oxygen species" (Bettina et al. 2017). Because of the potential to inhibitreactive oxygen species, especially during the biosynthesis of steroid hormones, antioxidant enzymes within granulosa cells, cumulus cells, and follicular fluid each play a pivotal role in oocyte protection (Bettina et al. 2017). The follicular fluid acts as a buffer for antioxidant because of close proximity of the cumulus oocytes cell complex which helps in maintaining redox balance. Oxidative stress-induced DNA damage in granulosa cells is inversely related to quality of embryo and fertilization rate.

Antioxidants are useful in reducing oxidative stress, as they scavenge free radicals and decrease the amount of reactive oxygen species. Antioxidants (enzymatic and non-enzymatic) are also used to decrease reactive oxygen species and therefore the degree of oxidative stress (Sharma and Agarwal 2004; Agarwal et al. 2012; Prasad et al. 2016)."In order to resolve this problem, antioxidants such as superoxide dismutase, catalase, glutathione peroxidase and glutathione oxidase, vit C, taurine, hypotaurine, vitE, Zn, selenium (Se), betacarotene, and carotene may be useful" (Prasad et al. 2016). Animal studies have demonstrated that supplementation with antioxidants is helpful in overcoming the disadvantageous effects of oxidative stressin oocytes (Sharma and Agarwal 2004; Agarwal et al. 2012; Prasad et al. 2016). Melatonin tends to play a significant role. To sustain the production of progesterone from ROS produced during the ovulation, melatonin protects granulosa cells (Taniguchi et al. 2009). "Melatonin (N-acetyl-5-methoxytryptamine) is a drug derived from the pineal glands and peripheral nerves". There are mainly two G protein–coupled receptors for melatonin, melatonin receptor-1 and -2. Melatonin's antioxidant effects may protect oocytes from oxidative stress harm. Action of melatonin on oocyte is mediated by its receptors. Follicular fluid concentrations of melatonin are closely correlated with both oocyte quantity and quality. In human granulosa cells, melatonin regulates the mRNA expression of vascular endothelial growth factor (VEGF) and down-regulate Inducible nitric oxide synthaseexpression. "Melatonin, antioxidant capacity, is five times more powerful thanGlutathione and eight times more powerful than mannitol". During the dry period under stress, melatonin enhanced production rate and decreased the rates of breeding dysfunction and disruption of pregnancy in animals.

During IVEP (in-vitro embryo production), melatonin decreases reactive oxygen species production and cellular apoptosis, improves the rate of blastocysts yield, increase the transcript expression of the super oxidase dismutase andB-cell lymphoma 2 (Bcl-2) and decrease transcript expression of p53. At the time of apoptosis cytochrome C (cyt-C) binds to Apaf-11 (has a caspase binding region) and the release of cyt- C inhibited by B-cell lymphoma 2. Reduced healthy follicles was reported in B-cell lymphoma2knockout animals and higher transcript expression of B-cell lymphoma 2gene in granulosa cells lower apoptosis rate. The existence of melatonin in follicular fluid was interrelated with oocyte number excellence and graafian follicles number, anti-Müllerian hormone (AMH) in serm, estradiol level in serum, and total number of embryos produced (Tiboni et al. 2004; Budani and Tiboni 2020). Melatonin enhanced the mitochondrial function of in vitro matured mice oocytes and protected in vitro matured oocytes against oxidative stress. Supplementation with melatonin increase mitochondrial function. Supplementation of melatonin during in-vitro maturation increases the copy number of the mitochondrial DNA (mtDNA), along with the potential of the mitochondrial membrane (Ochiai et al. 2019; Budani and Tiboni 2020). In mice, melatonin enhanced reproductive efficiency in an in vivo study conducted by Gao et al. (2012) steadily increasing melatonin concentrations in drinking water (0, 3, 30, 300 μg/mL) were exposed for 21 days to ICR mice, aged 7 weeks. A dosage of 30 μg/mL was linked to higher number of antral follicles in each region of the ovarian structure compared with control. In addition, 30 μg/mL dose of melatonin during thein- vitro fertilization significantly increased the hatching rate relative to the control group and other groups treated with melatonin. (Gaoet al. 2012; Budani and Tiboni 2020).

Vitamin E is commonly used in the field of assisted reproductive and can effectively reverse the harmful effects of oxidative stress on the reproductive system and the endocrine system. (Chen et al. 2020). Vitamin E, which is vitally important throughout the female reproductive process, can vilify oxidative damage by oxygen-free radicals and antioxidant deficiency by repressing the action of phospholipase A and lipoxygenase to sustain the cell membrane and guiding the usual physiological position of the reproductive system (Chen et al. 2020). Vitamin E can reduce the senile oxidative stress reaction with the anti-oxidant properties, which may have a negative impact on the number and quality of oocyte (Chen et al. 2020). Dairy cattle eating stored forages are often poor in vitamin E unless supplemented, and deficiencies in vitamin E are often found during the periparturial cycle leading to placenta retention and then anoestrus injection of diazinon (DZN) to rat decreases cells proliferation in secondary and Graafian follicles, supplementation of Vitamin E recover the toxic effect of diazinon. On oocytes, the impact of stress on maternal metabolic nutrition and antioxidant status is transmitted, resulting in embryos and the fertility state of the ruminant (Abdelatty et al. 2018). Long-term maternal diet programming is therefore the main factor in reducing the stress impact on the oocytes and the growth of embryos. Furthermore, adequate nutrition management of the high producing animal to preserve the energy balance can have a favorable outcome on the oocytes and the potential offspring (Abdelatty et al. 2018).

L-ascorbic acid (VitC) synthesized in the liver of many animals, with the exception of guinea pigs, human and other primates (Cantoni et al. 2017; Yu et al. 2018). Ascorbic acid was found to be strongly transported to cells by the high affinity sodium-dependent vitamin C transporters 1 and 2 to achieve a levels of 1 ~ 10 mM (Young et al. 2015;Cantoni et al. 2017;Yu et al. 2018). A complex molecular network coordinates the development of mammalian oocytes, and dynamic regulation of DNA methylation and histones is essential both for meiosis and embryonic development (Gu et al. 2010; Yamaguchi et al. 2012). Methylation (trimethylation) in histone-3(H3) at Lys-4 and Lys-36 linked with dynamicchromatin position while methylation at Lys-9 and Lys- 27 linked with suppressive chromatinstatus, which also have important evolutionary functions in the regulation of oogenesis and embryogenesis (Diao et al. 2014, 2016; Stewart et al. 2015; Yu et al. 2018.

According to outcomes of Chawalit et al. (2012) and Mallol et al. (2015) supplementation of ascorbic acid duringIVEPadvance blastocyst development rate in porcine hand-cloned embryosand mouse embryos made by SCNT (Somatic cell nuclear transfer). In porcine oocyte ascorbic acid supplementation increases meiotic maturation and developmental competence of porcine oocytes through epigenetic reprogramming (Yu et al. 2018). According to the finding of Yu et al. (2018) ascorbic acid reduces reactive oxygen species level and increase bone morphogenetic protein -15transcript level. Oxidative stress responsible for alter epigenetic status while supplementation of ascorbic acid responsible for modify epigenetic status in oocyte (McDonough et al. 2010; Young et al. 2015). During the periparturient period significant decline in concentration of vitamin C in serum was recorded in dairy cows. In lactating animal, heat stress reduced the serum Vit C concentration the serum Vit C concentration with regulated high ambient temperature decreased by 50%. In contrast, we further reported that the serum Vit C concentration in the summer was substantially lower than those in the autumn in milking animals. While H3K4me3 and H3K36me3 are typically associated with actively transcribed chromatin, H3K9me3 and H3K27me3 are associated with repressive chromatin. According to study of Hancock et al. (2015) ascorbic acid usage considerably reduced expression of H3K27me3 but higher the expression of H3K4me3 and H3K36me3 in matured oocyte.

Selenium (Se) is an effective antioxidant factor that has ruminant animal's functions, including reproductive activity of both sexes (Mehdi and Dufrasne 2016; Lizarraga et al. 2020). During early lactation cattle experienced reduced antioxidant defense capability, even with sufficient Se concentration in their rations. Selenium-yeast supplementation to Se-sufficient animals during late gestation increases serum selenium concentration, and it helps in improving antioxidant defense mechanism, and diminishes oxidative stress in early lactation. Supplementation with 10 ng/mL dose of selenium during in-vitro maturation of cattle (*Bosprimigenius Taurus)* improve embryo quality (Lizarraga et al. 2020). According to findings of several study in cow, reduced fertility, placental preservation, and augmented frequency of mastitis and metritis were linked with selenium deficiency (Spears and Weiss 2008; Hefnawy and Tórtora-Pérez 2010; Sordillo 2013). In human, supplementation of Se, calcium, and calcium ionophore to IVM medium increased the oocytes maturation rate (Makki et al. 2012). Selenium inhibits oxidative damage and also affect the transcipt expression of the follicles stimulating hormone receptor in granulosa cells, according to Basini and Tamanini (2000). It has been shown that selenoproteins such as glutathione peroxidase and thioredoxinreductase can exert their beneficial effects through selenium (Brigelius-Flohé and Maiorino 2013). Ceko et al. (2015) confirm the presence of selenium and glutathione peroxidase −1 in large follicles of bovine granulosa cellsand play a vibrant role as anti-oxidants throughout late follicular development. Coenzyme Q10 (CoQ10) acts as a scavenger against reactive oxygen species (Asensi-Fabado and Munné-Bosch 2010). According to Dai et al. (2017) treatment with coenzyme Q-10in mouse delayed ovarian reserve depletion, regained the mitochondrial gene expression of the oocyte, and enhanced mitochondrial activity. Several findings reviewed about the significance of antioxidants in oocyte growth and development as mentioned in Table 7.2, Fig. 7.2. The supplementation of antioxidant molecules in both (*In-vitro and In-vivo*) reduces oxidative stress and it's clearly required for optimum animal performance. Optimised antioxidant feeding may increase the therapeutic efficacy of the follicular environment, irrespective of the animal metabolic stress status, the above may offer a treatment to help' the oocyte that is metabolically affected.

7.5 Conclusion

Metabolic and nutritional stress undoubtedly place stress on all livestock species and will adversely affect their reproductive capacity. The effect of metabolic and nutritional stress was addressed in detail in this segment. This chapter also elaborated on improvement measures to be taken into account in order to avoid economic losses caused by metabolic and nutritional stress pressures on livestockreproduction. Metabolic and nutritional stress conditions negatively influencedoocyte and embryo development and quality. The difficulty in implementing this nutritional strategy is the lack of awareness of particular animals' individual antioxidant deficiency, the potential role of other nutrients, and their interactions with antioxidants. This first logical step towards enhanced fertility is to enhance metabolic health. The use of antioxidants *in vitro* embryo culture support development of embryos after a metabolic praise during oocyte maturation, which improve embryo quality. Although it is important to further elucidate the degree to which these antioxidant supplementation will potentially rescue the metabolically compromised oocyte by nutritional and metabolic stress in a combined in vivo and in vitro approach.

Table 7.2 Important antioxidants and their role in ovarian physiology

Animal model	Antioxidants	Experimental findings	References
Cattle	α-Tocopherol	Responsible forthe survival of pre-antral follicles and encourages the stimulation of primordial follicles in *in vitro* culture of cattle follicles.	Lisboa et al. (2009)
Bovine	Zinc	Zinc considerably affected intracellular GSH content and DNA integrity of cumulus cells during oocyte maturation and improved pre-implantation embryo development in bovine	Picco et al. (2010)
Ovine	α-Tocopherol and ascorbic acid	In combinations significantly increase the maturation rate of sheepoocyte	Miclea et al. (2012)
Porcine	Selenium and vitamin E	Increased the maturation rate of porcine oocytes and blastocysts.	Tareq et al. (2012)
Porcine	Rresveratrol	Increase blastocyst formation rates and total cells number. Reduced mRNA expression of apoptosis-related genes in COC treated with resveratrol	Kwak et al. (2012)
Bovine	Melatonin	Reduce the negative effect of stress on bovine oocytes.	Cebrian-et al. (2013)
Bovine	Resveratrol	Improved cumulus expansion, polar body formation, hatched blastocyst rate and mean number of cells in blastocysts	Wang et al. (2014)
Porcine	Ascorbioc acid	Significantly enhanced survival rates of blastocysts and reduced peroxide levels.	Castillo-Martín et al. (2014)
Camel	Selenium, melatonin, ascorbic acid, β-mercaptoethanol etc.	Maturation medium supplemented with different antioxidants had a beneficial impact on maturation rates.	Mayada et al. (2015)
Ovine	Fenugreek seed extract (FSE)	Significantly improved the maturation of oocytes.	Barakat et al. (2018)
Human	CoQ10	Improved oocytes retrieval, higher fertilization rate, and more high-quality embryos	Xu et al. (2018)
Human	Vitamin C	Treatment via oral route accelerates the level of Vit C in serum and follicular fluid improving the quality of oocytes and embryos in IVF-ET cycles	Lu et al. (2018)
Cattle	Selenium	Improve embryo quality.	Lizarraga et al. (2020)
Bovine	Resveratrol	Attenuated the increasing in active mitochondria in embryos cryopreserved	Gaviria et al. (2019)

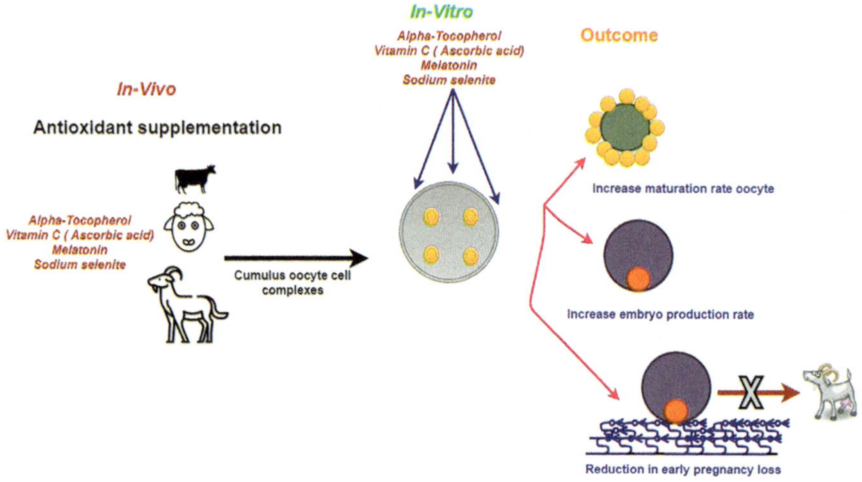

Fig. 7.2 Antioxidant administration strategies for overcoming Nutritional and metabolic stress

References

Aardema H, Lolicato F, van de Lest CH, Brouwers JF, Vaandrager AB, van Tol HT, Roelen BA, Vos PL, Helms JB, Gadella BM (2013) Bovine cumulus cells protect maturing oocytes from increased fatty acid levels by massive intracellular lipid storage. Biol Reprod. 88(6):164. https://doi.org/10.1095/biolreprod.112.106062

Abdelatty AM, Iwaniuk ME, Potts SB, Gad A (2018) Influence of maternal nutrition and heat stress on bovine oocyte and embryo development.Int. J Vet Sci Med 6(1):S1–S5. https://doi.org/10.1016/j.ijvsm.2018.01.005

Abedal-Majed MA, Cupp S (2019) Livestock animals to study infertility in women. Anim Front 9(3):28–33. https://doi.org/10.1093/af/vfz017

Adams GP, Singh J, Baerwald AR (2012) Large animal models for the study of ovarian follicular dynamics in women. Theriogenology 78:1733–1748. https://doi.org/10.1016/j.theriogenology.2012.04.010

Agarwal A, Aponte-Mellado A, Premkumar BJ, Shaman A, Gupta S (2012) The effects of oxidative stress on female reproduction: a review. Reprod Biol Endocrinol 10:49. https://doi.org/10.1186/1477-7827-10-49

Allan T, Ian C, O'Brien S (1998) Evidence-based fertility treatment. Royal College of Obstetricians and Gynaecologists Press, London

Andreas E, Pandey HO, Hoelker M, Salilew-Wondim D, Gebremedhn S, Schellander K, Tesfaye D (2021) The regulatory role of miR-20a in bovine cumulus cells and its contribution to oocyte maturation. Zygote 23:1–10. https://doi.org/10.1017/S0967199420000933

Asensi-Fabado MA, Munné-Bosch S (2010) Vitamins in plants: occurrence, biosynthesis and antioxidant function. Trends Plant Sci 15:582–592. https://doi.org/10.1016/j.tplants.2010.07.003

Baddela VS, Sharma A, Vanselow J (2020) Non-esterified fatty acids in the ovary: friends or foes? Reprod Biol Endocrinol 18:60. https://doi.org/10.1186/s12958-020-00617-9

Barakat IAH, Alajmi RA, Zoheir KMA, Salem LM, Al-Hemidiy AR (2018) Gene expression and maturation evaluation of sheep oocytes cultured in medium supplemented with natural antioxidant source. S Afr J Anim Sci 48(2):261–270. https://doi.org/10.4314/sajas.v48i2.7

Basini G, Tamanini C (2000) Selenium stimulates estradiol production in bovine granulosa cells: possible involvement of nitric oxide. Domest Anim Endocrinol 18:1–17. https://doi.org/10.1016/s0739-7240(99)00059-4
Berry DP, Buckley F, Dillon P (2007) Body condition score andlive-weight effects on milk production in Irish Holstein-Friesiandairy cows. Animal 1:1351–1359. https://doi.org/10.1017/S1751731107000419
Bettina PM, Redgrove KA, McLaughlin EA, Nixon B (2017) Molecular mechanisms responsible for increased vulnerability of the ageing oocyte to oxidative damage. Oxid Med Cell Longe. https://doi.org/10.1155/2017/4015874
Bilodeau-Goeseels S, Kastelic JP (2003) Factors affecting embryo survival and strategies to reduce embryonic mortality in cattle. Can J Anim Sci 83(4):659–671. https://doi.org/10.4141/A03-029
Bindari YR, Shrestha S, Shrestha N, Tara Gaire N (2013) Effects of nutrition on reproduction- a review. Adv Appl Sci Res 4(1):421–429
Bosco D, Haefliger JA, Meda P (2011) Connexins: key mediators of endocrine function. Physiol Rev 91:1393–1445. https://doi.org/10.1152/physrev.00027.2010
Bridges GA, Kruse SG, Funnell B, Perry GA, Gunn PJ, Arias RP, Lake SL (2012) Changes in body condition on oocyte quality andembryo survival. Proc App Reprod Strat Beef Cattle:269–284
Brigelius-Flohé R, Maiorino M (2013) Glutathione peroxidases. Biochim Biophys Acta 1830:3289–3303. https://doi.org/10.1016/j.bbagen.2012.11.020
Britt JH (1992) Impacts of early postpartum metabolism on follicular development and fertility. Bov Proc 24:39–43. https://doi.org/10.21423/aabppro19916706
Budani MC, Tiboni GM (2020) Effects of supplementation with natural antioxidants on oocytes and preimplantation embryos. Antioxidants (Basel, Switzerland) 9(7):612. https://doi.org/10.3390/antiox9070612
Butler WR (2003) Energy balance relationships with follicular development, ovulation and fertility in postpartum dairy cows. Livestock Prod Sci 83:211–218. https://doi.org/10.1016/S0301-6226(03)00112-X
Cantoni O, Guidarelli A, Fiorani M (2017) Mitochondrial uptake and accumulation of vitamin C: what can we learn from cell culture studies? Antioxid Redox Signal. https://doi.org/10.1089/ars.2017.7253
Carenzi C, Verga M (2009) Animal welfare: review of the scientific concept and definition. Ital J Anim Sci 8(Suppl.1):21–30. https://doi.org/10.4081/ijas.2009.s1.21
Carlsson F, Frykblom P, Lagerkvist CJ (2007) Farm animal welfare – testing for market failure. J Agric Appl Econ 39(1):61–73. https://doi.org/10.1017/S1074070800022756
Castillo-Martín M, Bonet S, Morató R, Yeste M (2014) Comparative effects of addingβ mercaptoethanol or L-ascorbic acid to culture or vitrification-warming media on IVF porcine embryos. Reprod Fertil Dev 26:875–882. https://doi.org/10.1071/RD13116
Cebrian-Serrano A, Salvador I, Raga E, Dinnyes A, Silvestre MA (2013) Beneficial effect of melatonin on blastocyst in vitro production from heat-stressed bovine oocytes. Reprod Domest Anim 48:738–746. https://doi.org/10.1111/rda.12154
Ceko MJ, Hummitzsch K, Hatzirodos N, Bonner WM, Aitken JB, Russell DL, Lane M, Rodgers RJ, Harris HH (2015) X-Ray fluorescence imaging and other analyses identify selenium and GPX1 as important in female reproductive function. Metallomics 7:71–82. https://doi.org/10.1039/c4mt00228h
Chawalit S, Nguyen NT, Tseng JK, Lo NW, Tu CF, Ju JC (2012) Trichostatin A and ascorbic acid assist in the development of porcine handmade cloned embryos via different physiologic pathways. Reprod Sci 19:976–986. https://doi.org/10.1177/1933719112440049
Chen J, Guo Q, Pei YH, Ren QL, Chi L, Hu RK, Tan Y (2020) Effect of a short-term vitamin E supplementation on oxidative stress in infertile PCOS women under ovulation induction: a retrospective cohort study. BMC Womens Health 20(1):69. https://doi.org/10.1186/s12905-020-00930-w
Christenson LK (2010) MicroRNA control of ovarian function. Anim Reprod 7:129–133

D'Occhio MJ, Baruselli PS, Campanile G (2019) Influence of nutrition, body condition, and metabolic status on reproduction in female beef cattle: a review. Theriogenology 125:277–284. https://doi.org/10.1016/j.theriogenology.2018.11.010

da Silveira JC, Veeramachaneni DNR, Winger QA, Carnevale EM, Bouma GJ (2012) Cell-secreted vesicles in equine ovarian follicular fluid contain miRNAs and proteins: a possible new form of cell communication within the ovarian follicle. Biol Reprod. 86:71. https://doi.org/10.1095/biolreprod.111.093252

Dai X, Lu Y, Zhang M, Miao Y, Zhou C, Cui Z, Xiong B (2017) Melatonin improves the fertilization ability of post-ovulatory aged mouse oocytes by stabilizing ovastacin and Juno to promote sperm binding and fusion. Hum Reprod 32:598–606. https://doi.org/10.1093/humrep/dew362

De Bie J, Marei WFA, Maillo V, Jordaens L, Gutierrez-Adan A, Bols PEJ, Leroy JLMR (2017) Differential effects of high and low glucose concentrations during lipolysis-like conditions on bovine in vitro oocyte quality, metabolism and subsequent embryo development. Reprod Fertil Dev 29:2284–2300. https://doi.org/10.1071/RD16474

Desmet KL, Van Hoeck V, Gagné D, Fournier E, Thakur A, O'Doherty AM, Walsh CP, Sirard MA, Bols PE, Leroy JL (2016) Exposure of bovine oocytes and embryos to elevated non-esterified fatty acid concentrations: integration of epigenetic and transcriptomic signatures in resultant blastocysts. BMC Genomics 17(1):1004. https://doi.org/10.1186/s12864-016-3366-y

Diao YF, Oqani RK, Li XX, Lin T, Kang JW, Jin DI (2014) Changes in histone H3 lysine 36 methylation in porcine oocytes and preimplantation embryos. PLoS One 9(6):e100205. https://doi.org/10.1371/journal.pone.0100205

Eler JP, Silva JA, Ferraz JB, Dias F, Oliveira HN, Evans JL, Golden BL (2002) Genetic evaluation of the probability of pregnancy at 14 months for Nellore heifers. J Anim Sci 80:951–954. https://doi.org/10.2527/2002.804951x

Elis S, Desmarchais A, Maillard V, Uzbekova S, Monget P, Dupont J (2015) Cell proliferation and progesterone synthesis depend on lipid metabolism in bovine granulosa cells. Theriogenology 83(5):840–853. https://doi.org/10.1016/j.theriogenology.2014.11.019

Emori C, Sugiura K (2014) Role of oocyte-derived paracrine factors in follicular development. Anim Sci J 85(6):627–633. https://doi.org/10.1111/asj.12200

Farman M, Tripathi SK, Nandi S, Girish Kumar V (2015) Follicular fluid concentrations of metabolic stressors in normal, obese, metabolic stressed and emaciated ewes. Asian J Anim Sci 9:466–470. https://doi.org/10.3923/ajas.2015.466.470

Farman M, Tripathi SK, Tej NK, Nandi S, Gupta PSP, Mondal S, Venkatesh GK (2018) Metabolic stress indicators in ewes (Ovisaries) under post-parturient and high protein diet conditions. Asian J Anim Vet Adv 13:360–368. https://doi.org/10.3923/ajava.2018.360.368

Ferreira RM, Chiaratti MR, Macabelli CH, Rodrigues CA, Ferraz ML, Watanabe YF et al (2016) The infertility of repeat-breeder cows during summer is associated with decreased mitochondrial DNA and increased expression of mitochondrial and apoptotic genes in oocytes. Biol Reprod. 94:66. https://doi.org/10.1095/biolreprod.115.133017

Gao C, Han HB, Tian XZ, Tan DX, Wang L, Zhou GB, Zhu SE, Liu GS (2012) Melatonin promotes embryonic development and reduces reactive oxygen species in vitrified mouse 2-cell embryos. J Pineal Res 52:305–311. https://doi.org/10.1111/j.1600-079X.2011.00944.x

Gaviria SM, Morado SA, López Herrera A, Betancur GR, Álvarez RAU, Zuluaga JE, Cética PD (2019) Resveratrol supplementation promotes recovery of lower oxidative metabolism after vitrification and warming of in vitro-produced bovine embryos. Reprod Fertil Dev 31:521–528. https://doi.org/10.1071/RD18216

Gebremedhn S, Ali A, Gad A, Prochazka R, Tesfaye D (2020) Extracellular vesicles as mediators of environmental and metabolic stress coping mechanisms during mammalian follicular development. Front Vet Sci 7:602043. https://doi.org/10.3389/fvets.2020.602043

Gu L, Wang Q, Sun QY (2010) Histone modifications during mammalian oocyte maturation: dynamics, regulation and functions. Cell Cycle 9:1942–1950. https://doi.org/10.4161/cc.9.10.11599

Gunaretnam I, Pretheeban T, Rajamahendran R (2013) Effects of ammonia and urea in vitro on mRNA of candidate bovine endometrial genes. Anim Reprod Sci 141(1–2):42–51. https://doi.org/10.1016/j.anireprosci.2013.07.001

Hailay T, Hoelker M, Poirier M, Gebremedhn S, Rings F, Saeed-Zidane M et al (2019a) Extracellular vesicle-coupled miRNA profiles in follicular fluid of cows with divergent post-calving metabolic status. Sci Rep 9:12851. https://doi.org/10.1038/s41598-019-49029-9

Hailay T, Hoelker M, Poirier M et al (2019b) Extracellular vesicle-coupled miRNA profiles in follicular fluid of cows with divergent post-calving metabolic status. Sci Rep 9:12851. https://doi.org/10.1038/s41598-019-49029-9

Halama A, Riesen N, Möller G, Hrabě de Angelis M, Adamski J (2013) Identification of biomarkers for apoptosis in cancer cell lines using metabolomics: tools for individualized medicine. J Intern Med 274:425–439. https://doi.org/10.1111/joim.12117

Hancock RL, Dunne K, Walport LJ, Flashman E, Kawamura A (2015) Epigenetic regulationby histone demethylases in hypoxia. Epigenomics 7:791–781. https://doi.org/10.2217/epi.15.24

Hefnawy AEG, Tórtora-Pérez JL (2010) The importance of selenium and the effects of its deficiency in animal health. Small Ruminant Res 89:185–192. https://doi.org/10.1016/j.smallrumres.2009.12.042

Hemmings KE, Leese HJ, Picton HM (2012) Amino acid turnover by bovine oocytes provides an index of oocyte developmental competence *in vitro*. Biol Reprod 86:1–12. https://doi.org/10.1095/biolreprod.111.092585

Hemmings KE, Maruthini D, Vyjayanthi S, Hogg JE, Balen AH, Campbell BK, Leese HJ, Picton HM (2013) Amino acid turnover by human oocytes is influenced by gamete developmental competence, patient characteristics and gonadotrophin treatment. Hum Reprod 28(4):1031–1044. https://doi.org/10.1093/humrep/des458

Hirshfield AN (1991) Development of follicles in the mammalian ovary. Int Rev Cytol 124:43–101. https://doi.org/10.1016/s0074-7696(08)61524-7

Hoedemaker M, Prange D, Gundelach Y (2009) Body condition change ante- and postpartum, health and reproductive performance in German Holstein cows. Reprod Domest Anim 44:167–173. https://doi.org/10.1111/j.1439-0531.2007.00992.x

Hung WT, Hong X, Christenson LK, McGinnis LK (2015) Extracellular vesicles from bovine follicular fluid support cumulus expansion. Biol Reprod 93:117. https://doi.org/10.1095/biolreprod.115.132977

Ibtisham F, Zhang L, Xiao M, An L, Bilal M (2017) Genomic selection and its application in animal breeding. Thai J Vet Med 47:301–310

Ibtisham F, Nawab A, Li G, Xiao M, An L, Naseer G (2018) Effect of nutrition on reproductive efficiency of dairy animals. Medycynaweterynaryjna. https://doi.org/10.21521/mw.6025

Keerthikumar S, Chisanga D, Ariyaratne D, Al Saffar H, Anand S, Zhao K, Samuel M, Pathan M, Jois M, Chilamkurti N, Gangoda L, Mathivanan S (2016) ExoCarta: a web-based compendium of exosomal cargo. J Mol Biol 428(4):688–692. https://doi.org/10.1016/j.jmb.2015.09.019

Kowsar R, Marey MA, Shimizu T, Miyamoto A (2016) Urea induces T helper 2 (Th2) type environment at transcriptional level and prostaglandin E2 secretion in bovine oviduct epithelial cells in culture. J Dairy Sci 99:5844–5850. https://doi.org/10.3168/jds.2016-10874

Kowsar R, Iranshahi VN, Sadeghi N, Ahmad R, Miyamoto A (2018) Urea influences amino acid turnover in bovine cumulus-oocyte complexes, cumulus cells and denuded oocytes, and affects *in vitro* fertilization outcome. Sci Rep 8:12191. https://doi.org/10.1038/s41598-018-30774-2

Kubovicova E, Makarevic A, Stadnik L, Holasek R, Hegedusova Z (2013) Effect of body condition and season on the yield and quality ofcattle embryos. J Micro Biotechnol Food Sci 2(1):1426–1435

Kumar D, Gupta D, Shankar S, Srivastava RK (2015) Biomolecular characterization of exosomes released from cancer stem cells: possible implications for biomarker and treatment of cancer. Oncotarget 6:3280–3291. https://doi.org/10.18632/oncotarget.2462

Kwak SS, Cheong SA, Jeon Y, Lee E, Choi KC, Jeung EB, Hyun SH (2012) The effects of resveratrol on porcine oocyte in vitro maturation and subsequent embryonic development after parthenogenetic activation and in vitro fertilization. Theriogenology 78:86–101. https://doi.org/10.1016/j.theriogenology.2012.01.024

Leroy JL, Vanholder T, Mateusen B, Christophe A, Opsomer G, de Kruif A, Genicot G, Van Soom A (2005) Non-esterified fatty acids in follicular fluid of dairy cows and their effect on developmental capacity of bovine oocytes in vitro. Reproduction 130:485–495. https://doi.org/10.1530/rep.1.00735

Leroy JL, Van Soom A, Opsomer G, Goovaerts IG, Bols PE (2008) Reduced fertility in high-yielding dairy cows: are the oocyte and embryo in danger? Part II. Mechanisms linking nutrition and reduced oocyte and embryo quality in high-yielding dairy cows. Reprod Domest Anim 43(5):623–632. https://doi.org/10.1111/j.1439-0531.2007.00961.x

Leroy JL, Valckx SD, Jordaens L, De Bie J, Desmet KL, Van Hoeck V, Britt JH, Marei WF, Bols PE (2015) Nutrition and maternal metabolic health in relation to oocyte and embryo quality: critical views on what we learned from the dairy cow model. Reprod Fertil Dev 27(4):693–703. https://doi.org/10.1071/RD14363

Lisboa LA, Andrade ER, Hertel MF, Melo-Sterza FA, Moreno K, Bracarense APFRL, Alfieri AA, Seneda MM (2009) Viability and growth of cattle preantral follicles after in vitro culture of ovarian fragments in α-tocopherol. Reprod Fertil Dev 22:318–319. https://doi.org/10.1071/RDv22n1Ab325

Lizarraga RM, Anchordoquy JM, Galarza EM, Farnetano NA, Carranza-Martin A, Furnus CC, Mattioli GA, Anchordoquy JP (2020) Sodium selenite improves in vitro maturation of Bosprimigeniustaurus oocytes. Biol Trace Elem Res. https://doi.org/10.1007/s12011-019-01966-2

Lu X, Wu Z, Wang M, Cheng W (2018) Effects of vitamin C on the outcome of in vitro fertilization-embryo transfer in endometriosis: a randomized controlled study. J Int Med Res 46(11):4624–4633. https://doi.org/10.1177/0300060518786918

Lucy MC (2019) Stress, strain, and pregnancy outcome in postpartum cows. Anim Reprod 16(3):455–464. https://doi.org/10.21451/1984-3143-AR2019-0063

Makki M, Saboori E, Sabbaghi MA, Aram R, Fallahian MH, Peyghambari F, Roustaei H, Ahmadi A (2012) Effects of selenium, calcium and calcium ionophore on human oocytes in vitro maturation in a chemically defined medium. Iran J Reprod Med 10(4):343–348. PMID: 25246896; PMCID: PMC4165952

Mallol A, Santaló J, Ibáñez E (2015) Improved development of somatic cell cloned mouse embryos by vitamin C and latrunculin A. PLoS One 10:e0120033

Mayada AE, Nadia AT,Abdel-Mohsen MH, Francois ARS (2015) Effect of antioxidant supplementation on in vitro maturation of Camelus Dromedaries oocytes. Nat Sci 13(2):17–24. http://www.sciencepub.net/nature

McDonough MA, Loenarz C, Chowdhury R, Clifton IJ, Schofield CJ (2010) Structural studies on human 2-oxoglutarate dependent oxygenases. Curr Opin Struct Biol 20(6):659–672. https://doi.org/10.1016/j.sbi.2010.08.006

Mehdi Y, Dufrasne I (2016) Selenium in cattle: a review. Molecules 21:545. https://doi.org/10.3390/molecules21040545

Miclea I, Pacală N, Hettig A, Zăhan M, Miclea V (2012) Alpha-tocopherol and ascorbic acid combinations influence the maturation of sheep oocytes. Sci Papers Anim Sci Biotechnol 45(1)

Mondou E, Dufort I, Gohin M, Fournier E, Sirard M-A (2012) Analysis of microRNAs and their precursors in bovine early embryonic development. Mol Hum Reprod 18:425–434. https://doi.org/10.1093/molehr/gas015

Morris CA (1980) A review of relationships between aspects of reproduction in beef heifers and their lifetime production. 2. Associations with relative calving date and with dystocia. Anim Breed Abstr 48:753–767

Nabenishi H, Ohta H, Nishimoto T, Morita T, Ashizawa K, Tsuzuki Y (2012) The effects of cysteine addition during *in vitro* maturation on the developmental competence, ROS, GSH and apoptosis level of bovine oocytes exposed to heat stress. Zygote 20(3):249–259. https://doi.org/10.1017/S0967199411000220

Nandi S, Shree USP, Kumar GV (2013) Metabolic stressors in ovine and caprine sera and ovarian follicular fluid. Appl Cell Biol 2:110–113

Nandi S, Mondal S, Pal DT, Gupta PSP (2015) Effect of ammonia-generating diet on ovine serum and follicular fluid ammonia and urea levels, serum oestrogen and progesterone concentrations and granulosa cell functions. J Anim Physiol Anim Nutr 100:309–331. https://doi.org/10.1111/jpn.12369. Epub 2015 Jul 27

Nandi S, Gupta PSP, Mondal S (2016) Ammonia concentrations in different size classes of ovarian follicles of sheep (*Ovisaries*): possible mechanisms of accumulation and its effect on oocyte and granulosa cell growth *in vitro*. Theriogenology 84(4):678–687. https://doi.org/10.1016/j.theriogenology.2015.10.007

Nandi S, Tripathi SK, Gupta PSP, Mondal S (2017) Effect of metabolic stressors on survival and growth of in vitro cultured ovine preantral follicles and enclosed oocytes. Theriogenology 104:80–86. https://doi.org/10.1016/j.theriogenology.2017.07.024

Nandi S, Tripathi SK, Gupta PSP, Mondal S (2018) Nutritional and metabolic stressors on ovine oocyte development and granulosa cell functions in vitro. Cell Stress Chaperones 23(3):357–371. https://doi.org/10.1007/s12192-017-0846-1

O'Callaghan D, Yaakub H, Hyttel P, Spicer LJ, Boland MP (2000) Effect of nutrition and superovulation on oocyte morphology, follicular fluid composition and systemic hormone concentrations in ewes. J Reprod Fertil 118:303–313

Ochiai A, Kuroda K, Ikemoto Y, Ozaki R, Nakagawa K, Nojiri S, Takeda S, Sugiyama R (2019) Influence of resveratrol supplementation on IVF-embryo transfer cycle outcomes. Reprod Biomed Online 39(2):205–210. https://doi.org/10.1016/j.rbmo.2019.03.205

Orisaka M, Tajima K, Tsang BK, Kotsuji F (2009) Oocyte-granulosa-theca cell interactions during preantral follicular development. J Ovarian Res 2(1):9. https://doi.org/10.1186/1757-2215-2-9

Pearson RA, Nengomasha E, Krecek RC (1997) The challenges in using donkeys for work in Africa. In: Meeting the challenges of animal traction. Proceedings of an ATNESA workshop. DebreZeit, Ethiopia, May 5–9, 1997, pp 190–198

Picco SJ, Anchordoquy JM, de Matos DG, Anchordoquy JP, Seoane A, Mattioli GA, Errecalde AL, Furnus CC (2010) Effect of increasing zinc sulphate concentration during in vitro maturation of bovine oocytes. Theriogenology 74(7):1141–1148. https://doi.org/10.1016/j.theriogenology.2010.05.015

Popescu S, Borda C, Lazar EA, Hegedüs CI (2009) Assessment of dairy cow welfare in farms from Transylvania. In: Proceeding of 4th Croatian & 4th international symposium on agriculture, pp 752–756

Prasad S, Tiwari M, Pandey AN, Shrivastav TG, Chaube SK (2016) Impact of stress on oocyte quality and reproductive outcome. J Biomed Sci 23:36. https://doi.org/10.1186/s12929-016-0253-4

Richani D (2014) Epidermal growth factor-like peptide signalling and oocyte in vitro maturation. PhD thesis, University of Adelaide

Roth Z (2017) Effect of heat stress on reproduction in dairy cows: insights into the cellular and molecular responses of the oocyte. Annu Rev Anim Biosci 5:151–170. https://doi.org/10.1146/annurev-animal-022516-022849

Roth Z, Hansen PJ (2005) Disruption of nuclear maturation and rearrangement of cytoskeletal elements in bovine oocytes exposed to heat shock during maturation. Reproduction 129:235–244. https://doi.org/10.1530/rep.1.00394

Rukkwamsuk T, Kruip TA, Wensing T (1999) Relationship between overfeeding and overconditioning in the dry period and the problems of high producing dairy cows during the postparturient period. Vet Q 21(3):71–77. https://doi.org/10.1080/01652176.1999.9694997

Sakatani M, Kobayashi S, Takahashi M (2004) Effects of heat shock on in vitro development and intracellular oxidative state of bovine preimplantation embryos. Mol Reprod Dev 67(1):77–82. https://doi.org/10.1002/mrd.20014

Santonocito M, Vento M, Guglielmino MR, Battaglia R, Wahlgren J, Ragusa M et al (2014) Molecular characterization of exosomes and their microRNA cargo in human follicular fluid:

bioinformatic analysis reveals that exosomal microRNAs control pathways involved in follicular maturation. Fertil Steril 102:1751–61.e1. https://doi.org/10.1016/j.fertnstert.2014.08.005
Schwab CG, Bozak CK, Whitehouse NL (1992) Amino acid limitation and flow to duodenum at four stages of lactation. 1. Sequence of lysine and methionine limitation. J Dairy Sci 75:3486–3502
Sharma RK, Agarwal A (2004) Role of reactive oxygen species in gynecologic diseases. Reprod Med Biol 3(4):177–199. https://doi.org/10.1111/j.1447-0578.2004.00068.x
Sharma A, Baddela VS, Becker F, Dannenberger D, Viergutz T, Vanselow J (2019) Elevated free fatty acids affect bovine granulosa cell function: a molecular cue for compromised reproduction during negative energy balance. Endocrine Connect 8(5):493–505. https://doi.org/10.1530/EC-19-0011
Silva M, Melo SA (2015) Non-coding RNAs in exosomes: new players in cancer biology. Curr Genomics 16(5):295–303. https://doi.org/10.2174/1389202916666150707154719
Smith OB, Akinbamijo OO (2000) Micronutrients and reproduction in farm animals. Anim Reprod Sci 60–61:549–560. https://doi.org/10.1016/s0378-4320(00)00114-7
Smith OB, Somade B (1994) Nutrition reproduction interactions in farm animals. In: Proceedings of the international foundation seminar on animal reproduction. Niamey, Niger, 17–21, pp 1–31
Sordillo LM (2013) Selenium-dependent regulation of oxidative stress and immunity in periparturient dairy cattle. Vet Med Int 2013:154045. https://doi.org/10.1155/2013/154045
Spears JW, Weiss WP (2008) Role of antioxidants and trace elements in health and immunity of transition dairy cows. Vet J 176(1):70–76. https://doi.org/10.1016/j.tvjl.2007.12.015
Starkey P, Jaiyesimi-Njobe F, Hanekom D (1995) Animal traction in South Africa: overview of the key issues. In: Starkey P (ed) Animal traction in South Africa – empowering rural communities. Gauteng, Development Bank of Southern Africa. A DBSA-SANAT, pp 17–33
Stewart KR, Veselovska L, Kim J, Huang J, Saadeh H, Tomizawa S, Smallwood SA, Chen T, Kelsey G (2015) Dynamic changes in histone modifications precede de novo DNA methylation in oocytes. Genes Dev 29(23):2449–2462. https://doi.org/10.1101/gad.271353.115
Sutton-McDowall ML, Gilchrist RB, Thompson JG (2010) The pivotal role of glucose metabolism in determining oocyte developmental competence. Reproduction 139(4):685–695. https://doi.org/10.1530/REP-09-0345
Sutton-McDowall ML, Wu LL, Purdey M, Abell AD, Goldys EM, MacMillan KL, Thompson JG, Robker RL (2016) Nonesterified fatty acid-induced endoplasmic reticulum stress in cattle cumulus oocyte complexes alters cell metabolism and developmental competence. Biol Reprod. 94(1):23. https://doi.org/10.1095/biolreprod.115.131862
Taniguchi K, Taketani T, Lee LM, Kizuka F, Tamura I, Sugino N (2009) Melatonin protects granulosa cells for progesterone production as an antioxidant in human ovarian follicles. Biol Reprod. 81(1):378
Tareq KM, Akter QS, Khandoker MA, Tsujii H (2012) Selenium and vitamin E improve the in vitro maturation, fertilization and culture to blastocyst of porcine oocytes. J Reprod Dev 58(6):621–628. https://doi.org/10.1262/jrd.2012-064
Tiboni GM, Bucciarelli T, Giampietro F, Sulpizio M, Di Ilio C (2004) Influence of cigarette smoking on vitamin E, vitamin A, beta-carotene and lycopene concentrations in human preovulatory follicular fluid. Int J Immunopathol Pharm 17:389–393
Tripathi SK, Farman M, Nandi S, Girish Kumar V, PSP G (2016a) Oviductal and uterine fluid analytes as biomarkers of metabolic stress in ewes (Ovisaries). Small Rumin Res 144:225–228. https://doi.org/10.1016/j.smallrumres.2016.09.022
Tripathi SK, Farman M, Nandi S, Mondal S, Gupta P, Kumar VG (2016b) In vitro culture of oocytes and granulosa cells collected from normal, obese, emaciated and metabolically stressed ewes. Anim Reprod Sci 170:83–89. https://doi.org/10.1016/j.anireprosci.2016.04.007
Valadi H, Ekström K, Bossios A, Sjöstrand M, Lee JJ, Lötvall JO (2007a) Exosome-mediated transfer of mRNAs and microRNAs is a novel mechanism of genetic exchange between cells. Nat Cell Biol 9:654–659. https://doi.org/10.1038/ncb1596

Valadi H, Ekström K, Bossios A, Sjöstrand M, Lee JJ, Lötvall JO (2007b) Exosome-mediated transfer of mRNAs and microRNAs is a novel mechanism of genetic exchange between cells. Nat Cell Biol 9(6):654–659. https://doi.org/10.1038/ncb1596

Valckx SD, Van Hoeck V, Arias-Alvarez M, Maillo V, Lopez-Cardona AP, Gutierrez-Adan A, Berth M, Cortvrindt R, Bols PE, Leroy JL (2014) Elevated non-esterified fatty acid concentrations during in vitro murine follicle growth alter follicular physiology and reduce oocyte developmental competence. Fertil Steril 102:1769–1776 e1761. https://doi.org/10.1016/j.fertnstert.2014.08.01

Van Hoeck V, Sturmey RG, Bermejo-Alvarez P, Rizos D, Gutierrez-Adan A, Leese HJ, Bols PE, Leroy JL (2011) Elevated non-esterified fatty acid concentrations during bovine oocyte maturation compromise early embryo physiology. PLoS One 6(8):e23183

Van Hoeck V, Leroy JL, Arias Alvarez M, Rizos D, Gutierrez-Adan A, Schnorbusch K, Bols PE, Leese HJ, Sturmey RG (2013) Oocyte developmental failure in response to elevated nonesterified fatty acid concentrations: mechanistic insights. Reproduction 145(1):33–44. https://doi.org/10.1530/REP-12-0174

Van Hoeck V, Rizos D, Gutierrez-Adan A, Pintelon I, Jorssen E, Dufort I, Sirard MA, Verlaet A, Hermans N, Bols PE, Leroy JL (2015) Interaction between differential gene expression profile and phenotype in bovine blastocysts originating from oocytes exposed to elevated non-esterified fatty acid concentrations. Reprod Fertil Dev 27:372–384. https://doi.org/10.1071/RD13263

Walters AH, Bailey TL, Pearson RE, Gwazdauskas FC (2002) Parity-related changes in bovine follicle and oocyte populations, oocyte quality, and hormones to 90 days postpartum. J Dairy Sci 85:824–832. https://doi.org/10.3168/jds.S0022-0302(02)74142-8

Wang F, Tian X, Zhang L, He C, Ji P, Li Y, Tan D, Liu G (2014) Beneficial effect of resveratrol on bovine oocyte maturation and subsequent embryonic development after in vitro fertilization. Fertil Steril 101(2):577–586. https://doi.org/10.1016/j.fertnstert.2013.10.041

Webb R, Garnsworthy PC, Gong JG, Armstrong DG (2004) Control of follicular growth: local interactions and nutritional influences. J Anim Sci 82(E-Suppl):E63–E74. https://doi.org/10.2527/2004.8213_supplE63x

Wilkanowska A, Kokoszyn'ski D (2015) Effect of diet and physical activity of farm animals on their health and reproductive performance. In: Handbook of fertility. https://doi.org/10.1016/B978-0-12-800872-0.00014-7

Wong SL, Wu LL, Robker RL, Thompson JG, McDowall ML (2015) Hyperglycaemia and lipid differentially impair mouse oocyte developmental competence. Reprod Fertil Dev 27(4):583–592. https://doi.org/10.1071/RD14328

Xu Y, Nisenblat V, Lu C, Li R, Qiao J, Zhen X, Wang S (2018) Pretreatment with coenzyme Q10 improves ovarian response and embryo quality in low-prognosis young women with decreased ovarian reserve: a randomized controlled trial. Reprod Biol Endocrinol 16:29

Yamaguchi S et al (2012) Tet1 controls meiosis by regulating meiotic gene expression. Nature 492:443–447. https://doi.org/10.1038/nature11709

Yáñez-Mó M, Siljander PRM, Andreu Z, BedinaZavec A, Borràs FE, Buzas EI et al (2015) Biological properties of extracellular vesicles and their physiological functions. J Extracell Vesicles 4:27066. https://doi.org/10.3402/jev.v4.27066

Yenuganti VR, Viergutz T, Vanselow J (2016) Oleic acid induces specific alterations in the morphology, gene expression and steroid hormone production of cultured bovine granulosa cells. Gen Comp Endocrinol 232:134–144. https://doi.org/10.1016/j.ygcen.2016.04.020

Young JI, Züchner S, Wang G (2015) Regulation of the epigenome by vitamin C. Annu Rev Nutr 35:545–564. https://doi.org/10.1146/annurev-nutr-071714-034228

Yu XX, Liu YH, Liu XM, Wang PC, Liu S, Miao JK, Du ZQ, Yang CX (2018) Ascorbic acid induces global epigenetic reprogramming to promote meiotic maturation and developmental competence of porcine oocytes. Sci Rep 8(1):6132. https://doi.org/10.1038/s41598-018-24395-y

Index

V. K. Yata et al. (eds.), *Sustainable Agriculture Reviews 57*, Sustainable Agriculture Reviews 57, https://doi.org/10.1007/978-3-031-07496-7

The manufacturer's authorised representative in the EU is Springer Nature Customer Service Centre GmbH, Europaplatz 3, 69115 Heidelberg, Germany. If you have any concerns regarding our products, please contact ProductSafety@springernature.com

Printed and bound by CPI Group (UK) Ltd, Croydon, CR0 4YY
01/10/2026
02225872-0003